AF898913

Control of Axially Moving Systems

Keum-Shik Hong · Li-Qun Chen ·
Phuong-Tung Pham · Xiao-Dong Yang

Control of Axially Moving Systems

Springer

Keum-Shik Hong
School of Mechanical Engineering
Pusan National University
Busan, Korea (Republic of)

Phuong-Tung Pham
School of Mechanical Engineering
Pusan National University
Busan, Korea (Republic of)

Li-Qun Chen
School of Mechanics and Engineering Science
Shanghai University
Shanghai, China

Xiao-Dong Yang
School of Mechanical Engineering
Beijing University of Technology
Beijing, China

ISBN 978-981-16-2914-3 ISBN 978-981-16-2915-0 (eBook)
https://doi.org/10.1007/978-981-16-2915-0

© The Editor(s) (if applicable) and The Author(s), under exclusive license to Springer Nature Singapore Pte Ltd. 2022
This work is subject to copyright. All rights are solely and exclusively licensed by the Publisher, whether the whole or part of the material is concerned, specifically the rights of translation, reprinting, reuse of illustrations, recitation, broadcasting, reproduction on microfilms or in any other physical way, and transmission or information storage and retrieval, electronic adaptation, computer software, or by similar or dissimilar methodology now known or hereafter developed.
The use of general descriptive names, registered names, trademarks, service marks, etc. in this publication does not imply, even in the absence of a specific statement, that such names are exempt from the relevant protective laws and regulations and therefore free for general use.
The publisher, the authors and the editors are safe to assume that the advice and information in this book are believed to be true and accurate at the date of publication. Neither the publisher nor the authors or the editors give a warranty, expressed or implied, with respect to the material contained herein or for any errors or omissions that may have been made. The publisher remains neutral with regard to jurisdictional claims in published maps and institutional affiliations.

This Springer imprint is published by the registered company Springer Nature Singapore Pte Ltd.
The registered company address is: 152 Beach Road, #21-01/04 Gateway East, Singapore 189721, Singapore

Preface

This book provides a complete in-depth understanding of the derivation of the equations of motion and control of various axially moving systems (AMSs) described by string, beam, belt, and plate models. Additionally, approximate models discretized from partial differential equations (PDEs) using approximate methods are introduced in this book. The relationship between the system parameters of the AMSs and the dynamic behaviors of the AMSs is emphasized. Also, the design of control strategies for axially moving systems is presented. Based on a subsequent discussion, conclusions are drawn, and future directions on axially moving systems are proposed. The contents of this book are briefly described as follows.

Chapter 1 introduces industrial axially moving systems and the motivation to study the axially moving systems. This chapter also gives a broad overview of the dynamics and control of axially moving systems.

Chapter 2 covers the dynamics of axially moving strings. The PDEs that describe the vibration of traveling strings (with constant or time-varying velocity) are developed using the Hamilton principle. The effects of nonlinear factors, such as axial deformation and viscoelasticity of materials, are also considered. Various boundary conditions are further investigated in this chapter. In addition to the exact model described by PDEs, approximate models characterized by ordinary differential equations (ODEs) are also derived. Subsequently, the dynamic behaviors of axially moving strings are analyzed.

Chapter 3 provides a complete discussion on the dynamics of the axially moving beam. The equations of motion of moving beams are derived based on diverse beam theories: the Euler–Bernoulli, Rayleigh, and Timoshenko beam theories. Besides, other aspects of axially moving beams, such as the cantilever beam, viscoelastic beam, fluid-conveying pipe, nanoscale beam, and composite beam, are also covered. The approximate models and dynamic behaviors of moving beams are further analyzed in this chapter.

Chapter 4 discusses the design of the control schemes for axially moving strings and beams. The control methods that use an ODE model, such as model-based feedback control and variable structure control, are presented in the first half of this chapter. The remainder of this chapter focuses on designing PDE-based control schemes: transfer function method, wave cancelation approach, boundary controls based on the Lyapunov method, adaptive control, and hybrid control methods. Numerical simulations are also shown to verify the effectiveness of the control schemes.

Chapters 5 and 6 are concerned with axially moving belts. Chapter 5 introduces the dynamics of various axially moving belts, whereas Chap. 6 covers boundary controls of moving belts based on the Lyapunov method.

Chapter 7 is the first of two sections on axially moving plates, which provides the derivations of the mathematical model of orthotropic, isotropic, and composite moving plates. The remainder of Chap. 7 discusses the approximate models and dynamic behaviors of axially moving plates. Chapter 8 is concerned with the control of axially moving plates based on velocity regulation.

Chapter 9 presents several conclusions and finally makes future research directions for axially moving systems.

Busan, Korea (Republic of) Keum-Shik Hong
Shanghai, China Li-Qun Chen
Busan, Korea (Republic of) Phuong-Tung Pham
Beijing, China Xiao-Dong Yang

Acknowledgments In preparing this book, Keum-Shik Hong and Phuong-Tung Pham were supported by the National Research Foundation (NRF) of Korea under the auspices of the Ministry of Science and ICT, Korea (grant no. NRF-2020R1A2B5B03096000).

Contents

Abbreviations

AMS	Axially moving system
DIQM	Differential and integral quadrature method
IPDE	Integro-partial differential equation
ODE	Ordinary differential equation
PDE	Partial differential equation
SMC	Sliding mode control
VSC	Variable structure control

Fundamental Notations

A	The cross-sectional area of a string/beam/belt
A_f	The cross-sectional area of fluid
a_{nl}	Internal characteristic length in the non-local constitutive relation of a nanobeam
b	The width of a plate
$\mathbf{C}$	Damping matrix of an ODE model for the lateral vibration
$\mathbf{C}_{u\eta}$	Damping matrix of an ODE model for the longitudinal and transverse vibrations
c	Damping coefficient
c_a	Damping coefficient of a rotating accessory
c_b	Damping coefficient of friction between a belt and a pulley
c_d	Damping coefficient of a wrap spring
c_e	External damping coefficient
c_i	Internal damping coefficient
c_h	Damping coefficient of a hydraulic actuator
D	Flexural rigidity of a plate
$d(t)$	External disturbance
$d(x, t)$	Distributed external disturbance
d_b	Bound of unknown disturbance
E	Elastic modulus
E_1, E_2	Elastic modulus of a viscoelastic model
$E_{1X}, E_{2X}, E_{1Z}, E_{2Z}$	Young's moduli of a viscoelastic plate
$E_B(t)$	Relaxation function
E_s	Mechanical energy of a string
E_X, E_Z	Elastic modulus of an elastic plate along with *i*- and *k*-axes
e	Tracking error
e_0	Constant appropriate to each material in the non-local constitutive relation of a nanobeam
$F(x, s)$	Laplace transforms of $f(x, t)$
F_f	Pressure between a fluid and a structure
$f(t)$	Control force
$\mathbf{f}(t)$	Control input vector

G	Shear modulus
$\tilde{G}$	Gyroscopic operator
$G(s)$	Transfer function of a feedback controller
$G_0(s, \xi, t)$	Open-loop transfer function of a linear string model (PDE)
$G_c(s, \xi, t)$	Closed-loop transfer function of a linear string model (PDE)
$\mathbf{G}$	Control gain matrix in a model-based feedback control scheme
g	Gravity acceleration
H_0^k	Sobolev space
h	Thickness of a plate
h_n	Distance from the external boundary of the *n*th layer to the *i*-axis of a laminated composite material
I	Inertia moment
$\mathbf{i}$	Unit vector along the *i*-axis
J	Rotational inertia of a pulley/roller
J_a	Rotational inertia of an accessory
$\mathbf{j}$	Unit vector along the *j*-axis
K	Kinetic energy
K_d	Stiffness of a wrap spring
$\tilde{K}$	Stiffness operator
$\mathbf{K}$	Stiffness matrix of an ODE model of the lateral motion
$\mathbf{K}_{u\eta}$	Stiffness matrix of an ODE model of the longitudinal and transverse motions
$\mathbf{k}$	Unit vector along the *k*-axis
L^2	Lebesgue space
l	Length of a string/beam/belt/plate
$M(x, t)$	Bending moment
M_1	Preload between a driven pulley and an accessory
$\tilde{M}$	Mass operator
M_X, M_Z	Sectional bending moments with respect to the *i*- and *k*-axes
M_{XZ}	Sectional torsional moment
M_{nl}	Non-local bending moment of a nanobeam
$\mathbf{M}$	Mass matrix of an ODE model of the lateral motion
$\mathbf{M}_{u\eta}$	Mass matrix of an ODE model of the longitudinal and transverse motions
m_a	Mass of a roller (an actuator)
m_e	End mass
m_f	Linear density of fluid
m_p	Mass of the payload of a crane
m_t	Mass of the trolley of a crane
$P(x, t)$	Axial force generated by the gravitational acceleration in an axially moving system with varying length
P_f	Fluid pressure
p_k^0	Open-loop poles ($i = 1, 2, 3, \ldots$)
p_k^c	Closed-loop poles ($i = 1, 2, 3, \ldots$)

N_X, N_Z	Sectional plate forces along with the *i*- and *k*-axes
N_{XZ}	Sectional shear force
Q_{ij}^n	Stiffness elements of the material used in the *n*th layer of a laminated composite material
$\hat{q}$	Shear force between a fluid and a pipe wall
$q_i(t)$	Generalized displacements
$q_{ij}^w(t), q_{ij}^u(t), q_{ij}^\eta(t)$	Generalized displacements of the lateral, longitudinal, and transverse vibrations of a plate/membrane
$\mathbf{q}$	Vector of the lateral displacement of an axially moving system
$\mathbf{q}_{u\eta}$	Vector of the longitudinal and transverse displacements of an axially moving system
$\mathbf{R}$	Reuter matrix
r	Radius of a pulley/roller
r_a	Radius of an accessory
$\mathbf{r}_d$	Displacement vector of a unit element of a string/beam/belt/plate
$\mathbf{r}_v$	Velocity vector of a unit element of a string/beam/belt/plate
S_c	Circumference of the cross section
s	Sliding surface
s	Laplace variable (complex variable)
T_0	Initial tension
T_d	Dynamic tension
$T(x, t)$	Varying tension
$\mathbf{T}_L$	Transformation matrix in the strain–stress relationship of a lamina
$\mathbf{T}_a, \mathbf{T}_s$	Matrices corresponding to the locations of the actuator and sensors in a model-based feedback control scheme
t	Time variable
t_d	Time delay
$\bar{t}$	Dimensionless time
U	Potential energy
$u(x, t)$	Longitudinal vibration (i.e., vibration along the *i*-axis)
$\bar{u}(\bar{x}, \bar{t})$	Dimensionless longitudinal vibration
$u(x, y, t)$	Longitudinal vibration of a plate
$V(t)$	Lyapunov's function candidate
$v(t)$	Time-varying axial velocity
v_0	Constant axial velocity
$v_f(t)$	Fluid velocity
$\bar{v}(t)$	Dimensionless time-varying velocity
$\bar{v}_0$	Dimensionless constant velocity
$W(x, s)$	Laplace transform of $w(x, t)$
$w(x, t)$	Lateral vibration (i.e., vibration along the *j*-axis)
$\bar{w}(\bar{x}, \bar{t})$	Dimensionless lateral vibration
$w_\infty(x)$	Equilibrium solution of an axially moving material
$w(x, y, t)$	Lateral vibration of a plate

$X_i(x)$	Basis functions in the i-axis of a two-dimensional model
x	Spatial variable in the i-axis
x_a	Location of an actuator
x_s	Location of a sensor
$\bar{x}$	Dimensionless term of x
y	Spatial variable in the j-axis
y_{tr}	Position of the trolley of a crane
$\bar{y}$	Dimensionless term of y
$Z_i(z)$	Basis functions in the k-axis of a two-dimensional model
z	Spatial variable in the k-axis
$\bar{z}$	Dimensionless term of z
$\gamma(x, t)$	Shear strain of a Timoshenko beam
ε	Bookkeeping device used in the multiple scale method
$\varepsilon(x, t)$	Strain of a string/beam/belt
$\varepsilon(t)$	Approximated strain based on the quasi-static stretch assumption
$\varepsilon_X, \varepsilon_Y, \varepsilon_{XY}$	Strain components of a plate
$\varepsilon_X^0, \varepsilon_Z^0, \varepsilon_{XZ}^0$	Strain components for the mid-surface of a plate
$\eta(x, t)$	Transverse vibration (i.e., vibration along the k-axis)
$\bar{\eta}$	Dimensionless longitudinal vibration
$\eta(x, y, t)$	Transverse vibration of a plate
$\theta(x, t)$	Angle of rotation of the cross section of a Timoshenko beam
$\theta_1(t), \theta_2(t), \theta_3(t)$	Angular displacements of the driven pulley, the driving pulley, and the accessory, respectively, in a pulley–belt system
$\boldsymbol{\theta}$	Vector of unknown parameters
$\hat{\theta}$	Estimated vector of unknown parameters
$\tilde{\theta}$	Parameter estimation error
θ_a	Angle from the i-axis to the normal direction of the fibers of a lamina
κ	Shape factor of a Timoshenko beam
κ_X, κ_Y	Curvature modifications of the mid-surface of a plate
κ_{XY}	Torsion of the mid-surface of a plate
μ	Viscosity coefficient
μ_X, μ_Z	Viscosity coefficient of a viscoelastic plate
μ_1, μ_2	Viscosity coefficients of a Burger model
ρ	Mass density
σ	Detuning parameter used for analyzing instability regions
$\sigma(x, t)$	Stress of a string/beam/belt
$\sigma_X, \sigma_Y, \sigma_{XY}$	Stress resultants in i- and k-axes
σ_{nl}	Non-local stress of a nanobeam
σ^n	Strain of the nth layer of a laminate composite material
τ_0, τ_1	Timescales used in the multiple scale method
$\tau(x, t)$	Shear stress of a Timoshenko beam
υ	Poisson's number of an isotropic plate

$\upsilon_{\mathrm{XZ}}, \upsilon_{\mathrm{ZX}}$	Poisson's numbers for the main orthotropic directions of an orthotropic plate
$\Phi(x, z, t)$	Airy stress function
$\varphi_i(x)$	Basis functions of the vibration of a string/beam/belt ($i = 1, 2, 3, \ldots$)
$\varphi_i(x, z)$	Basis functions of the vibration of a plate/membrane ($i = 1, 2, 3, \ldots$)
$\phi_{ij}(t)$	Stress variables of the approximation of the Airy stress function of a plate
$\phi_{\mathrm{X}}, \phi_{\mathrm{Z}}$	Creep functions of a viscoelastic plate
χ	Orthotropy factor of a plate
$\psi(x)$	Weight function
ψ_{XZ}	Relaxation function of a Zener-type viscoelastic element
$\psi_{\mathrm{H}}(x)$	Heaviside step function
Ω_T	Angular frequency of a varying tension
Ω_V	Angular frequency of a varying velocity
ω	Natural frequency of the vibration of a string/beam/belt

Chapter 1
Introduction

Abstract This chapter introduces industrial examples and applications of axially moving systems (AMSs), motivating the research on AMSs. A brief overview of the dynamics and control of AMSs is presented.

Keywords Industrial application · Axially moving system · Dynamics · Vibration control

Axially moving systems (AMSs) that constitute a gyroscopic distributed parameter system have an essential role in various engineering systems, particularly high-speed continuous manufacturing processes. Since the 1950s of the twentieth century, these systems' vibrations have been a fascinating research subject: Dynamic behaviors of diverse axially moving systems have been discovered, and control strategies for vibration suppression have been developed. With the development of complicated hybrid systems and nonhomogeneous materials, investigations on the dynamics and control of axially moving systems are more demanding. This chapter first introduces applications of axially moving systems and motivation to investigate AMSs. Additionally, a brief overview of studies on the dynamics and control of AMSs is presented.

1.1 Industrial Axially Moving Systems

Axially moving systems are noteworthy components widely applied in various industrial sectors (Figs. 1.1, 1.2, 1.3, 1.4, 1.5, 1.6, and 1.7), such as continuous manufacturing processes, transport systems, power transmission systems, and lifting/lowing systems. In the manufacturing fields, the roll-to-roll system, which is regarded as an axially moving system, is an essential fabricating technology platform used in numerous packaging, coating, printing, and laminating applications. The manufacturing processes for fabrics (Fig. 1.1) and rods (Fig. 1.2), zinc galvanization lines (Fig. 1.3), and nanoscale metal printing processes for marking electronic devices are

© The Author(s), under exclusive license to Springer Nature Singapore Pte Ltd. 2022
K.-S. Hong et al., *Control of Axially Moving Systems*,
https://doi.org/10.1007/978-981-16-2915-0_1

Fig. 1.1 Fabric manufacturing process (https://www.piqsels.com/en/search?q=production+line&page=3)

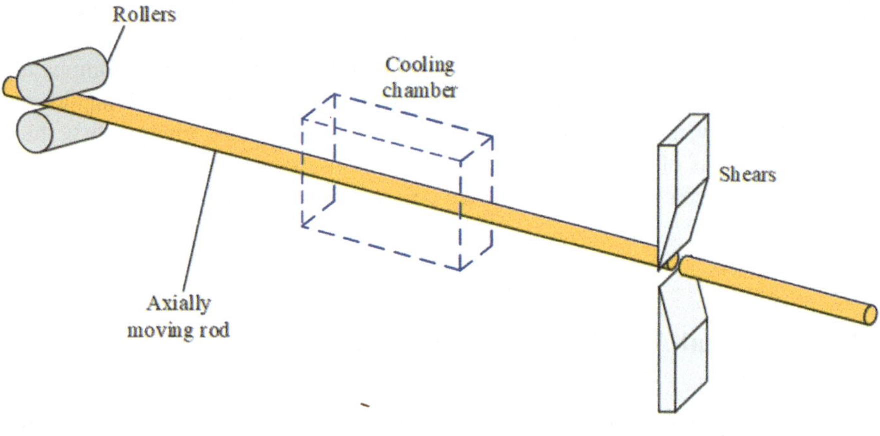

Fig. 1.2 Rod manufacturing process

typical examples. In the transport sector, cables of the aerial tramway and cables in elevator systems (Fig. 1.4) can be considered AMSs. The AMS can also exist as belts in power transmission systems or linear guides with belt transmission (Fig. 1.5). In addition to these examples of AMSs with two supported boundaries, an AMS model of one free boundary can also be used to describe the hosting cables in cranes

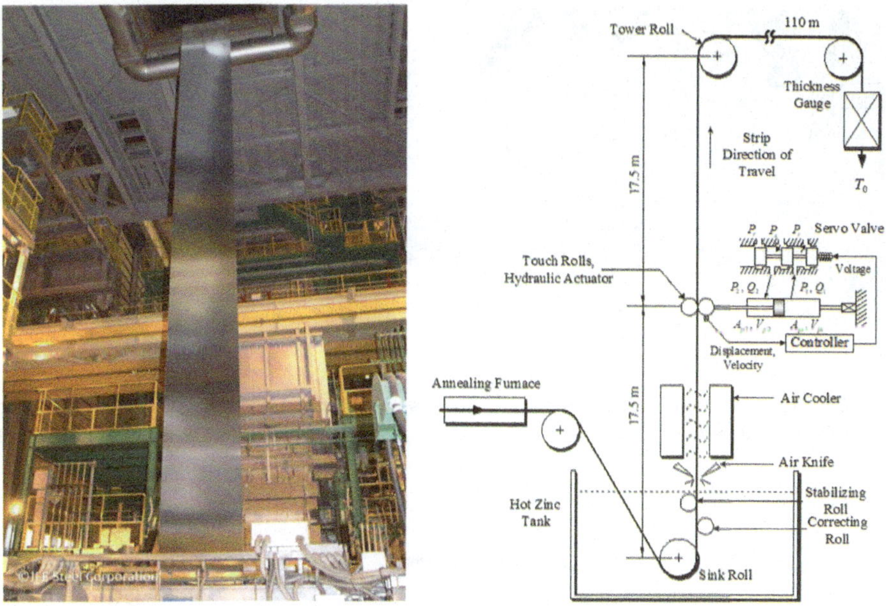

a b

Fig. 1.3 Hot-dip galvanizing line: **a** hot-dip galvanizing line inside a factory (https://tc.ifac-control.org/6/2/gallery/jfe-steel-corporation) **b** schematic of a hot-dip galvanizing line (Hong et al. 2004)

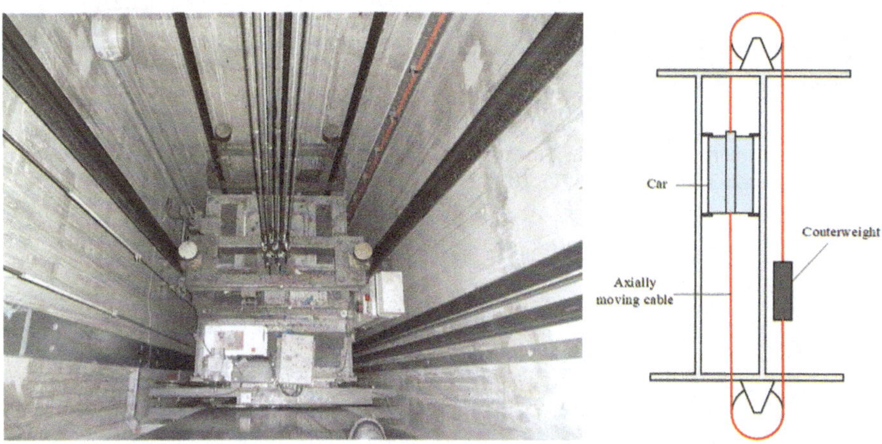

a b

Fig. 1.4 Lift elevator system: **a** cable system of an elevator (https://commons.wikimedia.org/wiki/File:Traction_lift_design.jpg), **b** schematic of an elevator system

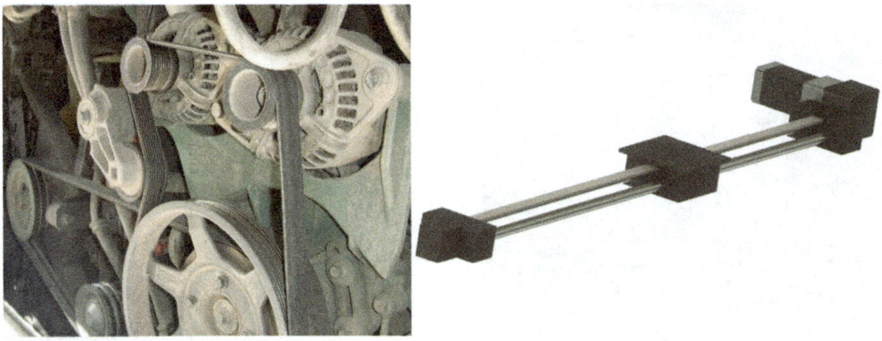

a b

Fig. 1.5 Axially moving systems in power transmission systems: **a** belt driver system (https://commons.wikimedia.org/wiki/File:Belt_drive_systen_01.JPG), **b** a linear guide with belt transmission

a b

Fig. 1.6 Construction tower crane: **a** tower cranes operating at a construction site (https://www.flickr.com/photos/26344495@N05/50125013768), **b** schematic of a construction tower crane

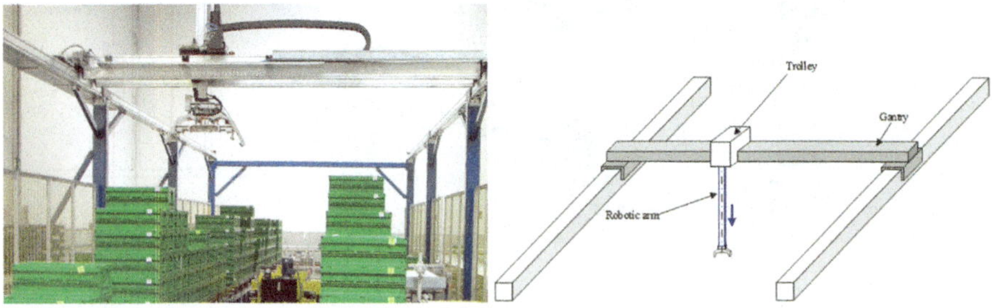

a b

Fig. 1.7 Gantry robot: **a** gantry robot operating inside a factory (https://commons.wikimedia.org/wiki/File:Robot_Portico_tecno-840.jpg), **b** schematic of a gantry robot during the lifting/lowing processes (Fig. 1.6) and prismatic joints in a robotic system (Fig. 1.7).

In many applications, mechanical vibrations that occur within AMSs are perceived as the principal factors that degrade the systems' performance and productivity. For instance, the vibrations of strips in the zinc galvanization line (Fig. 1.3) reduce the uniformity of the zinc layer coating on the strip surface and waste zinc consumption; the oscillations of the hosting cables of a crane (Fig. 1.6) produce errors in the positioning process of the payload. Accordingly, to optimize the design and improve the efficiency of AMSs, various investigations on the vibrations of these systems have been conducted.

1.2 Dynamics of Axially Moving Systems

Axially moving material constitutes a continuous element under tension, as depicted in Fig. 1.8, wherein the vibrations along the *i*-axis, *j*-axis, and *k*-axis are defined as the longitudinal vibration, lateral vibration, and transverse vibration, respectively. Depending on the system's flexibility and geometric parameters, the mathematical models of AMSs can be categorized into four models: string model, beam model, belt model, and plate model (Pham and Hong 2020).

The string is the most fundamental continuous element that is utilized to model an AMS without considering the bending stiffness of the material (i.e., fibers in technique textile-manufacturing processes, cables in aerial tramways, etc.) (Sack 1954; Mahalingam 1957; Zhu and Mote 1994; Zhao et al. 2008; Yang et al. 2017). In contrast to the string model, the beam model addresses the bending stiffness [i.e., Euler–Bernoulli beam (Wickert 1992; Fung et al. 1998; Özkaya and Pakdemirli 2002)], and further includes the rotary inertia effect [i.e., Rayleigh beam (Ghayesh and Balar 2008; Rezaee and Lotfan 2015)], and more also examines the shear deformation [i.e., Timoshenko beam (Lee et al. 2004; Dung et al. 2016)]. The steel rod in the continuous manufacturing process (Fig. 1.2) and the gantry robot with a flexible robotic arm in a robotic system (Fig. 1.7) are examples of axially moving beams. The belt model accounts for the lateral and longitudinal vibrations (Thurman and Mote 1969; Hong et al. 2004; Ding and Chen 2011) by considering the longitudinal inertia force. An

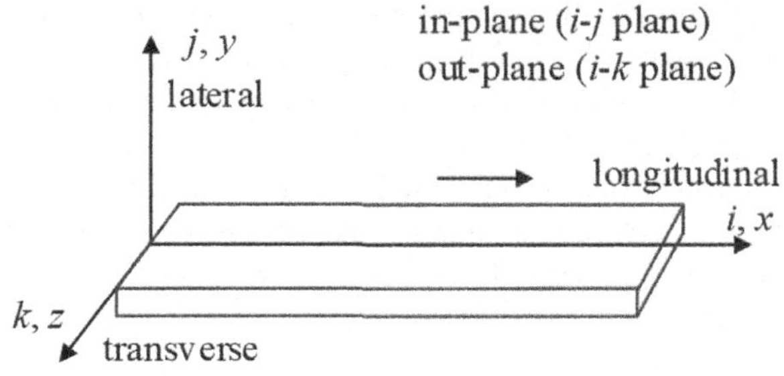

Fig. 1.8 Definition of the vibrations in the three-dimensional coordinate system

Table 1.1 Characteristics of mathematical models for axially moving systems

Axially moving systems	Number of independent spatial variables	Number of PDEs	Differential order
String	1	1	2
Beam			
Euler–Bernoulli beam	1	1	4
Rayleigh beam	1	1	4
Timoshenko beam	1	2	2
Belt	1	2	4
Plate	2	3	4

axially moving belt model can be developed based on the Euler–Bernoulli, Rayleigh, or Timoshenko beam theories. The belt model is utilized to describe systems such as the steel strip in a zinc galvanization line (Fig. 1.3) and the belt used in the linear guide (Fig. 1.5b). The string/beam/belt models are one-dimensional models from the sense of one independent spatial variable x in the mathematical model. In contrast, the plate model consisting of two independent spatial variables (e.g., x and z) is two dimensional. The plate model is regarded as a two-dimensional generalization of the beam model, including the bending and shear stiffness. In addition to the plate, the membrane is a two-dimensional model that considers only tension. The axially moving plate model is generally used to model materials with significant widths (e.g., steel strip in hot-dip galvanizing lines (Fig. 1.4), metal layers in nanoscale metal printing processes) (Marynowski and Kolakowski 1999; Marynowski 2004; Hatami et al. 2006). The mathematical properties of string/beam/belt/plate models are shown in Table 1.1.

From a physical point of view, the AMS is a distributed parameter system with an infinite number of vibration modes. The AMS can be described as an infinite-dimensional system via partial differential equations (PDEs). In contrast to the static distributed parameter system, the AMS has gyroscopic properties because of the Coriolis acceleration. An early study of the dynamics of AMSs was conducted by Sack (1954), who presented the governing equation describing the lateral vibration of a traveling string. Mahalingam (1957) reported a study concerning the natural frequencies of a moving string three years later. These studies are a prerequisite for further works on AMSs. Recently, various investigations on the dynamics of AMSs have been published. Figure 1.9 illustrates the distribution of AMS models, where the dynamics of the string and beam models have received the most attention.

During the early stages of developing studies on AMSs, researchers focused on elastic materials. Subsequently, viscoelastic models were employed to examine the influences of material damping in the AMS, and dynamic models that describe translating viscoelastic materials were derived by Zhao and Chen (2002), Marynowski and Kapitaniak (2002), Li et al. (2003) and Tang et al. (2020). With the development of

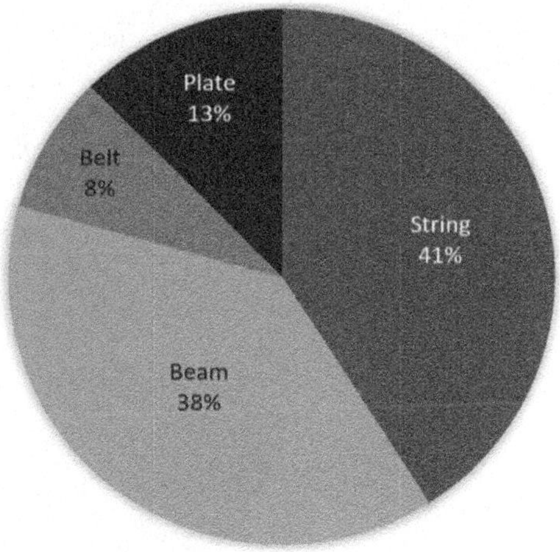

Fig. 1.9 Modeling of axially moving systems (total: 365 papers including dynamics papers)

various materials, the vibrations of moving laminate composite materials or traveling nanoscale systems have also been analyzed; refer to Ghayesh (2011), Arani et al. (2017), and Li et al. (2018). Furthermore, the dynamics of AMSs subject to environmental factors, such as magnetic effects (Hu et al. 2015; Arani et al. 2016), hydrodynamic forces (Lin and Qiao 2008; Wang et al. 2016; Li et al. 2019), and thermal effects (Marynowski and Grabski 2013; Wu and Qi 2020) have also been undertaken. The PDE model is an exact model that is appropriate for precisely describing the dynamics of AMSs. However, the use of the PDE model to analyze the vibrations of AMSs is a challenge due to their infinite number of vibration modes. To overcome this problem, approximate models characterized by a finite-dimensional set of ordinary differential equations (ODEs) were employed; refer to Wickert and Mote (1988), Chen and Yang (2006), and Marynowski and Kolakowski (1999).

1.3 Control of Axially Moving Systems

Control of axially moving systems focuses on suppressing undesired vibrations via passive control or active control: Refer to Hong and Pham (2019) for a comprehensive review. Compared with belt and plate models, vibration control of string and beam models has attracted more research interest; refer to Fig. 1.10. Control techniques using external forces can be classified into two: "in-domain" and "boundary" control techniques. The in-domain control approaches attenuate oscillations of the AMS by using actuators and sensors that are distributed along the material (i.e., distributed control) or located at several points in the middle of the component (i.e., pointwise

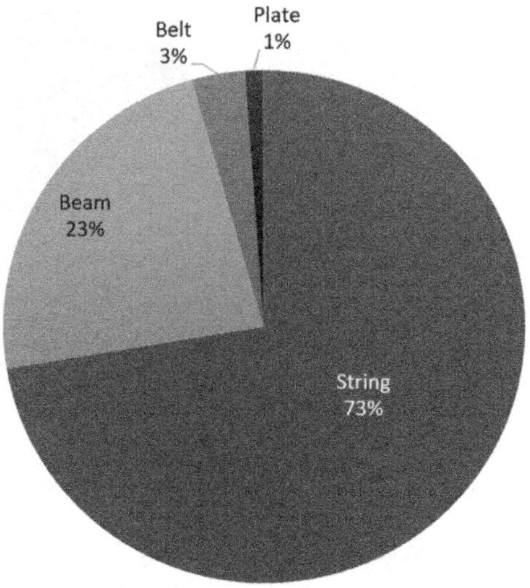

Fig. 1.10 Models used to control axially moving systems (total: 106 papers)

control) (Ulsoy 1984; Fung and Liao 1995; Tan and Ying 2000; Zhu et al. 2001). Conversely, boundary control approaches use actuators and sensors located at the boundary of a system, which does not interfere with its operation (Hong et al. 2004; Kim and Hong 2009; Nguyen and Hong 2012; Liu et al. 2019; Zhao et al. 2019; Ali et al. 2020). Therefore, boundary control is regarded as a practical control solution for the vibration suppressing problem of AMSs. In several situations, the control technique that implements external forces can impair the surface of a material. Thus, vibration attenuation techniques by regulating the axial velocity have been developed; refer to Nguyen and Hong (2011), Nguyen et al. (2015).

Early studies concerning vibration suppression of AMSs were conducted based on ODE models. Ulsoy (1984) employed the model-based control strategy to design a feedback control scheme for a traveling string. Later, Fung and Liao (1995) reported a robust control law based on variable structure control. Control strategies using an ODE model, unfortunately, can lead to the spillover phenomenon. Therefore, schemes that directly use a PDE in control system design have been developed to improve the control performance and avoid the spillover phenomenon. Control methods such as wave cancelation methods (Tan and Ying 2000; Zhang and Chen 2006), the Lyapunov method (Lee and Mote 1996; Yang et al. 2010; Liu et al. 2018; Zhao et al. 2018, 2019), computed torque control (Fung et al. 2002), and adaptive control (Ngo et al. 2009; Nguyen et al. 2018; Guo et al. 2018; Yue 2021) have been implemented for the vibration suppression of AMSs. The distribution of the control methods applied to AMSs is depicted in Fig. 1.11.

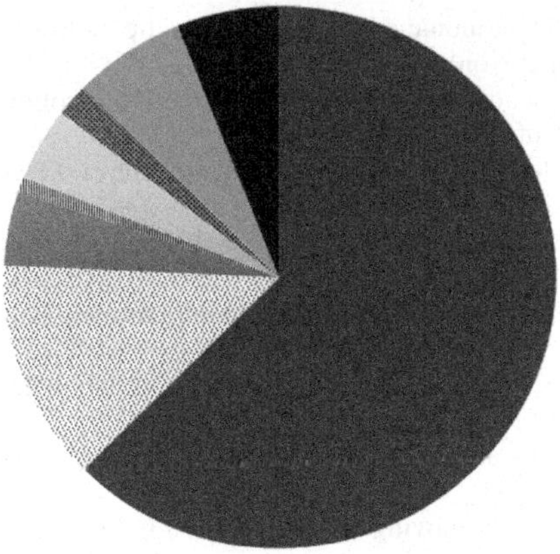

Fig. 1.11 Control methods applied to axially moving systems (106 papers)

References

Ali S, Hawwa MA, Hardt DE (2020) Vibration suppression of an axially moving web in a multi-span roll-to-roll microcontact printing system. J Vib Eng Technol 8(1):35–46

Arani AG, Haghparast E, Zarei HB (2016) Nonlocal vibration of axially moving graphene sheet resting on orthotropic visco-Pasternak foundation under longitudinal magnetic field. Phys B 495:35–49

Arani AG, Haghparast E, Zarei HB (2017) Vibration analysis of functionally graded nanocomposite plate moving in two directions. Steel Compos Struct 23(5):529–541

Chen LQ, Yang XD (2006) Transverse nonlinear dynamics of axially accelerating viscoelastic beams based on 4-term Galerkin truncation. Chaos Soliton Fract 27(3):748–757

Ding H, Chen LQ (2011) Equilibria of axially moving beams in the supercritical regime. Arch Appl Mech 81(1):51–64

Ding H, Tan X, Zhang GC, Chen LQ (2016) Equilibrium bifurcation of high-speed axially moving Timoshenko beams. Acta Mech 227(10):3001–3014

Fung RF, Liao CC (1995) Application of variable-structure control in the nonlinear string system. Int J Mech Sci 37(9):985–993

Fung RF, Lu PY, Tseng CC (1998) Non-linearly dynamic modelling of an axially moving beam with a tip mass. J Sound Vib 218(4):559–571

Fung RF, Wu JW, Lu PY (2002) Adaptive boundary control of an axially moving string system. J Vib Acoust 124(3):435–440

Ghayesh MH (2011) On the natural frequencies, complex mode functions, and critical speeds of axially traveling laminated beams: parametric study. Acta Mech Solida Sin 24(4):373–382

Ghayesh MH, Balar S (2008) Non-linear parametric vibration and stability of axially moving visco-elastic Rayleigh beams. Int J Solids Struct 45(25–26):6451–6467

Guo F, Luo F, Liu Y, Wu Y (2018) Adaptive output feedback boundary control for a class of axially moving system. ET Contr Theory Appl 13(2):213–221

Hatami S, Azhari M, Saadatpour MM (2006) Exact and semi-analytical finite strip for vibration and dynamic stability of traveling plates with intermediate supports. Adv Struct Eng 9(5):639–651

Hong K-S, Pham P-T (2019) Control of axially moving systems: A review. Int J Control Autom 17(12):2983–3008

Hong K-S, Kim CW, Hong KT (2004) Boundary control of an axially moving belt system in a thin-metal production line. Int J Control Autom 2(1):55–67

Hu YD, Hu P, Zhang JZ (2015) Strongly nonlinear subharmonic resonance and chaotic motion of axially moving thin plate in magnetic field. J Comput Nonlin Dyn 10(2):021010

Kim CS, Hong K-S (2009) Boundary control of container cranes from the perspective of controlling an axially moving string system. Int J Control Autom 7(3):437–445

Lee SY, Mote CD (1996) Vibration control of an axially moving string by boundary control. J Dyn Syst-T Asme 118(1):66–74

Lee U, Kim JH, Oh HM (2004) Spectral analysis for the transverse vibration of an axially moving Timoshenko beam. J Sound Vib 271(3–5):685–703

Li YH, Gao Q, Jian KL, Yin XG (2003) Dynamic responses of viscoelastic axially moving belt. Appl Math Mech-Engl 24(11):1348–1354

Li YH, Wang L, Yang EC (2018) Nonlinear dynamic responses of an axially moving laminated beam subjected to both blast and thermal loads. Int J Nonlinear Mech 101:56–67

Li H, Lang T, Liu Y, Li J (2019) Nonlinear vibrations and stability of an axially moving plate immersed in fluid. Acta Mech Solida Sin 32(6):737–753

Lin W, Qiao N (2008) Vibration and stability of an axially moving beam immersed in fluid. Int J Solids Struct 45(5):1445–1457

Liu Y, Guo F, He X, Hui Q (2018) Boundary control for an axially moving system with input restriction based on disturbance observers. IEEE Trans Syst Man Cybern Syst 49(11):2242–2253

Mahalingam S (1957) Transverse vibrations of power transmission chains. Br J Appl Phys 8(4):145

Marynowski K (2004) Non-linear vibrations of an axially moving viscoelastic web with time-dependent tension. Chaos Soliton Fract 21(2):481–490

Marynowski K, Grabski J (2013) Dynamic analysis of an axially moving plate subjected to thermal loading. Mech Res Commun 51:67–71

Marynowski K, Kapitaniak T (2002) Kelvin-Voigt versus Burgers internal damping in modeling of axially moving viscoelastic web. Int J Nonlinear Mech 37(7):1147–1161

Marynowski K, Kolakowski Z (1999) Dynamic behaviour of an axially moving thin orthotropic plate. J Theor Appl Mech 1(37):109–128

Ngo QH, Hong K-S, Jung IH (2009) Adaptive control of an axially moving system. J Mech Sci Techno l 23(11):3071–3078

Nguyen QC, Hong K-S (2011) Stabilization of an axially moving web via regulation of axial velocity. J Sound Vib 330(20):4676–4688

Nguyen QC, Hong K-S (2012) Simultaneous control of longitudinal and transverse vibrations of an axially moving string with velocity tracking. J Sound Vib 331(13):3006–3019

Nguyen QC, Le TH, Hong K-S (2015) Transverse vibration control of axially moving web systems by regulation of axial tension. Int J Control Autom 13(3):689–696

Nguyen QC, Piao M, Hong K-S (2018) Multivariable adaptive control of the rewinding process of a roll-to-roll system governed by hyperbolic partial differential equations. Int J Control Autom 16(5):2177–2186

Ozkaya E, Pakdemirli M (2002) Group—theoretic approach to axially accelerating beam problem. Acta Mech 155(1–2):111–123

Pham P-T, Hong K-S (2020) Dynamic models of axially moving systems: a review. Nonlinear Dynam 100(1):315–349

Rezaee M, Lotfan S (2015) Non-linear nonlocal vibration and stability analysis of axially moving nanoscale beams with time-dependent velocity. Int J Mech Sci 96-97:36–46

Sack RA (1954) Transverse oscillations in travelling strings. Br J Appl Phys 5(6):224

Tan CA, Ying S (2000) Active wave control of the axially moving string: theory and experiment. J Sound Vib 236(5):861–880

Tang JL, Liu JK, Huang JL (2020) Nonlinear dynamics of high-dimensional models of in-plane and out-of-plane vibration in an axially moving viscoelastic beam. Appl Math Model 79:161–179

Thurman AL, Mote CD (1969) Free, periodic, nonlinear oscillation of an axially moving strip. J Appl Mech 36(1):83–91

Ulsoy AG (1984) Vibration control in rotating or translating elastic-systems. Mech Eng 106(2):90–90

Wang YQ, Huang XB, Li J (2016) Hydroelastic dynamic analysis of axially moving plates in continuous hot-dip galvanizing process. Int J Mech Sci 110:201–216

Wickert JA (1992) Nonlinear vibration of a traveling tensioned beam. Int J Nonlinear Mech 27(3):503–517

Wickert JA, Mote CD (1988) Linear transverse vibration of an axially moving string-particle system. J Acoust Soc Am 84(3):963–969

Wu Q, Qi G (2020) Homoclinic bifurcations and chaotic dynamics of non-planar waves in axially moving beam subjected to thermal load. Appl Math Model 83:674–682

Yang XD, Lim CW, Liew KM (2010) Vibration and stability of an axially moving beam on elastic foundation. Adv Struct Eng 13(2):241–247

Yang XD, Wu H, Qian YJ, Zhang W, Lim CW (2017) Nonlinear vibration analysis of axially moving strings based on gyroscopic modes decoupling. J Sound Vib 393:308–320

Yue X (2021) Adaptive fault-tolerant control of an axially moving system with time-varying constraints. Asian J Control 23(1):331–340

Zhang W, Chen LQ (2006) Vibration control of an axially moving string system: wave cancellation method. Appl Math Comput 175(1):851–863

Zhao WJ, Chen LQ (2002) A numerical algorithm for non-linear parametric vibration analysis of a viscoelastic moving belt. Int J Nonlinear Sci 3(2):139–144

Zhao WJ, Chen LQ (2008) Iterative algorithm for axially accelerating strings with integral constitutive law. Acta Mech Solida Sin 21(5):449–456

Zhao Z, He X, Ren Z, Wen G (2018) Output feedback stabilization for an axially moving system. IEEE Trans Syst Man Cybern Syst 49(12):2374–2383

Zhao Z, Ma Y, Liu G, Zhu D, Wen G (2019) Vibration control of an axially moving system with restricted input. Complexity. https://doi.org/10.1155/2019/2386435

Zhu WD, Mote CD (1994) Free and forced response of an axially moving string transporting a damped linear-oscillator. J Sound Vib 177(5):591–610

Zhu WD, Ni J, Huang J (2001) Active control of translating media with arbitrarily varying length. J Vib Acoust 123(3):347–358

Chapter 2
String Model

Abstract This chapter discusses the mathematical modeling of axially moving strings. Various partial differential equations of moving strings, including linear and nonlinear equations, are derived. An approximated model based on the Galerkin method is also explored. The dynamics issues, including stability, bifurcation, and chaos for axially moving strings, are thoroughly analyzed.

Keywords Axially moving string · Modeling · Lateral vibration · Dynamics · Approximate model · Stability · Vibration analysis · Bifurcation · Chaos

This chapter discusses the mathematical modeling of axially moving strings. Understanding the dynamics of a moving string is the first step in understanding axially moving systems (AMSs). Various partial differential equations of moving strings, including linear and nonlinear equations, are derived. An approximated model representing axially moving string is then explored based on the Galerkin method. Additionally, solutions for lateral vibration (both approximate solutions and closed-form solutions) are presented. Other dynamical issues, including stability, bifurcation, and chaos, are also discussed.

2.1 Introduction

The string model is recognized as the most fundamental model for describing AMSs. The string model considers the lateral vibration of a material under tension with negligible bending stiffness. This model is characterized by one partial differential equation that predicts the dynamics of the lateral vibrations of AMSs. In early studies on strings, Sack (1954) and Mahalingam (1957) employed a linear string model to model traveling strings' dynamics. Later, Mote (1966) and Bapat and Srinivasan (1967) developed a nonlinear partial differential equation, wherein the large deflection was considered, whereas Pakdemirli et al. (1994) addressed strings with axial acceleration. An axially moving string with varying length was also considered in Fung et al. (1998a, b) and a series of studies by Zhu and Ni (2000) and Zhu et al. (2001). In addition to elastic strings, the dynamics of axially moving viscoelastic

© The Author(s), under exclusive license to Springer Nature Singapore Pte Ltd. 2022
K.-S. Hong et al., *Control of Axially Moving Systems*,
https://doi.org/10.1007/978-981-16-2915-0_2

strings were investigated based on various viscoelastic models (Li et al. 2003; Zhang et al. 2007; Ghayes and Balar 2008; Ghayesh and Moradian 2011).

Although the string model considers only tension, it is useful for describing axially moving materials that are entirely flexible concerning bending. In engineering, the string model can be utilized for modeling axially moving components in many systems. There are technical textiles, papers, and fibers in continuous manufacturing processes, cables in winding machines, and narrow belts and chains in power transmission systems. We can consider axially moving materials with minor bending stiffness as axially moving strings by disregarding the flexural rigidity: Hoisting cable(s) in container crane systems and elevator systems are typical examples.

2.2 Dynamic Models of Axially Moving Strings

2.2.1 Linear Model

Figure 2.1 depicts an axially moving string model, wherein l is the string length, v denotes the axial velocity, and $w(x, t)$ indicates the lateral vibration. The length l and the axial velocity v_0 are assumed to be constant, whereas the lateral vibration is a function of both the spatial variable and the temporal variable, x and t, respectively. The string is stretched to the initial tension T_0 and simply supported at both ends by rollers.

The mathematical model of the axially moving string can be obtained based on D'Alembert's principle or the extended Hamilton principle. In this chapter, we use the extended Hamilton principle to derive the equation of motion of the string. Notably, $(\cdot)_t$ and $(\cdot)_x$ are employed to denote the partial derivative with respect to time t and spatial variable x, respectively, of spatiotemporal functions (e.g., $w(x, t)$). To simplify the notation, we use w instead of $w(x, t)$. The displacement vector and velocity vector (i.e., $\mathbf{r}_{\mathrm{d}}(x, t)$ and $\mathbf{r}_{\mathrm{v}}(x, t)$, respectively) of a unit element of the string at x are given by

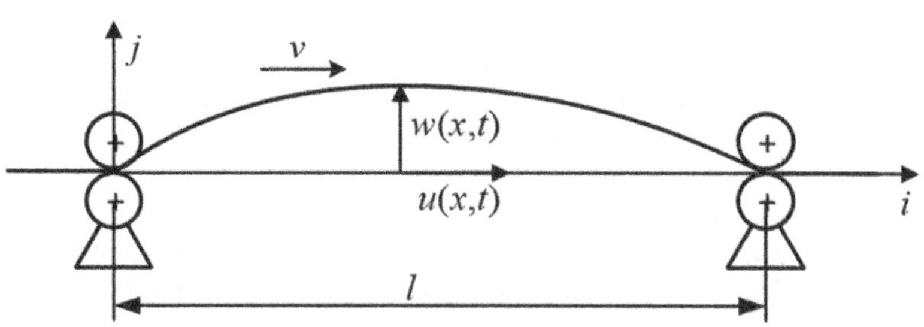

Fig. 2.1 Axially moving string

$$\mathbf{r}_{\mathrm{d}}(x,t) = x\mathbf{i} + w\mathbf{j}, \tag{2.1}$$

$$\mathbf{r}_{\mathrm{v}}(x,t) = \frac{\mathrm{D}\mathbf{r}_{\mathrm{d}}(x,t)}{\mathrm{D}t} = v_0\mathbf{i} + \frac{\mathrm{D}w}{\mathrm{D}t}\mathbf{j}, \tag{2.2}$$

where **i** and **j** are unit vectors in the directions of the *i*-axis and the *j*-axis, respectively. Notably, the material derivative, i.e., $\mathrm{D}(\cdot)/\mathrm{D}t = (\cdot)_t + v_0(\cdot)_x$, is used instead of the time derivative to describe the time rate of change of the displacement of a material element because of the gyroscopic property of the moving string. The kinetic energy K and potential energy U of the considered system are given by

$$K = \frac{1}{2}\rho A \int_0^l \mathbf{r}_{\mathrm{v}} \cdot \mathbf{r}_{\mathrm{v}} \mathrm{d}x = \frac{1}{2}\int_0^l \rho A\left[v_0^2 + \left(\frac{\mathrm{D}w}{\mathrm{D}t}\right)^2\right]\mathrm{d}x, \tag{2.3}$$

$$U = \int_0^l T(x,t)\varepsilon(x,t)\mathrm{d}x \tag{2.4}$$

where ρ is the mass density, A denotes the string's cross-sectional area, $T(x, t)$ is the tension, and $\varepsilon(x, t)$ indicates the distributed strain. To derive the distributed strain, we consider the deflection of an element on the string. Due to the lateral and longitudinal displacements, an element with initial length Δx is deformed into

$$\Delta s = \sqrt{\left(\frac{\partial w}{\partial x}\Delta x\right)^2 + \left(\Delta x + \frac{\partial u}{\partial x}\Delta x\right)^2}, \tag{2.5}$$

wherein u is the longitudinal vibration. The strain, therefore, is expressed as

$$\begin{aligned}\varepsilon(x,t) &= \frac{\Delta s - \Delta x}{\Delta x} = \frac{\sqrt{\left(\frac{\partial w}{\partial x}\Delta x\right)^2 + \left(\Delta x + \frac{\partial u}{\partial x}\Delta x\right)^2} - \Delta x}{\Delta x} \\ &= \sqrt{w_x^2 + (1+u_x)^2} - 1.\end{aligned} \tag{2.6}$$

If the longitudinal vibration is ignored, the strain is rewritten as follows:

$$\varepsilon(x,t) = \sqrt{w_x^2 + 1} - 1. \tag{2.7}$$

We assume that the initial tension T_0 is significant, and the lateral vibration w is sufficiently small. It follows that the tension $T(x, t)$ can be approximated by the initial tension during the motion, i.e., $T(x, t) = T_0$. Applying the Taylor series expansion for $\varepsilon(x, t)$, i.e.,

$$\varepsilon(x,t) = -1 + \sqrt{w_x^2 + 1} \approx -1 + 1 + \frac{1}{2}w_x^2 - \frac{1}{8}w_x^4 + \frac{1}{16}w_x^6 - \ldots = \frac{1}{2}w_x^2 + O(w_x^4), \tag{2.8}$$

and eliminating the higher-order terms, the potential energy in (2.4) can be simplified further to

$$U = \frac{1}{2}\int_0^l T_0 w_x^2 \mathrm{d}x. \tag{2.9}$$

For the free vibration problem, the work done δW is zero. According to the extended Hamilton principle,

$$\int_{t_1}^{t_2} (\delta K - \delta U + \delta W)\mathrm{d}t = 0, \tag{2.10}$$

the following equation is obtained

$$\begin{aligned} 0 &= \int_{t_1}^{t_2} (\delta K - \delta U)\mathrm{d}t \\ &= \frac{1}{2}\int_{t_1}^{t_2} \delta\left(\int_0^l [\rho A v_0^2 + \rho A(w_t + v_0 w_x)^2 - T_0 w_x^2]\mathrm{d}x\right)\mathrm{d}t \\ &= \frac{1}{2}\int_{t_1}^{t_2}\int_0^l \delta L_s \mathrm{d}x\mathrm{d}t \end{aligned} \tag{2.11}$$

where $L_s = \rho A v_0^2 + \rho A(w_t + v_0 w_x)^2 - T_0 w_x^2$. Taking the variation of L_s,

$$0 = \frac{1}{2}\int_{t_1}^{t_2}\int_0^l \left[\frac{\partial L_s}{\partial w}\delta w + \frac{\partial L_s}{\partial w_x}\delta w_x + \frac{\partial L_s}{\partial w_t}\delta w_t\right]\mathrm{d}x\mathrm{d}t, \tag{2.12}$$

and using the procedure for integration by parts yields

$$0 = \frac{1}{2}\int_{t_1}^{t_2}\int_0^l \left[\frac{\partial L_s}{\partial w} - \frac{\partial}{\partial x}\frac{\partial L_s}{\partial w_x} - \frac{\partial}{\partial t}\frac{\partial L_s}{\partial w_t}\right]\delta w \mathrm{d}x\mathrm{d}t$$

$$+\frac{1}{2}\int_{t_1}^{t_2}\frac{\partial L_s}{\partial w_x}\delta w\Big|_{x=0}^{x=l}\mathrm{d}t+\frac{1}{2}\int_0^l\frac{\partial L_s}{\partial w_t}\delta w\Big|_{t=t_1}^{t=t_2}\mathrm{d}x$$

$$=-\int_{t_1}^{t_2}\int_0^l\left[\rho A\left(w_{tt}+2v_0w_{xt}+v_0^2w_{xx}\right)-T_0w_{xx}\right]\delta w\mathrm{d}x\mathrm{d}t$$

$$+\frac{1}{2}\int_{t_1}^{t_2}\frac{\partial L_s}{\partial w_x}\delta w\Big|_{x=0}^{x=l}\mathrm{d}t+\frac{1}{2}\int_0^l\frac{\partial L_s}{\partial w_t}\delta w\Big|_{t=t_1}^{t=t_2}\mathrm{d}x. \tag{2.13}$$

Because the variations at two endpoints t_1 and t_2 are zero, i.e., $\delta w(x, t_1) = \delta w(x, t_2) = 0$, the last term vanishes. Based on the remainder of (2.13), the equation of motion describing the lateral vibration of the string is obtained as

$$\rho A\left(w_{tt}+2v_0w_{xt}+v_0^2w_{xx}\right)-T_0w_{xx}=0, \tag{2.14}$$

and the boundary conditions are

$$w(0,t)=0,\quad w(l,t)=0. \tag{2.15}$$

The partial differential equation in (2.14) constitutes the fundamental equation of motion that describes an axially moving string. The terms on the left side of (2.14) (from left to right) are associated with the lateral acceleration, Coriolis force (due to concurrent longitudinal and lateral motions), centrifugal acceleration, and the unit element's tension of the string at x. Equation (2.14) can be rewritten in dimensionless form as follows:

$$\overline{w}_{\overline{t}\overline{t}}+2\overline{v}_0\overline{w}_{\overline{x}\overline{t}}-(1-\overline{v}_0^2)\overline{w}_{\overline{x}\overline{x}}=0, \tag{2.16}$$

where

$$\overline{w}=\frac{w}{l},\ \overline{x}=\frac{x}{l},\ \overline{t}=t\sqrt{\frac{T_0}{\rho Al^2}},\ \overline{v}_0=v_0\sqrt{\frac{\rho A}{T_0}}. \tag{2.17}$$

2.2.2 *Strings with Time-Varying Velocity*

The previously considered problem assumes that the axial velocity is constant. Particularly, accelerating and decelerating motions tend to affect the vibrational behaviors of axially moving systems severely. Therefore, the cases in which the moving velocity of the string is prescribed as a function of time should be considered. There is not much difference in deriving the mathematical model of the string with varying velocity and deriving that with constant velocity. With the velocity depending on time $v(t)$, the term inside the first integral in (2.13) is expressed as follows:

$$\begin{aligned}\frac{\partial L_s}{\partial w}-\frac{\partial}{\partial x}\frac{\partial L_s}{\partial w_x}-\frac{\partial}{\partial t}\frac{\partial L_s}{\partial w_t} &= -2\frac{\partial}{\partial x}[\rho A v(t)(w_t+v(t)w_x)-T_0 w_x] \\ &\quad -2\frac{\partial}{\partial t}[\rho A(w_t+v(t)w_x)] \\ &= -2\rho A v(t)(w_{xt}+v(t)w_{xx})+2T_0 w_{xx} \\ &\quad -2\rho A(w_{tt}+\dot{v}(t)w_x+v(t)w_{xt}) \\ &= 2\left[-\rho A\left(w_{tt}+\dot{v}(t)w_x+2v(t)w_{xt}+v(t)^2 w_{xx}\right)\right. \\ &\quad \left.+T_0 w_{xx}\right]. \end{aligned} \tag{2.18}$$

Hence, the string's equation of motion with time-varying velocity becomes

$$\rho A\left(w_{tt}+\dot{v}(t)w_x+2v(t)w_{xt}+v(t)^2 w_{xx}\right)-T_0 w_{xx}=0, \tag{2.19}$$

where $\dot{(\cdot)}$ denotes the time derivative of a temporal function.

2.2.3 *Large Amplitude Vibration*

In the previously discussed mathematical models, tension T is approximated by the initial tension (i.e., a constant value), which is the result of assuming that the lateral vibration is sufficiently small. Now, we assume that the lateral displacement is significant. This assumption follows the influence of axial deformation on tension. With this assumption, the potential energy of the string becomes

$$U=\int_0^l\left(T_0+\frac{1}{2}A\sigma(x,t)\right)\varepsilon(x,t)\mathrm{d}x \tag{2.20}$$

where $\sigma(x,t)=E\varepsilon(x,t)$ is the strain, E is the Young modulus, and the second term is known as dynamic tension. This constitutive model is based on Hooke's law for continuous media. The term $A\sigma(x,t)$ in (2.20) constitutes the effect of the axial deformation on the potential energy of the string. The use of the Taylor series expansion for $\varepsilon(x,t)$, i.e., (2.8), allows the replacement of the potential energy in (2.20) by

$$U=\int_0^l\left(\frac{1}{2}T_0 w_x^2+\frac{EA}{8}w_x^4\right)\mathrm{d}x. \tag{2.21}$$

Substituting the kinetic energy in (2.3) and potential energy in (2.21) into the extended Hamilton principle, the axially moving string's dynamic model, wherein axial deformation is considered, is obtained as follows:

$$\rho A\left(w_{tt}+2v_0 w_{xt}+v_0^2 w_{xx}\right)-T_0 w_{xx}-\frac{3\mathrm{EA}}{2}w_x^2 w_{xx}=0. \tag{2.22}$$

Equation (2.22) is a nonlinear PDE owing to the presence of the last term. Mathematically, this model is more complicated than the linear model in (2.14). However, (2.22) is useful when considering lateral vibrations with large amplitudes.

2.2.4 *Varying-Length Strings*

The strings with constant length have been considered in the previous sections. In practice, many systems, such as the hoisting cable(s) in the container crane system and elevator system, can be modeled by axially moving strings with varying length. We now consider the case wherein the string's length depends on time, as shown in Fig. 2.2. We consider an axially moving string that has one simply supported end, and another end is attached to end mass m_e. The string's length $l(t)$ changes during the lifting and lowering processes of the mass. It should be noted that the axial velocity and the axial acceleration of the string are $\dot{l}(t)$ and $\ddot{l}(t)$, respectively.

The axial force along the string $P(x, t)$, which arises from the axial acceleration $\ddot{l}(t)$ and gravitational acceleration g, is given as follows (Zhu and Ni 2000):

$$P(x,t)=(m_\mathrm{e}+\rho(l(t)-x))\left(g-\ddot{l}(t)\right). \tag{2.23}$$

The kinetic energy of the considered system is given by

$$K=\frac{1}{2}\int_0^{l(t)}\rho A\left[\dot{l}(t)^2+\left(\frac{\mathrm{D}w}{\mathrm{D}t}\right)^2\right]\mathrm{d}x+\frac{1}{2}m_\mathrm{e}\left[\dot{l}(t)^2+\left(\frac{\mathrm{D}w(l(t),t)}{\mathrm{D}t}\right)^2\right]. \tag{2.24}$$

In (2.24), the first term on the right-hand side is associated with the kinetic energy of the string, whereas the second term represents one of the end masses. The potential energy is given as follows:

$$U=\frac{1}{2}\int_0^{l(t)}P(x,t)w_x^2\mathrm{d}x. \tag{2.25}$$

Note that the integration domain for spatial variable x in the string's kinetic energy and potential energy depends on time due to the time-varying length. Substituting (2.24) and (2.25) into the extended Hamilton principle yields the following equation:

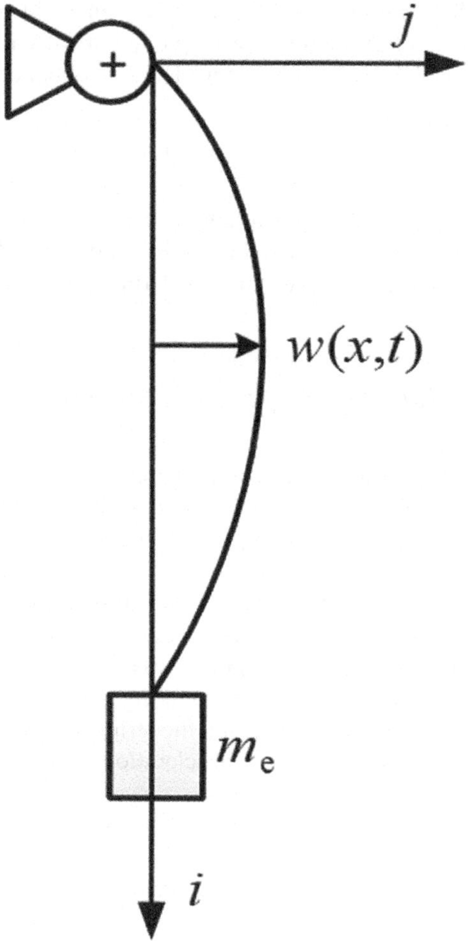

Fig. 2.2 Axially moving string with varying length

$$
\begin{aligned}
0 &= \int_{t_1}^{t_2} (\delta K - \delta U + \delta W)\mathrm{d}t \\
&= \int_{t_1}^{t_2} \left\{ \delta \int_0^{l(t)} \left[\frac{1}{2}\rho A\left(\dot{l}(t)^2 + \left(w_t + \dot{l}(t)w_x\right)^2\right) - \frac{1}{2}P(x,t)w_x^2 \right]\mathrm{d}x \right. \\
&\quad \left. + \delta\left[\frac{1}{2}m_\mathrm{e}\left(\dot{l}(t)^2 + \left(w_t(l(t),t) + \dot{l}w_x(l(t),t)\right)^2\right)\right]\right\}\mathrm{d}t \\
&= \int_{t_1}^{t_2} \left(\delta \int_0^{l(t)} L_\mathrm{s}(x,t)\mathrm{d}x + \delta L_\mathrm{e}(t) \right)\mathrm{d}t
\end{aligned}
\tag{2.26}
$$

where

$$L_s(x,t) = \frac{1}{2}\rho A\left[\dot{l}(t)^2 + w_t^2 + 2\dot{l}(t)w_t w_x + \dot{l}(t)^2 w_x^2\right] - \frac{1}{2}P(x,t)w_x^2, \quad (2.27)$$

$$L_e(t) = \frac{1}{2}m_e\left[\dot{l}(t)^2 + \left(\frac{\mathrm{D}w(l(t),t)}{\mathrm{D}t}\right)^2\right]. \quad (2.28)$$

Owing to the time-varying domain of integration for the spatial variable in (2.26), we must apply the Leibniz integral rule to (2.26). We now take the variation in (2.26) by using the partial integration technique and Leibniz integral rule. Equation (2.26) is expressed as follows:

$$\begin{aligned}
0 &= \int_{t_1}^{t_2}\left(\delta\int_0^{l(t)} L_s \mathrm{d}x + \delta L_e\right)\mathrm{d}t \\
&= \int_{t_1}^{t_2}\left(\int_0^{l(t)} \delta L_s \mathrm{d}x + L_s(l(t),t)\delta l + \delta L_e\right)\mathrm{d}t \\
&= \int_{t_1}^{t_2}\left(\int_0^{l(t)}\left[\frac{\partial L_s}{\partial w}\delta w + \frac{\partial L_s}{\partial w_x}\delta w_x + \frac{\partial L_s}{\partial w_t}\delta w_t\right]\mathrm{d}x + L_s(l(t),t)\delta l\right)\mathrm{d}t \\
&\quad + \int_{t_1}^{t_2}\delta L_e \mathrm{d}t. \quad (2.29)
\end{aligned}$$

Substituting (2.27) and (2.28) into (2.29) yields

$$\begin{aligned}
0 &= \int_{t_1}^{t_2}\int_0^{l(t)}\left\{\left[\rho A\dot{l}(t)\left(w_t + \dot{l}(t)w_x\right) - P(x,t)w_x\right]\delta w_x + \rho A\left(w_t + \dot{l}(t)w_x\right)\delta w_t\right\}\mathrm{d}x\mathrm{d}t \\
&\quad + \int_{t_1}^{t_2} L_s(l(t),t)\delta l \mathrm{d}t - \int_{t_1}^{t_2}\left[m_e\frac{\mathrm{D}^2 w}{\mathrm{D}t^2}\delta w\right]\bigg|_{x=l(t)}\mathrm{d}t. \quad (2.30)
\end{aligned}$$

Using the Leibniz rule gives the following equation:

$$\begin{aligned}
\int_0^{l(t)}\rho A\left(w_t + \dot{l}(t)w_x\right)\delta w_t \mathrm{d}x &= \frac{\partial}{\partial t}\int_0^{l(t)}\rho A\left(w_t + \dot{l}(t)w_x\right)\delta w \mathrm{d}x \\
&\quad - \dot{l}(t)\left[\rho A\left(w_t + \dot{l}(t)w_x\right)\delta w\right]\Big|_{x=l(t)}
\end{aligned}$$

$$-\int_0^{l(t)} \rho A \frac{\partial}{\partial t}\left(w_t + \dot{l}(t) w_x\right) \delta w \mathrm{d}x. \tag{2.31}$$

Integrating (2.31) from t_1 to t_2 and the vanishing variation δw at the two endpoints t_1 and t_2 yields

$$\int_{t_1}^{t_2}\int_0^{l(t)} \rho A\left(w_t + \dot{l}(t) w_x\right)\delta w_t \mathrm{d}x\mathrm{d}t = -\int_{t_1}^{t_2}\left[\rho A \dot{l}(t)\left(w_t + \dot{l}(t) w_x\right)\delta w\right]\Big|_{x=l(t)} \mathrm{d}t$$
$$-\int_{t_1}^{t_2}\int_0^{l(t)} \rho A \frac{\partial}{\partial t}\left(w_t + \dot{l}(t) w_x\right)\delta w \mathrm{d}x\mathrm{d}t. \tag{2.32}$$

By substituting (2.32) into (2.30) and applying the procedure for integration by parts, (2.30) can be rewritten as follows:

$$0 = \int_{t_1}^{t_2} L_{\mathrm{s}}(l(t), t)\delta l \mathrm{d}t$$
$$-\int_{t_1}^{t_2}\int_0^{l(t)} \left[\rho A\left(w_{tt} + \ddot{l}(t) w_x + 2\dot{l}(t) w_{xt} + \dot{l}(t)^2 w_{xx}\right) - (P(x,t) w_x)_x\right]\delta w \mathrm{d}x\mathrm{d}t$$
$$-\int_{t_1}^{t_2} P(x,t) w_x \delta w|_{x=l(t)} \mathrm{d}t + \int_{t_1}^{t_2}\left[\rho A \dot{l}(t)\left(w_t + \dot{l}(t) w_x\right) - P(x,t) w_x\right]\delta w\big|_{x=0} \mathrm{d}t$$
$$-\int_{t_1}^{t_2}\left[m_{\mathrm{e}}\left(w_{tt} + \ddot{l}(t) w_x + 2\dot{l}(t) w_{xt} + \dot{l}(t)^2 w_{xx}\right)\delta w\right]\Big|_{x=l(t)} \mathrm{d}t$$
$$= \int_{t_1}^{t_2} L_{\mathrm{s}}(l(t), t)\delta l \mathrm{d}t$$
$$-\int_{t_1}^{t_2}\int_0^{l(t)} \left[\rho A\left(w_{tt} + \ddot{l}(t) w_x + 2\dot{l}(t) w_{xt} + \dot{l}(t)^2 w_{xx}\right) - (P(x,t) w_x)_x\right]\delta w \mathrm{d}x\mathrm{d}t$$
$$-\int_{t_1}^{t_2}\left[\left[m_{\mathrm{e}}\left(w_{tt} + \ddot{l}(t) w_x + 2\dot{l}(t) w_{xt} + \dot{l}(t)^2 w_{xx}\right) + P(x,t) w_x\right]\delta w\right]_{x=l(t)} \mathrm{d}t$$

$$+\int_{t_1}^{t_2}\left[\rho A\dot{l}(t)\left[w_t+\dot{l}(t)w_x\right]-P(x,t)w_x\right]\delta w\big|_{x=0}\mathrm{d}t. \tag{2.33}$$

It is noted that $\delta w(x, t)$ and $\delta l(t)$ vanish at the endpoints (i.e., at t_1 and t_2). From (2.33), the equations of motion of the string and payload are obtained as follows:

$$\rho A\left(w_{tt}+\ddot{l}(t)w_x+2\dot{l}(t)w_{xt}+\dot{l}(t)^2w_{xx}\right)-\left(P(x,t)w_x\right)_x=0, \tag{2.34}$$

$$\begin{aligned}&m_e\left[w_{tt}(l(t),t)+\ddot{l}(t)w_x(l(t),t)+2\dot{l}(t)w_{xt}(l(t),t)+\dot{l}(t)^2w_{xx}(l(t),t)\right]\\&+P(l(t),t)w_x(l(t),t)=0,\end{aligned} \tag{2.35}$$

and the boundary condition is $w(0, t) = 0$. The dynamic model of the considered system includes a PDE describing the axially moving string and an ODE representing the mass's dynamics. Equations (2.34) and (2.35) describe the free vibration of an axially moving string with varying length. Furthermore, if the string is affected by external forces, such as the damping force, the work done of the system can be given by

$$\delta W=-\int_0^{l(t)}c\left(w_t+\dot{l}(t)w_x\right)\mathrm{d}x \tag{2.36}$$

where c is the damping coefficient. By substituting (2.36) into the extended Hamilton principle, the PDE describing the damped vibration of the string with varying length is written as follows.

$$\rho A\left(w_{tt}+\ddot{l}(t)w_x+2\dot{l}(t)w_{xt}+\dot{l}(t)^2w_{xx}\right)-\left(P(x,t)w_x\right)_x+c\left(w_t+\dot{l}(t)w_x\right)=0. \tag{2.37}$$

2.2.5 *Viscoelastic Strings*

Viscoelastic damping has been broadly used for noise and vibration absorption in diverse industries, such as automobiles, aircraft, propulsion systems, and computer peripherals. Viscoelastic materials can provide a wide range of possibilities for producing desired damping levels in structures and machines. By utilizing the Kelvin viscoelastic model of axially translating strings and the multiple scale method, it has been found that the damping described by the viscoelastic model has no meaningful effect on nonlinear natural frequencies. In contrast, it has a significant effect on the magnitude. Thus, materials with substantial viscoelastic properties can effectively reduce the string's vibration without suffering from significantly reduced natural frequencies (Zhang and Zu 1999).

On the other hand, the contribution of viscosity to the parametrically excited moving string is somewhat complicated. Hou and Zu (2001) conducted vibration analyses of a parametrically excited moving string expressed by the standard linear viscoelastic solid model. It was found that the first non-trivial solution's amplitude decreases with the increase of the viscoelastic parameter, while the second non-trivial solution's amplitude increases with the increase of the viscoelastic parameter. Also, they found that the first non-trivial solution was consistently stable while the second one was unstable, and viscoelasticity revealed the existence of the upper boundary for non-trivial solutions.

The simplest technique to characterize the energy dissipative mechanism due to internal and external friction is the first-order damping, by introducing the velocity-related linear viscous force

$$f(x,t) = c\frac{\mathrm{d}w}{\mathrm{d}t} \tag{2.38}$$

where c is the damping coefficient. The external friction occurred by the viscosity of the surrounding medium plays a vital role in the study of axially moving materials. In (2.38), it is assumed that the surrounding medium applies a resistive force on each string element that is proportional to the rate of change of the string's shape, though not necessarily to the material's actual velocity. If the surrounding medium is in axial motion, independent of the traveling string's axial velocity, one needs to consider the material derivative of (2.38), i.e.,

$$f_\mathrm{e}(x,t) = c_\mathrm{e}\frac{\mathrm{D}w}{\mathrm{D}t} = c_\mathrm{e}(w_t + v_0 w_x) \tag{2.39}$$

where c_e is the external damping coefficient. If the traveling material's internal friction is considered, the material differentiation does not work since the material operates in the co-moving coordinate system. Hence, it is assumed that

$$f_\mathrm{i}(x,t) = c_\mathrm{i}\frac{\partial w}{\partial t} = c_\mathrm{i} w_t \tag{2.40}$$

where c_i is the internal damping coefficient. By adding both external and internal viscous terms into (2.22), one obtains

$$\rho A\left(w_{tt} + 2v_0 w_{xt} + v^2 w_{xx}\right) - T_0 w_{xx} + c_\mathrm{e}(w_t + v_0 w_x) + c_\mathrm{i} w_t - \frac{3}{2}\mathrm{EA}w_x^2 w_{xx} = 0. \tag{2.41}$$

The above linear viscoelastic properties give us a simple way to show the work of dissipation. However, the viscous parameter can only be determined by experiments, since such model is based on the linear assumption. If one considers the viscoelastic effect by axial elongation, then a new one-dimensional viscoelastic model needs to be developed.

When the viscoelastic property of the material denoted in the axial deformation is considered, one differential constitutive relation based on the Kelvin viscoelastic model is

$$\sigma(x, t) = E^* \varepsilon(x, t),$$
$$E^* = E + \mu \frac{\partial}{\partial t} \tag{2.42}$$

where μ is the viscosity coefficient. Replacing Young's modulus E in (2.21) by the differential operator E^* leads to the following governing equation of the axially moving string's lateral vibration (Zhang and Zu 1999)

$$\rho A\left(w_{tt} + 2v_0 w_{xt} + v^2 w_{xx}\right) - T_0 w_{xx} - \frac{3}{2}\mathrm{EA} w_x^2 w_{xx} - \frac{3}{2}\mu\left[w_x^2 w_{xx}\right]_t = 0. \tag{2.43}$$

A general differential linear constitutive law can be expressed as follows.

$$P_\mathrm{V} \sigma(x, t) = Q_\mathrm{V} \varepsilon(x, t) \tag{2.44}$$

where P_V and Q_V are linear differential operators in time (Christensen 1982). For example, Hou and Zu (2001) investigated the standard linear solid model for viscoelasticity formulated by

$$P_\mathrm{S} = 1 + \frac{\mu}{E_1 + E_2}\frac{\partial}{\partial t},\ Q_S = \frac{E_1}{E_1 + E_2}\left(E_2 + \mu\frac{\partial}{\partial t}\right). \tag{2.45}$$

The standard viscoelastic model can be converted to the Kelvin solid model (Hou and Zu 2002) by defining

$$P_\mathrm{K} = 1,\ Q_\mathrm{K} = E + \mu\frac{\partial}{\partial t}. \tag{2.46}$$

Letting $E_1 \to \infty$ and $E_2 = E$, Eq. (2.44) becomes the Maxwell fluid model given by

$$P_\mathrm{M} = 1 + \frac{\mu}{E}\frac{\partial}{\partial t},\ Q_\mathrm{M} = \mu\frac{\partial}{\partial t}. \tag{2.47}$$

Perturbation analysis on axially moving viscoelastic strings shows that either bigger stiffness constant E_1 or lesser stiffness constant E_2 leads to larger vibration amplitude, and increasing the dynamic viscosity μ to a particular value benefits vibration elimination. Another differential constitutive model is the Maxwell–Kelvin fluid model given by

$$P_{\mathrm{MK}} = 1 + \left(\frac{\mu_1}{E_1} + \frac{\mu_2}{E_1 + E_2}\right)\frac{\partial}{\partial t} + \frac{\mu_1\mu_2}{E_1E_2}\frac{\partial^2}{\partial t^2},$$
$$Q_{\mathrm{MK}} = \mu_1\frac{\partial}{\partial t} + \frac{\mu_1\mu_2}{E_2}\frac{\partial^2}{\partial t^2} \tag{2.48}$$

where P_{V} and Q_{V} in (2.44) are replaced by P_{MK} and Q_{MK}, respectively. Zu and Hou (2000) derived a general solution of the nonlinear axially moving string's free nonlinear vibrations and compared the similarities and differences of the various viscoelastic models given by (2.45)–(2.48) via numerical simulations. The magnitudes and frequencies obtained from all four models are close to each other, with neglectable discrepancies. For big stiffness constants, dynamic viscosity, and initial tension, the standard linear solid model of (2.45) forecasts the largest magnitude, which is slightly bigger than that obtained by the Maxwell model or the Maxwell–Kelvin model, while the Kelvin model gives the smallest amplitude. At most moving speeds, the frequency obtained by the Maxwell–Kelvin model is more significant than those projected by the other three models.

The widely used integral constitutive relation characterizing the material viscoelastic property is the Boltzmann superposition principle (Chen et al. 2003; Chen 2005), i.e.,

$$\sigma(x,t) = E_0\varepsilon(x,t) + \int_0^t E_{\mathrm{B}}(t-\tau)\varepsilon(x,\tau)\mathrm{d}\tau \tag{2.49}$$

where $E_{\mathrm{B}}(t)$ is the relaxation function. E_0 is the value at $t = 0$, that is, the material's initial Young's modulus. Replacing the elastic relation $\sigma(x,t) = E\varepsilon(x,t)$ in (2.20) by the integral viscoelastic relation of (2.49), (2.22) becomes the following governing equation.

$$\rho A\left(w_{tt} + 2v_0w_{xt} + v_0^2w_{xx}\right) - T_0w_{xx}$$
$$-\frac{3}{2}E_0Aw_x^2w_{xx} - \frac{3}{2}\left[Aw_x\int_0^t E_{\mathrm{B}}(t-\tau)w_x^2d\tau\right] = 0. \tag{2.50}$$

Chen et al. (2003a, b) and Chen (2005) studied the axially moving viscoelastic string, described by the Boltzmann superposition principle. It is found that viscosity has a significant effect on the amplitude in the steady-state response and the existence of a boundary. For certain viscoelastic materials such as synthetic rubber and synthetic fiber, the Abelian-type relaxed function with a weak singularity depicts the viscoelasticity's features well. Therefore, the fractional differentiation law (an integral formula with a weakly singular kernel) is used to describe those materials. Experiments have illustrated that it is a valuable model for diverse materials such as synthetic rubbers and fibers (Drozdov and Kalamkarov 1996). For such materials, the stress–strain relation is

$$\sigma(x,t) = E_0\varepsilon(x,t) + \mu_0 D_T^\alpha(\varepsilon(x,t)) \tag{2.51}$$

where E_0 is the stiffness constant, μ_0 denotes the dynamic viscosity, and the Riemann–Liouville α order fractional differentiation operator with respect to t is given by

$$D_\mathrm{T}^\alpha(f) = \frac{1}{\Gamma(1-\alpha)}\frac{\mathrm{d}}{\mathrm{d}t}\int_0^t \frac{f(\tau)}{(1-\tau)^\alpha}\mathrm{d}\tau \tag{2.52}$$

where Γ is the gamma function. Substituting (2.51) into (2.20), the following governing equation is derived.

$$\begin{aligned}&\rho A\left(w_{tt} + 2v_0 w_{xt} + v^2 w_{xx}\right) - T_0 w_{xx}\\&- w_{xx}\left[\frac{3}{2}E_0 A w_x^2 + \frac{\mu_0}{2}D_T^\alpha w_x^2\right] - \mu_0 w_x D_T^\alpha(w_x w_{xx}) = 0.\end{aligned} \tag{2.53}$$

Chen et al. (2004) have computationally demonstrated the effects of viscoelastic parameters and frequencies of parametric excitations on the axially moving string's transient responses. Numerical simulations revealed that the transient response's amplitudes decreased with the increase of both the viscoelastic power parameter and the viscoelastic coefficient. In contrast, the transient response's periods decreased with the viscoelastic coefficient's increase.

It should be noted that the damping due to the string's viscoelasticity in both differential and integral models exists only in nonlinear terms, and thus there are no linear contributions. The contributions of the viscoelasticity have been widely investigated in both regular and chaotic motions of axially moving strings (Chen et al. 2004a, b; Chen and Zhang 2005). Adopting one of the above viscoelastic models, the bifurcation and chaos can be detected in the lateral motions of axially moving strings.

2.2.6 *Strings with Non-homogenous Boundaries*

Thus far, most axially moving strings had simply supported boundaries, wherein the boundary conditions were homogeneous. In practice, the boundary conditions, however, are non-homogenous due to the operations of other machine elements or the effects of external excitations. This subsection focuses on the non-homogenous boundary problems of axially moving strings.

In some cases, the lateral motion of the medium is controlled through contact between its edge and one or more stationary guides. Vibrating supports belong to typical non-homogenous boundary problems. Such non-homogenous boundary can be formulated as follows:

$$w(0, t) = a(t), w(l, t) = b(t). \tag{2.54}$$

If $a(t)$ and $b(t)$ can be cast into sinusoidal functions, the linear response of the axially moving string can be obtained by substituting (2.54) into the general linear solutions (Mahalingam 1957).

Pakdemirli and Boyaci (2002) assumed such boundaries undergoing small excitations. Therefore, perturbation methods, such as Lindstedt–Poincaré's method and multiple scale method, could be used in getting the responses of both linear and nonlinear moving materials. Based on perturbation methods, nonlinear vibrations of axially moving multi-supported strings having non-homogenous support conditions have been studied by Yurddas et al. (2013).

In optical and magnetic tape libraries for storing data, the tape's lateral motions arise from the excitation sources, including the wear of packs and rollers or bumps occurring between the tape and guiding flanges. Lateral motion is undesirable because it results in the misalignment of the data tracks and the read/write heads. Kartik and Wickert (2006) have investigated the steady-state forced response of moving media in the presence of edge guides and an imperfect sinusoidal motion. The response has been found by employing a modal analysis procedure specifically for dynamic gyroscopic systems.

The lateral oscillations of traveling strings due to lateral vibrations of support rollers have been studied by van Horssen (2003). A homogeneous boundary condition at $x = 0$ and an inhomogeneous boundary condition at $x = l$ have been investigated. It has been shown that the oscillations can be exactly solved using the Laplace transform method. It should be noted that even if both boundary conditions are inhomogeneous, the Laplace transform method still works.

Since no time variable is involved, axially moving strings with moving particle, elastic or viscous guide, or on elastic foundations belong to homogeneous systems (Wickert and Mote, 1988; Lakshmikumaran and Wickert, 1996; Wickert, 1994; Ghayesh, 2008).

2.3 Approximate Model

In many works about the dynamics of axially moving systems, approximation methods have been pursued to convert the partial differential equations (PDEs), describing the vibrations of the continua, into a low-dimensional system of ordinary differential equations (ODEs) to facilitate the use of specific techniques employed for solving discrete problems. In this section, approximate methods, particularly the Galerkin method, finite element method, finite differential method, and differential quadrature method, will be explored. Approximated models of axially moving strings corresponding to the use of each of the above methods are derived.

The PDE of an axially moving string based on the linear non-damped model with distributed disturbance becomes

$$\rho A\left(w_{tt}+2v_0w_{xt}+v_0^2w_{xx}\right)-T_0w_{xx}=d(x,t) \tag{2.55}$$

or its dimensionless equation becomes

$$\overline{w}_{\overline{tt}}+2\overline{v}\overline{w}_{\overline{xt}}-\left(1-\overline{v}_0^2\right)\overline{w}_{\overline{xx}}=\overline{d}\left(\overline{x},\overline{t}\right). \tag{2.56}$$

Equation (2.56) has an exact explicit solution, contrary to the implicit ones to axially moving beams and plates, which involves bending stiffness. Wickert and Mote (1990) first treated the lateral vibration of a moving string using modal analysis. They obtained an exact closed-form expression for the response to arbitrary excitation and initial conditions.

If nonlinear effects are considered, see the governing Eqs. (2.43), (2.50), and (2.53), the exact explicit solutions cannot be obtained. Wickert and Mote (1990) solved the linear vibration problem of a translating string completely and laid the foundation for studying vibrations of translating strings under complicated constraints or coupling and studying nonlinear vibrations of translating strings by approximate analytical approaches. The most common method, the Galerkin method, is to truncate the PDE into a set of ODEs. Therefore, the Galerkin method is also referred to as the Galerkin truncation or the Galerkin discretization. The Galerkin method is a special form of the weighted residual method with trial functions the same as the weight functions. In the applications to axially moving strings, the eigenfunctions of linear non-translating strings usually serve as both the trial functions and the weight functions.

Now, Eq. (2.22) is revisited to show the truncation procedure of the classical nonlinear axially moving string model. Introduce the dimensionless terms and parameters as follows.

$$\overline{w}=\frac{w}{l},\ \overline{x}=\frac{x}{l},\ \overline{t}=t\sqrt{\frac{T_0}{\rho Al^2}},\ \overline{v}_0=v_0\sqrt{\frac{\rho A}{T_0}},\ k_{\mathrm{E}}=\frac{EA}{T_0}. \tag{2.57}$$

Then, nonlinear Eq. (2.22) can be written as follows.

$$\overline{w}_{\overline{tt}}+2\overline{v}_0\overline{w}_{\overline{xt}}+\left(\overline{v}_0^2-1\right)\overline{w}_{\overline{xx}}-\frac{3}{2}k_{\mathrm{E}}(\overline{w}_{\overline{x}})^2\overline{w}_{\overline{xx}}=0. \tag{2.58}$$

The eigenfunctions of the static string are adopted as trial functions, and the solution of (2.58) is expanded as

$$\overline{w}\left(\overline{x},\overline{t}\right)=\sum_{n=1}^{N}q_n\left(\overline{t}\right)\sin(n\pi\overline{x}), \tag{2.59}$$

where $q_n\left(\overline{t}\right)$ are the generalized displacements. Generally, higher truncation order N leads to more exact result.

Taking the appropriate derivatives and substituting (2.59) into (2.58), one obtains the residual function as follows.

$$\begin{aligned} R(\bar{x},\bar{t}) = & \sum_{n=1}^{N} \ddot{q}_n(\bar{t})\sin(n\pi\bar{x}) + 2\pi\bar{v}_0\sum_{n=1}^{N} n\dot{q}_n(\bar{t})\cos(n\pi\bar{x}) \\ & - \pi^2(\bar{v}_0^2 - 1)\sum_{n=1}^{N} n^2 q_n(\bar{t})\sin(n\pi\bar{x}) \\ & + \frac{3}{2}\pi^4 k_{\mathrm{E}}\left[\sum_{n=1}^{N} n q_n(\bar{t})\cos(n\pi\bar{x})\right]^2 \sum_{n=1}^{N} n^2 q_n(\bar{t})\sin(n\pi\bar{x}). \end{aligned} \tag{2.60}$$

Choose again the eigenfunctions of the static string as weighting functions. Application of the Galerkin method requires

$$\int_0^1 R(\bar{x},\bar{t})\sin(k\pi\bar{x}) = 0, \ (k = 1, 2, \ldots, N). \tag{2.61}$$

Substituting (2.60) into (2.61) yields

$$\ddot{q}_k + 8\bar{v}\sum_{k+n \text{ is odd}}^{N} \frac{nk}{k^2 - n^2}\dot{q}_n + \left(1 - \bar{v}^2\right)k^2\pi^2 q_k = 3\pi^4 k_{\mathrm{E}}\sum_{n=1}^{N}\sum_{j=1}^{N}\sum_{m=1}^{N} d_{knjm}q_n q_j q_m$$
$$(k = 1, 2, \cdots, N) \tag{2.62}$$

where

$$d_{knjm} = \begin{cases} \frac{njmk}{8}, & n + j = \pm(m + k),\ \pm(m - k) \text{ or } n - j = \pm(m + k),\ \pm(m - k), \\ 0, & \text{other}. \end{cases} \tag{2.63}$$

Several investigations for linear systems were reported by choosing a different truncation order N. The Galerkin truncation procedure can also be applied to other types of nonlinear or non-homogeneous equations of axially translating strings.

Fung et al. (1998a, b, 1997) investigated two-term and four-term Galerkin truncations to compute the transient motion of a translating viscoelastic string based on the Boltzmann superposition principle. They pointed out that the vibration frequency depends on the moving velocity and demonstrated the potential occurrence of instability resulting in the parametric resonance. In Chen et al. (2004b), an axially moving viscoelastic string based on a fractional differentiation relation was investigated. In this work, a four-term Galerkin truncation was utilized to study the influences of the viscoelastic parameters and the longitudinal speed on the frequency and the amplitude of transient response. Chen et al. (2003a, 2009) analyzed bifurcation and chaos in the lateral vibrations of AMSs with different models of viscoelasticity via exploiting Galerkin's discretization. It has been demonstrated that even order term truncation is better in the truncation procedure when stationary string eigenfunctions are used as both trial and weight functions. In another study on axially translating nonlinear Kirchhoff elastic strings, Pellicano et al. (2004) utilized an eight-term Galerkin truncation to determine the amplitude–frequency characteristics and the bifurcation diagrams of the considered system.

Besides the abovementioned studies using the stationary string's eigenfunctions, the Galerkin method can be implemented based on other functions, such as the eigenfunctions of moving strings (Zhang et al. 2002; Zhang 2008) and the Hermite functions (Chen and Zhao 2004). With the emergence of nonlinear normal modes, more options for trial and weight functions have been adopted. Yang et al. (2017) used the gyroscopic modes of the linear axially moving string to truncate the nonlinear PDE. Further, the nonlinear modes have been used to obtain their approximate solutions (Liang et al. 2017, 2018). Generally speaking, the sophisticated choices of trial and weight functions yielded superior performance.

Wickert and Mote (1991) further used their speed-adapting modes to improve the effectiveness of the Ritz discretization. Another investigation published by Jia and Parker (2000), however, revealed that the approach proposed by Wickert and Mote cannot assure upper bound approximations and suffers from poor convergence at supercritical speeds. Hence, it is clear that good approximate models make further possible investigations of the axially translating strings with complicated constraints, coupling, nonlinear effect, or other non-homogenous conditions. For example, when treating the axially moving strings with pulsating length, tension, or velocity, the Galerkin truncated model fails to describe the lateral vibrations and stability, especially when the moving speed is high. A translating string, wherein the string's length harmonically varies about a constant mean length, was investigated by Sandilo and van Horssen (2014). This study showed that the Galerkin method could not be utilized to achieve asymptotic approximations on long timescales.

Malookani and van Horssen (2015) suggested that Galerkin truncation cannot be implemented in all cases of moving strings at speed depending harmonically on time. In the relatively high constant speed, the approximate analytical solution was obtained by directly using the multiple scale method and the Laplace transform method. On the contrary, in low speed, the authors first discretized the PDE by using the Fourier series and then utilized the multiple scale method.

Numerical approaches are a valuable tool to investigate nonlinear or non-homogenous problems. In 1976, a numerical method to solve PDEs describing the nonlinear vibrations of an axially translating string was proposed by Moustafa and Salma. They developed an iterative algorithm based on the discretization of spatial variable and temporal variable by the central difference, wherein the error is bounded to any desired value, and the convergence is guaranteed. The simulation results pointed out that the string's maximum tension increases with the longitudinal speed, whereas the fundamental frequency decreased with the longitudinal speed. Chen et al. (2005) investigated frequencies and magnitudes of free nonlinear vibrations of an axially translating string. In this study, a novel finite difference scheme was presented, wherein the stress was assumed to be an unknown auxiliary parameter used to discretize spatial variables in the dynamic model and the viscoelastic model. Accordingly, the PDE model is discretized into a set of differential–algebraic equations that is solved by numerical methods. Numerical simulations showed that vibration's magnitudes and frequencies decreased with the string's longitudinal speed increase. Although only the linear elastic strings with homogeneous boundaries were investigated in their works, the proposed finite difference scheme can be extended to more complicated viscoelastic models and boundary conditions.

A number of approximated analytic methods for nonlinear vibrations have been implemented to analyze free nonlinear vibrations of longitudinally translating strings. Bapat and Srinivasan (1971) applied a direct linearization method to a translating string. Later, Bapat and Srinivasan (1976) used the harmonic balance method to derive the approximated period–tension relationship of one-term Galerkin truncated systems. The harmonic balance method was also utilized to obtain the approximate analytical form of the period–tension relationship in Ghangrekar (1975). In Mote and Thurman (1971), hybrid discretization and perturbation methods were used to present linear and nonlinear modes of an axially translating string. Accordingly, the authors pointed out that the effect of nonlinearities on the fundamental mode increases with the translating speed. In Korde (1985), a time transformation method was implemented to describe the period–amplitude–tension relationships, wherein only the fundamental mode was considered.

Without truncation, there has an alternative direct perturbation approach dealing with PDEs describing axially moving systems directly. This approach has been used in Nayfeh et al. 1992, Pakdemirli 1994, Pakdemirli et al. 1995, and Pakdemirli and Boyaci 1995. According to this approach, these studies also found the discrepancies between schemes for principal, subharmonic, and superharmonic resonances of some other nonlinear systems instead of axially translating strings. In Wickert (1993), the author reported the response of the frictionally damped string and validated the results via simulation results. Additionally, the author pointed out that self-excitation vibration appears below a specific value of the string's moving speed. In another study, Wickert (1993) used the asymptotic method of Krylov, Bogoliubov, and Mitropolsky to investigate a translating string's vibration.

2.4 Dynamic Analysis of String Model

The equilibrium and its stability can be explored based on the linear governing equation. Substituting the solution

$$\overline{w}(\overline{x}, \overline{t}) = \varphi(\overline{x})e^{i\omega\overline{t}} \tag{2.64}$$

into the homogeneous part of (2.56) or the linear part of PDE (2.58) yields the second-order ODE

$$\omega^2\varphi - 2\mathrm{i}\omega\overline{v}_0\varphi_{\overline{x}} + \left(1 - \overline{v}_0^2\right)\varphi_{\overline{x}\overline{x}} = 0. \tag{2.65}$$

By using $\varphi(\overline{x}) = \mathrm{e}^{\lambda\overline{x}}$, the algebraic characteristic equation can be obtained as follows.

$$\left(\overline{v}_0^2 - 1\right)\lambda^2 + 2\mathrm{i}\omega\overline{v}_0\lambda - \omega^2 = 0 \tag{2.66}$$

which yields two eigenvalues

$$\lambda_1 = -\frac{\mathrm{i}\omega}{1 + \overline{v}_0}, \quad \lambda_2 = \frac{\mathrm{i}\omega}{1 - \overline{v}_0}. \tag{2.67}$$

The general solution to (2.66) can be expressed as

$$\varphi(\overline{x}) = C_1\mathrm{e}^{\lambda_1\overline{x}} + C_2\mathrm{e}^{\lambda_2\overline{x}}. \tag{2.68}$$

The fixed supporting conditions of two ends at $\overline{x} = 0$ and $\overline{x} = 1$ require

$$C_1 + C_2 = 0, \tag{2.69}$$

$$C_1\mathrm{e}^{\lambda_1} + C_2\mathrm{e}^{\lambda_2} = 0. \tag{2.70}$$

Substituting (2.69) into (2.70) yields

$$\mathrm{e}^{\lambda_1} = \mathrm{e}^{\lambda_2}. \tag{2.71}$$

Further substituting (2.67) into (2.71) leads to the equation in complex domain

$$\cos\left(\frac{\omega}{1 + \overline{v}_0}\right) = \cos\left(\frac{\omega}{1 - \overline{v}_0}\right), \quad \sin\left(\frac{\omega}{1 + \overline{v}_0}\right) = -\sin\left(\frac{\omega}{1 - \overline{v}_0}\right). \tag{2.72}$$

The above two equations can be satisfied under the following relation

$$\frac{\omega}{1 + \overline{v}_0} = -\frac{\omega}{1 - \overline{v}_0} + 2k\pi \tag{2.73}$$

and the natural frequencies are

$$\omega = \omega_k = k\pi\left(1 - \overline{v}_0^2\right), \quad (k = 1, 2, \ldots). \tag{2.74}$$

The above equation shows that the natural frequencies may decrease with the increasing speed of the axially moving string. Substituting (2.74) back into (2.67) yields

$$\lambda_1 = -\mathrm{i}k\pi(1 - \overline{v}_0), \quad \lambda_2 = \mathrm{i}k\pi(1 + \overline{v}_0). \tag{2.75}$$

Inserting (2.75) into (2.68) and considering (2.69), one can derive the modal functions of the axially moving string as follows.

$$\begin{aligned}\varphi(\overline{x}) &= C_1 e^{\lambda_1 \overline{x}} + C_2 e^{\lambda_2 \overline{x}} = C_1 \cos(k\pi(1-\overline{v}_0)\overline{x}) - C_1 \cos(k\pi(1+\overline{v}_0)\overline{x}) \\ &\quad - \mathrm{i}C_1(\sin k\pi(1+\overline{v}_0)\overline{x}) - \mathrm{i}C_1(\sin k\pi(1-\overline{v}_0)\overline{x}) \\ &= 2C_1 \sin(k\pi\overline{x})\sin(k\pi\overline{v}_0\overline{x}) - 2C_1\mathrm{i}\sin(k\pi\overline{x})\cos(k\pi\overline{v}_0\overline{x}) \\ &= -2C_1\mathrm{i}\sin(k\pi\overline{x})[\mathrm{i}\sin(k\pi\overline{v}_0\overline{x}) + \cos(k\pi\overline{v}_0\overline{x})] \\ &= C_k e^{\mathrm{i}k\pi\overline{v}_0\overline{x}}\sin(k\pi\overline{x})\end{aligned} \tag{2.76}$$

where $C_k = -2\mathrm{i}C_1$.

Natural frequencies (2.74) and modal functions (2.76) both recover to the static string case when $\overline{v}_0 = 0$. Substituting (2.74) and (2.76) into (2.64) yields the solution

$$\overline{w}(\overline{x}, \overline{t}) = C_k e^{\mathrm{i}(k\pi\overline{v}_0\overline{x} + \omega\overline{t})}\sin(k\pi\overline{x}). \tag{2.77}$$

Hence, the exponential modal functions show the traveling wave modes in the axially moving string instead of the standing wave modes in a static string.

2.4.1 Equilibrium Solution

This subsection discusses equilibrium configurations of axially moving strings and the critical speed. The subcritical and the supercritical regimes are defined. The obvious equilibrium is the trivial solution $\overline{w}(\overline{x}, \overline{t}) = \varphi(\overline{x}) = 0$, which satisfies the governing Eqs. (2.58) and (2.65) for both linear and nonlinear cases. The stiffness operator $\widetilde{K} = (1 - v^2)\frac{\partial^2}{\partial \overline{x}^2}$, where $\overline{v} = v\sqrt{\rho A/T_0}$, is positive definite if $\overline{v} < 1$ and is negative definite if $\overline{v} > 1$. The velocity $\overline{v}_{\mathrm{cr}} = 1$ is called the critical value. Hence, the zero equilibrium is stable in the subcritical domain and becomes unstable in the supercritical domain. The detailed stability of the supercritical regime can be found in Parker (1999).

2.4.2 Stability Analysis

Axially moving strings may become unstable due to external excitations or parametric excitations (i.e., periodical variation of their parameter resulting from the tension variation and the axial transport acceleration). In this subsection, the stability conditions of axially moving strings are investigated. The method of multiple scales is applied to the approximate model to analyze the stability of an axially moving string. The multiscale analysis is then directly performed via the PDE model. Consequently, the stability boundaries for principal parametric and combination resonances are determined and demonstrated via the simulation results.

By introducing the dimensionless variables and parameters (2.57), the linear equation of motion of the string with time-varying speed (2.19) becomes

$$\overline{w}_{\overline{t}\overline{t}} + 2\overline{v}\overline{w}_{\overline{x}\overline{t}} + \dot{\overline{v}}\overline{w}_{\overline{x}} + \left(1 - \overline{v}^2\right)\overline{w}_{\overline{x}\overline{x}} = 0. \tag{2.78}$$

The axial tension is assumed time-varying as follows.

$$T = T_0 + T_1 \cos(\Omega_T t). \tag{2.79}$$

Then, the linear equation in dimensionless form of the string with time-varying tension becomes

$$\overline{w}_{\overline{t}\overline{t}} + 2\overline{v}\overline{w}_{\overline{x}\overline{t}} + \left(1 - \overline{v}^2 - \delta\cos\left(\omega\overline{t}\right)\right)\overline{w}_{\overline{x}\overline{x}} = 0 \tag{2.80}$$

where

$$\overline{w} = \frac{w}{l},\ \overline{x} = \frac{x}{l},\ \overline{t} = t\sqrt{\frac{T_0}{\rho A l^2}},\ \overline{v} = v\sqrt{\frac{\rho A}{T_0}},\ \delta = \frac{T_1}{T_0},\ \omega = \Omega_T l\sqrt{\frac{\rho A}{T_0}}. \tag{2.81}$$

If the velocity is assumed a sinusoid variation with a mean value, the governing Eq. (2.78) shows the similar behaviors with that of (2.80). The time-varying tension model (2.80) will be considered in this section.

There are two approaches to study analytically the stability and corresponding boundaries of the vibration. Perturbation methods are applied to the truncated ODE system or directly to the PDE model.

In the truncation approach, the Galerkin method is employed to simplify (2.80). The solution of (2.80) may be expanded into the series of eigenfunctions of the stationary string with two fixed ends as expressed in (2.59). Applying the Galerkin procedure introduced in Sect. 2.3, the one-term truncated ordinary differential equation becomes

$$\ddot{q}_1 + \pi^2\left(1 - \overline{v}^2 - \delta\cos\left(\omega\overline{t}\right)\right)q_1 = 0 \tag{2.82}$$

which is a typical Mathieu equation. The stability boundaries can be obtained by traditional perturbation methods.

Similarly, the two-term truncated coupled ordinary differential equations are

$$\begin{cases} \ddot{q}_1 - \dfrac{16\overline{v}}{3}\dot{q}_2 + \pi^2\left(1 - \overline{v}^2 - \delta\cos\left(\omega\overline{t}\right)\right)q_1 = 0, \\ \ddot{q}_2 + \dfrac{16\overline{v}}{3}\dot{q}_1 + 4\pi^2\left(1 - \overline{v}^2 - \delta\cos\left(\omega\overline{t}\right)\right)q_2 = 0 \end{cases} \tag{2.83}$$

which can be investigated by perturbation methods to locate the stability boundaries.

The axially moving string undergoes flutter if the excitation parameters fall in the unstable region. If only linear terms are considered, the flutter makes the amplitude infinite until broken. However, if the nonlinear terms are involved, the response amplitude in the unstable region becomes finite. If nonlinear terms are caused by the axial deformation and the Kelvin viscoelastic damping, (2.80) becomes

$$\begin{aligned}&\overline{w}_{\overline{t}\overline{t}}+2\overline{v}\overline{w}_{\overline{x}\overline{t}}+c\overline{w}_{\overline{t}}+\left(1-\overline{v}^2-\varepsilon\delta\cos\left(\omega\overline{t}\right)\right)\overline{w}_{\overline{x}\overline{x}}\\&+\varepsilon\frac{3}{2}\overline{k}_{\mathrm{E}}\overline{w}_{\overline{x}\overline{x}}\overline{w}_{\overline{x}}^2+\mu\overline{w}_{\overline{x}\overline{x}}\overline{w}_{\overline{x}}\overline{w}_{\overline{x}\overline{t}}=0\end{aligned}\tag{2.84}$$

where $\overline{k}_{\mathrm{E}}$ is the nonlinear elastic coefficient, c is the material viscosity, μ is the damping coefficient, and ε is a bookkeeping device to denote that the speed variation and the nonlinear term are small. Equation (2.84) can be truncated into ODEs by the Galerkin method and then studied by perturbation methods. It can be treated directly by perturbation methods. The method of multiple scales, as a direct perturbation technique, is applied to obtain the steady-state response of the axially moving string under parametric excitation.

The solutions to (2.84) can be assumed as follows.

$$\overline{w}\left(\overline{x},\overline{t};\varepsilon\right)=w_0(\overline{x},\tau_0,\tau_1)+\varepsilon w_1(\overline{x},\tau_0,\tau_1)+O\left(\varepsilon^2\right)\tag{2.85}$$

where $\tau_0=\overline{t}$ and $\tau_1=\varepsilon\overline{t}$ are, respectively, the fast and the slow timescales. Substitution of (2.85) and the following relationship

$$\frac{\partial}{\partial\overline{t}}=\frac{\partial}{\partial\tau_0}+\varepsilon\frac{\partial}{\partial\tau_1},\ \frac{\partial^2}{\partial\overline{t}^2}=\frac{\partial^2}{\partial\tau_0^2}+2\varepsilon\frac{\partial^2}{\partial\tau_0\partial\tau_1}+O\left(\varepsilon^2\right)\tag{2.86}$$

into (2.84) and then equalization of coefficients of ε^0 and ε^1 in the resulting equations lead to, at order ε^0,

$$w_{0\tau_0\tau_0}+2\overline{v}w_{0\overline{x}\tau_0}+\left(\overline{v}^2-1\right)w_{0\overline{x}\overline{x}}=0,\tag{2.87}$$

and, at order ε^1,

$$\begin{aligned}&w_{1\tau_0\tau_0}+2\overline{v}w_{1\overline{x}\tau_0}+\left(\overline{v}^2-1\right)w_{1\overline{x}\overline{x}}=\\&-\left[2\left(w_{0\tau_0\tau_1}+\overline{v}w_{0\overline{x}\tau_1}\right)+cw_{0\tau_0}\delta\cos(\omega t)w_{0\overline{x}\overline{x}}\right.\\&\left.-\frac{3}{2}\overline{k}_{\mathrm{E}}w_{0\overline{x}\overline{x}}w_{0\overline{x}}^2-\mu w_{0\overline{x}\overline{x}}w_{0\overline{x}}w_{0\overline{x}\tau_0}\right].\end{aligned}\tag{2.88}$$

Since the damping term due to viscoelasticity does not appear in (2.87), the linear system (2.87) is a gyroscopic continuous system with pure imaginary eigenvalues. The solution becomes

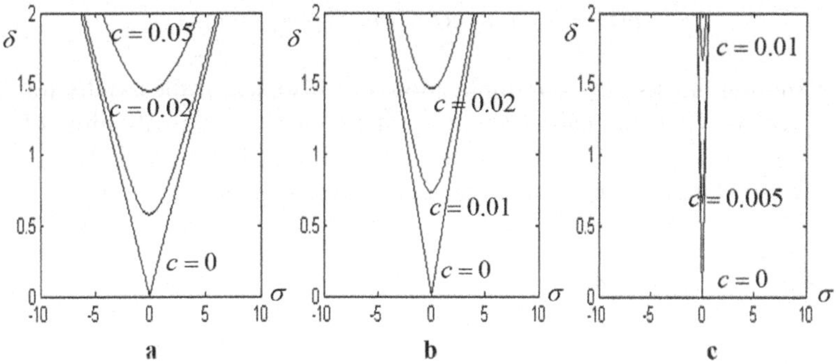

Fig. 2.3 Instability regions caused by parametric resonances: **a** first principle, **b** second principle, **c** combination of addition type

$$w_0(\overline{x}, \tau_0, \tau_1) = \sum_{n=1}^{\infty} A_n(\tau_1)\varphi_n(x)e^{i\omega_n\tau_0} + cc \tag{2.89}$$

where cc stands for the complex conjugates, and the parameters can be obtained by (2.74) and (2.77). Substitution of (2.89) into (2.87) leads to the solvability conditions determining the stability of parametric resonances.

Numerical demonstrations of the analytical solutions reveal three types of parametric resonances illustrated in Fig. 2.3. Figure 2.3a shows the instability region caused by the first principle parametric resonance with $\omega = 2\omega_1 + \varepsilon\sigma$, where ω_1 is the first-order natural frequency and σ is the detuning parameter. It can be seen that in the $\sigma - \delta$ plane, the unstable triangle area implies the wider range of instability occurs with higher perturbation amplitude. With the introduction of damping c, the instability regions will be narrowed, which means a higher excitation amplitude is needed to trigger the energy accumulation mechanism. The instability of the second principle parametric resonance, $\omega = 2\omega_2 + \varepsilon\sigma$, can be found in a similar way, as shown in Fig. 2.3b. The instability of combination resonance of summation type, $\omega = \omega_1 + \omega_2 + \varepsilon\sigma$, is shown in Fig. 2.3c. However, no instability was found for the difference-type combination resonance $\omega = \omega_2 - \omega_1 + \varepsilon\sigma$. It should be noted that the nonlinear terms do not contribute to the stability analysis of the trivial solutions. If only linear terms are considered, the system loses stability in the instability region with infinite amplitude, while the nonlinear terms make the response amplitude finite in the unstable region.

2.4.3 Bifurcation and Chaotic Dynamics

In the investigations on bifurcation and chaotic dynamics of axially moving strings, it is assumed that the transport speed is characterized as a simple harmonic variation about the constant mean speed, i.e.,

$$v(t) = v_0 + v_1 \cos \Omega_V t \quad (v_0, v_1 > 0). \tag{2.90}$$

The assumption has its physical implication. For instance, if the axially translating string represents a belt on a pair of rotating pulleys, the rotational vibrations of pulleys will result in a small fluctuation in the translational speed of the belt. The governing equation of the lateral motions in dimensionless form is then

$$\begin{aligned}
&\frac{\partial^2 \overline{w}(\overline{x}, \overline{t})}{\partial \overline{t}^2} + 2(\gamma_0 + \gamma_1 \cos \omega \overline{t}) \frac{\partial^2 \overline{w}(\overline{x}, \overline{t})}{\partial \overline{x} \partial \overline{t}} \\
&+ \left(\gamma_0^2 + \frac{\gamma_1^2}{2} + 2\gamma_0 \gamma_1 \cos \omega \overline{t} + \frac{\gamma_1^2}{2} \cos 2\omega \overline{t} - 1 \right) \frac{\partial^2 \overline{w}(\overline{x}, \overline{t})}{\partial \overline{x}^2} \\
&- \omega \gamma_1 \sin \omega \overline{t} \frac{\partial \overline{w}(\overline{x}, \overline{t})}{\partial \overline{x}} \\
&- \frac{\partial}{\partial x} \left\{ \left[\frac{k_E}{2} \left(\frac{\partial \overline{w}(\overline{x}, \overline{t})}{\partial \overline{x}} \right)^2 + \frac{k_\mu}{2} \frac{\partial}{\partial \overline{t}} \left(\frac{\partial \overline{w}(\overline{x}, \overline{t})}{\partial \overline{x}} \right)^2 \right] \frac{\partial \overline{w}(\overline{x}, \overline{t})}{\partial \overline{x}} \right\} = 0
\end{aligned} \tag{2.91}$$

where

$$\begin{aligned}
&\overline{w} = \frac{w}{l}, \overline{x} = \frac{x}{l}, \overline{t} = t\sqrt{\frac{P}{\rho A}}, \omega = \Omega_V l \sqrt{\frac{\rho A}{T_0}}, k_E = \frac{EA}{T_0}, \\
&k_\mu = \frac{\mu b}{l} \sqrt{\frac{T_0}{\rho A}}, \gamma_0 = v_0 \sqrt{\frac{\rho A}{T_0}}, \gamma_1 = v_1 \sqrt{\frac{\rho A}{T_0}}.
\end{aligned} \tag{2.92}$$

Equation (2.91) is in the form of a continuous nonlinear gyroscopic system with Kelvin viscoelastic damping and parametric excitation terms.

The Galerkin method is employed to simplify (2.91). Under the fixed-end boundary conditions

$$\overline{w}(0, t) = 0, \overline{w}(1, t) = 0, \tag{2.93}$$

the solution of (2.91) is expanded into the series of the string's eigenfunctions with two fixed boundary conditions as follows.

$$\overline{w}(\overline{x}, \overline{t}) = \sum_{n=1}^{m} q_n(\overline{t}) \sin(n\pi \overline{x}) \tag{2.94}$$

where $q_n(\overline{t})$ are the generalized displacements. Taking the appropriate derivatives and substituting (2.94) into (2.91), one obtains the residual function as follows.

$$R_m(\overline{x}, \overline{t}) = \sum_{n=1}^{m} \ddot{q}_n(\overline{t}) \sin(n\pi \overline{x}) + 2\pi(\gamma_0 + \gamma_1 \cos \omega \overline{t}) \sum_{n=1}^{m} n\dot{q}_n(\overline{t}) \cos(n\pi \overline{x})$$

$$
\begin{aligned}
&- \pi^2\left(\gamma_0^2 + \frac{\gamma_1^2}{2} + 2\gamma_0\gamma_1 \cos\omega\bar{t} + \frac{\bar{v}_1^2}{2}\cos 2\omega\bar{t} - 1\right)\sum_{n=1}^{m} n^2 q_n(\bar{t})\sin(n\pi\bar{x}) \\
&- \omega s \gamma_1 \pi \sin\omega t \sum_{n=1}^{m} n q_n(\bar{t})\cos(n\pi\bar{x}) \\
&+ \frac{3}{2}\pi^4 k_{\mathrm{E}}\left(\sum_{n=1}^{m} n\dot{q}_n(\bar{t})\cos(n\pi\bar{x})\right)^2 \sum_{n=1}^{m} n^2 q_n(\bar{t})\sin(n\pi\bar{x}) \\
&+ 2\pi^4 k_{\mu}\left(\sum_{n=1}^{m} n q_n(\bar{t})\cos(n\pi\bar{x})\right)\left(\sum_{n=1}^{m} n\dot{q}_n(\bar{t})\cos(n\pi\bar{x})\right)\sum_{n=1}^{m} n^2 q_n(\bar{t})\sin(n\pi\bar{x}) \\
&+ \pi^4 k_{\mu}\left(\sum_{n=1}^{m} n q_n(\bar{t})\cos(n\pi\bar{x})\right)^2 \sum_{n=1}^{m} n^2 \dot{q}_n(\bar{t})\sin(n\pi\bar{x}).
\end{aligned}
\tag{2.95}
$$

If the weighting functions are selected as the eigenfunctions of a stationary string, the implementation of the Galerkin method requires that the residual function (2.95) satisfies

$$
\int_0^1 R_m(\bar{x}, \bar{t})\sin(i\pi\bar{x})\mathrm{d}\bar{x} = 0 \quad (i = 1, 2, \cdots, m).
\tag{2.96}
$$

For given m, inserting (2.96) into (2.95) and integrating the resulting terms yield the m-term Galerkin approximation of the PDE describing a translating string. The one-term Galerkin approximation becomes

$$
\ddot{q}_1 + \frac{3}{4}k_{\mu}\pi^4 q_1^2\dot{q}_1 + \frac{\pi^2}{2}(2 - 2\gamma_0^2 - \gamma_1^2 - 4\gamma_0\gamma_1\cos\omega\bar{t} - \gamma_1^2\cos 2\omega\bar{t})q_1 + \frac{3k_{\mathrm{E}}\pi^4}{8}q_1^3 = 0.
\tag{2.97}
$$

The two-term Galerkin approximation is

$$
\begin{aligned}
&\ddot{q}_1 + \frac{k_{\mu}\pi^4}{4}(3q_1^2 + 8q_2^2)\dot{q}_1 + \frac{4}{3}(3k_{\mu}\pi^4 q_1 q_2 - 4\gamma_0 - 4\gamma_1\cos\omega t)\dot{q}_2 + \frac{8\omega\gamma_1\sin\omega\bar{t}}{3}q_2 \\
&\quad + \frac{\pi^2}{2}(2 - 2\gamma_0^2 - \gamma_1^2 - 4\gamma_0\gamma_1\cos\omega\bar{t} - \gamma_1^2\cos 2\omega\bar{t})q_1 + \frac{3k_{\mathrm{E}}\pi^4}{8}(q_1^2 + 8q_2^2)q_1 = 0, \\
&\ddot{q}_2 + \frac{4}{3}(3k_{\mu}\pi^4 q_1 q_2 + 4\gamma_0 + 4\gamma_1\cos\omega\bar{t})\dot{q}_1 + 2k_{\mu}\pi^4(q_1^2 + 6q_2^2)\dot{q}_2 - \frac{8\omega\gamma_1\sin\omega\bar{t}}{3}q_1 \\
&\quad + 2\pi^2(2 - 2\gamma_0^2 - \gamma_1^2 - 4\gamma_0\gamma_1\cos\omega\bar{t} - \gamma_1^2\cos 2\omega\bar{t})q_2 + 3k_{\mathrm{E}}\pi^4(q_1^2 + 2q_2^2)q_2 = 0,
\end{aligned}
\tag{2.98}
$$

which contains a gyroscopically coupled linear term $-16\bar{v}_0\dot{q}_2/3$ in the first formula and $16\bar{v}_0\dot{q}_1/3$ in the second formula. The three-term Galerkin approximation is

$$
\begin{aligned}
&\ddot{q}_1 + \frac{k_\mu \pi^4}{4}\left(3q_1^2 + 8q_2^2 + 18q_3^2 + 6q_1q_3\right)\dot{q}_1 + \frac{2}{3}\left(-8\gamma_0 - 8\gamma_1\cos\omega\bar{t} + 6k_\mu\pi^4 q_1q_2 + 9k_\mu\pi^4 q_2q_3\right)\dot{q}_2\\
&\quad + \frac{3k_\mu\pi^4}{4}\left(q_1^2 + 4q_2^2 + 12q_1q_3\right)\dot{q}_3 + \frac{\pi^2}{2}\left(2 - 2\gamma_0^2 - \gamma_1^2 - 4\gamma_0\gamma_1\cos\omega\bar{t} - \gamma_1^2\cos 2\omega\bar{t}\right)q_1\\
&\quad + \frac{3k_E\pi^4}{8}\left(q_1^3 + 8q_1q_2^2 + 3q_1^2q_3 + 12q_2^2q_3 + 18q_1q_3^2\right) + \frac{8\omega\gamma_1\sin\omega\bar{t}}{3}q_2 = 0,\\
&\ddot{q}_2 + \frac{2}{3}\left(8\gamma_0 + 8\gamma_1\cos\omega t + 6k_\mu\pi^4 q_1q_2 + 9k_\mu\pi^4 q_2q_3\right)\dot{q}_1 + 2k_\mu\pi^4\left(q_1^2 + 6q_2^2 + 3q_1q_3 + 9q_3^2\right)\dot{q}_2\\
&\quad + \frac{2}{5}\left(-24\gamma_0 - 24\gamma_1\cos\omega t + 15k_\mu\pi^4 q_1q_2 + 90k_\mu\pi^4 q_2q_3\right)\dot{q}_3 + 2\pi^2\left(2 - 2\gamma_0^2 - \gamma_1^2 - 4\gamma_0\gamma_1\cos\omega\bar{t}\right.\\
&\quad \left.-\gamma_1^2\cos 2\omega\bar{t}\right)q_2 + 3k_E\pi^4\left(q_1^2q_2 + 2q_2^3 + 3q_1q_2q_3 + 9q_2q_3^2\right) + \frac{8\omega\gamma_1\sin\omega\bar{t}}{15}(9q_3 - 5q_1) = 0,\\
&\ddot{q}_3 + \frac{3k_\mu\pi^4}{4}\left(q_1^2 + 4q_2^2 + 12q_1q_3\right)\dot{q}_1 + \frac{6}{5}\left(8\gamma_0 + 8\gamma_1\cos\omega\bar{t} + 5k_\mu\pi^4 q_1q_2 + 5k_\mu\pi^4 q_2q_3\right)\dot{q}_2\\
&\quad + \frac{9k_\mu\pi^4}{4}\left(2q_1^2 + 8q_2^2 + 27q_3^2\right)\dot{q}_3 + \frac{9\pi^2}{8}\left(8 - 8\gamma_0^2 - 4\gamma_1^2 - 16\gamma_0\gamma_1\cos\omega\bar{t} - 4\gamma_1^2\cos 2\omega\bar{t}\right)q_3\\
&\quad + \frac{3k_E\pi^4}{8}\left(q_1^3 + 12q_1q_2^2 + 18q_1^2q_3 + 72q_2^2q_3 + 81q_3^3\right) - \frac{24\omega\gamma_1\sin\omega\bar{t}}{5}q_2 = 0.
\end{aligned}
\tag{2.99}
$$

The four-term Galerkin approximation becomes

$$
\begin{aligned}
&\ddot{q}_1 + \frac{k_\mu\pi^4}{4}\left(3q_1^2 + 8q_2^2 + 18q_3^2 + 32q_4^2 + 6q_1q_3 + 16q_2q_4\right)\dot{q}_1 + \frac{2}{3}(-8\gamma_0 - 8\gamma_1\cos\omega\bar{t}\\
&\quad +6k_\mu\pi^4 q_1q_2 + 9k_\mu\pi^4 q_2q_3 + 6k_\mu\pi^4 q_1q_4 + 18k_\mu\pi^4 q_3q_4\Big)\dot{q}_2 + \frac{3k_\mu\pi^4}{4}\Big(q_1^2 + 4q_2^2 + 12q_1q_3\\
&\quad + 16q_2q_4)\dot{q}_3 + \frac{4}{15}\left(-8\gamma_0 - 8\gamma_1\cos\omega\bar{t} + 15k_\mu\pi^4 q_1q_2 + 45k_\mu\pi^4 q_2q_3 + 60k_\mu\pi^4 q_1q_4\right)\dot{q}_4\\
&\quad + \frac{\pi^2}{2}\left(2 - 2\gamma_0^2 - \gamma_1^2 - 4\gamma_0\gamma_1\cos\omega\bar{t} - \gamma_1^2\cos 2\omega\bar{t}\right)q_1 + \frac{8\omega\gamma_1\sin\omega\bar{t}}{15}(5q_2 + 2q_4)\\
&\quad + \frac{3k_E\pi^4}{8}\left(q_1^3 + 8q_1q_2^2 + 3q_1^2q_3 + 12q_2^2q_3 + 18q_1q_3^2 + 16q_1q_2q_4 + 48q_2q_3q_4 + 32q_1q_4^2\right) = 0,\\
&\ddot{q}_2 + \frac{2}{3}\left(8\gamma_0 + 8\gamma_1\cos\omega t + 6k_\mu\pi^4 q_1q_2 + 9k_\mu\pi^4 q_2q_3 + 6k_\mu\pi^4 q_1q_4 + 18k_\mu\pi^4 q_3q_4\right)\dot{q}_1\\
&\quad + 2k_\mu\pi^4\left(q_1^2 + 6q_2^2 + 3q_1q_3 + 9q_3^2 + 16q_4^2\right)\dot{q}_2 + \frac{6}{5}\Big(-8\gamma_0 - 8\gamma_1\cos\omega\bar{t} + 5k_\mu\pi^4 q_1q_2\\
&\quad +30k_\mu\pi^4 q_2q_3 + 10k_\mu\pi^4 q_1q_4 + 30k_\mu\pi^4 q_3q_4\Big)\dot{q}_3 + 2k_\mu\pi^4\left(q_1^2 + 6q_1q_3 + 9q_3^2 + 32q_2q_4\right)\dot{q}_4\\
&\quad + 2\pi^2\left(2 - 2\gamma_0^2 - \gamma_1^2 - 4\gamma_0\gamma_1\cos\omega\bar{t} - \gamma_1^2\cos 2\omega\bar{t}\right)q_2 + \frac{8\omega\gamma_1\sin\omega\bar{t}}{15}(9q_3 - 5q_1)\\
&\quad + 3k_E\pi^4\left(q_1^2q_2 + 2q_2^3 + 3q_1q_2q_3 + 9q_2q_3^2 + q_1^2q_4 + 6q_1q_3q_4 + 9q_3^2q_4 + 16q_2q_4^2\right) = 0,\\
&\ddot{q}_3 + \frac{3k_\mu\pi^4}{4}\left(q_1^2 + 4q_2^2 + 12q_1q_3 + 16q_2q_4\right)\dot{q}_1 + \frac{6}{5}\Big(8\gamma_0 + 8\gamma_1\cos\omega t + 5k_\mu\pi^4 q_1q_2\\
&\quad + 30k_\mu\pi^4 q_2q_3 + 10k_\mu\pi^4 q_1q_4 + 30k_\mu\pi^4 q_3q_4\Big)\dot{q}_2 + \frac{9k_\mu\pi^4}{4}\Big(2q_1^2 + 8q_2^2 + 27q_3^2 + 16q_2q_4\\
&\quad +32q_4^2\Big)\dot{q}_3 + \frac{12}{7}\left(-8\gamma_0 - 8\gamma_1\cos\omega\bar{t} + 7k_\mu\pi^4 q_1q_2 + 21k_\mu\pi^4 q_2q_3 + 84k_\mu\pi^4 q_3q_4\right)\dot{q}_4\\
&\quad + \frac{9\pi^2}{2}\left(2 - 2\gamma_0^2 - \gamma_1^2 - 4\gamma_0\gamma_1\cos\omega\bar{t} - \gamma_1^2\cos 2\omega\bar{t}\right)q_3 + \frac{24\omega\gamma_1\sin\omega\bar{t}}{35}(10q_4 - 7q_2)
\end{aligned}
$$

$$+\frac{3k_E\pi^4}{8}\left(q_1^3+12q_1q_2^2+18q_1^2q_3+72q_2^2q_3+81q_3^3+48q_1q_2q_4+144q_2q_3q_4+288q_3q_4^2\right)=0,$$

$$\ddot{q}_4+\frac{4}{15}\left(8\gamma_0+8\gamma_1\cos\omega t+15k_\mu\pi^4q_1q_2+45k_\mu\pi^4q_2q_3+60k_\mu\pi^4q_1q_4\right)\dot{q}_1$$
$$+2k_\mu\pi^4\left(q_1^2+6q_1q_3+9q_3^2+32q_2q_4\right)\dot{q}_2+\frac{12}{7}\left(8\gamma_0+8\gamma_1\cos\omega t+7k_\mu\pi^4q_1q_2\right.$$
$$\left.+21k_\mu\pi^4q_2q_3+84k_\mu\pi^4q_3q_4\right)\dot{q}_3+8k_\mu\pi^4\left(q_1^2+4q_2^2+9q_3^2+24q_4^2\right)\dot{q}_4$$
$$+8\pi^2\left(2-2\gamma_0^2-\gamma_1^2-4\gamma_0\gamma_1\cos\omega\bar{t}-\gamma_1^2\cos 2\omega\bar{t}\right)q_4-\frac{16\omega\gamma_1\sin\omega\bar{t}}{105}(7q_1+45q_3)$$
$$+3k_E\pi^4\left(q_1^2q_2+6q_1q_2q_3+9q_2q_3^2+4q_1^2q_4+16q_2^2q_4+36q_3^2q_4+32q_4^3\right)=0 \quad (2.100)$$

which contains two gyroscopically coupled linear terms $-16\gamma_0\dot{q}_2/3-32\gamma_0\dot{q}_4/15$ in the first formula, $16\gamma_0\dot{q}_1/3-48\gamma_0\dot{q}_3/5$ in the second formula, $48\gamma_0\dot{q}_2/5-96\gamma_0\dot{q}_4/7$ in the third formula, and $32\gamma_0\dot{q}_1/15+96\gamma_0\dot{q}_3/7$ in the fourth formula.

The Poincaré map is a convenient and powerful tool to investigate dynamic behaviors. In the following, the Poincaré maps are presented by sampling the dimensionless displacement and the dimensionless speed of the string's center. To examine the effects of truncation on longtime dynamic behaviors, we construct the Poincaré maps for different term Galerkin truncations, namely

$$u_1(nT,1/2)=q_1(nT),\ v_1(nT,1/2)=\dot{q}_1(nT), \quad (2.101)$$

$$u_2(nT,1/2)=q_1(nT),\ v_2(nT,1/2)=\dot{q}_1(nT), \quad (2.102)$$

$$u_3(nT,1/2)=q_1(nT)-q_3(nT),\ v_3(nT,1/2)=\dot{q}_1(nT)-\dot{q}_3(nT), \quad (2.103)$$

$$u_4(nT,1/2)=q_1(nT)-q_3(nT),\ v_4(nT,1/2)=\dot{q}_1(nT)-\dot{q}_3(nT) \quad (2.104)$$

where $T=2\pi/\omega$ and $n=1,2,3,\ldots$.

The bifurcation analysis is a modern technique used to explore nonlinear systems. The dynamical behaviors may be globally observed over a range of parameter values and thereby simultaneously compare different types of vibrations. The bifurcation analysis shows a summary of fundamental dynamics via bifurcation diagrams; accordingly, the overall dynamics can be examined. In the bifurcation analysis, numerical integration is handled by using the fourth-order Runge–Kutta routine. The bifurcation diagrams are portrayed by varying system parameters, such as the amplitude of axial speed fluctuation γ_1, the frequency of axial speed fluctuation, the mean axial speed γ_0, the dynamic viscosity k_μ, and the string stiffness k_E.

The bifurcation diagrams for the mean axial speed γ_0 are depicted in Fig. 2.4, where $\omega=0.125$, $\gamma_1=0.6$, $k_E=400$, and $k_\mu=120$. All cases show a common feature: The equilibrium (or minimal oscillation) disappears its stability after γ_0

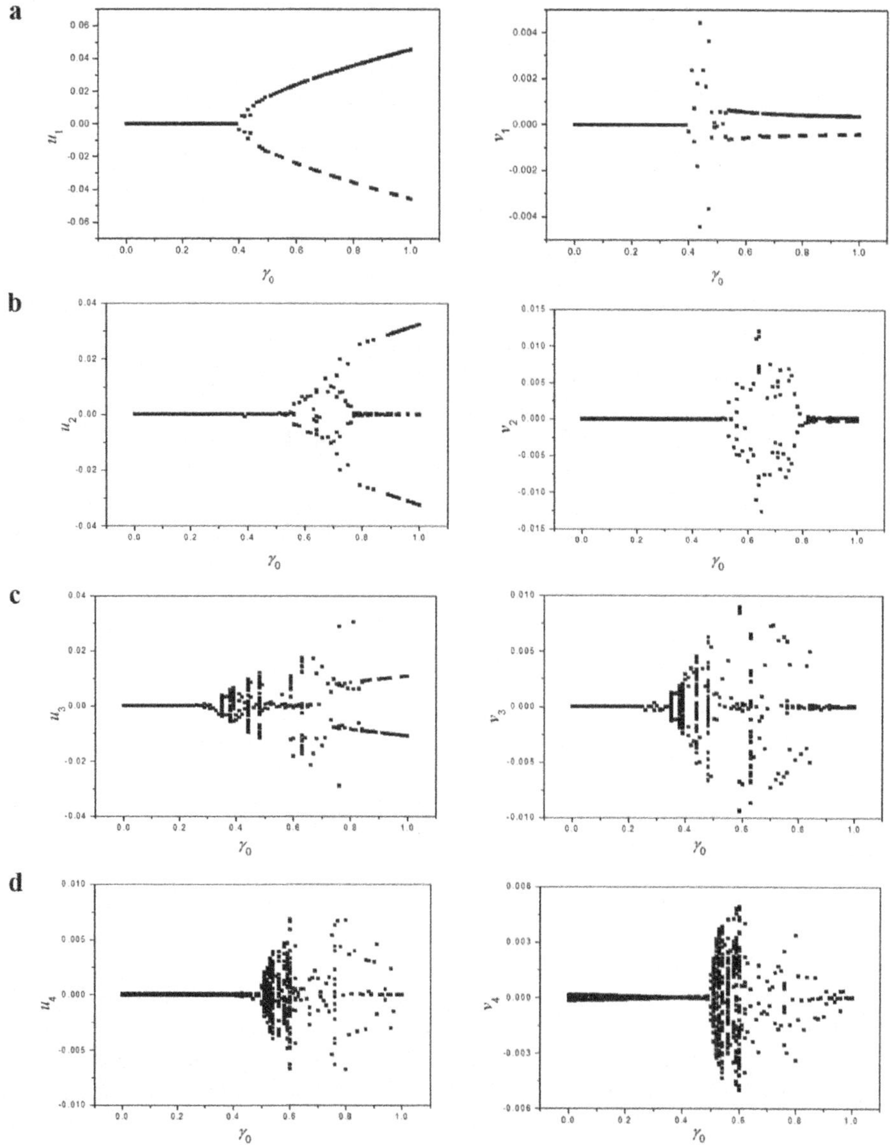

Fig. 2.4 Bifurcation versus the mean axial speed: **a** one-term Galerkin truncation, **b** two-term Galerkin truncation, **c** three-term Galerkin truncation, **d** four-term Galerkin truncation

exceeds a certain value. In Fig. 2.4a, a typical period-doubling bifurcation with a short transient process exists in the one-term Galerkin truncation. In the two-term Galerkin truncation, a tangent bifurcation and two period-doubling bifurcations appear successively (Fig. 2.4b). Figure 2.4c shows that the vibration's magnitude increases gradually in the three-term Galerkin truncation. Additionally, the periodic motion and complicated motion occur alternately. In the four-term Galerkin truncation, the explosive phenomena occur at a certain value of γ_0, approximately 0.5 (Fig. 2.4d). Besides that, the periodic motion replaces the complicated motion when the mean axial speed increases further.

Figure 2.5 depicts the bifurcation diagrams for the magnitude of the longitudinal speed fluctuation γ_1, where $\omega = 0.125$, $\gamma_0 = 0.7$, $k_E = 400$, and $k_\mu = 120$. All cases show a common feature: The equilibrium disappears its stability after the magnitude of γ_1 gets more significant than a particular value, approximately 0.4 for even-term truncations. In the one-term Galerkin truncation case, a period-doubling bifurcation with a rather long transient process exists; see Fig. 2.5a. For the two-term Galerkin truncation, Fig. 2.5b shows that the complicated transient process interrupts the period-doubling bifurcations. Besides that, for a more considerable longitudinal speed fluctuation, the lateral motion becomes the equilibrium again. In the three-term Galerkin truncation, the periodic motion and the complicated motion exchange alternately with any explicit patterns (Fig. 2.5c). Figure 2.5d reveals that the amplitude of the vibration increases gradually in the four-term Galerkin truncation. Additionally, for more significant longitudinal speed fluctuation, the original equilibrium is achieved again.

The bifurcation diagrams for the frequency of axial speed fluctuation are demonstrated in Fig. 2.6, where $\gamma_0 = 0.5$, $\gamma_1 = 0.8$, $k_E = 400$, and $k_\mu = 120$. The periodic motion and the complicated motion occur alternately in all cases. It seems that the bifurcation diagrams shown in Fig. 2.6 do not follow any known patterns. It reveals that the influence of the excitation frequency on dynamics has not been well understood.

The bifurcation diagrams for the string stiffness k_E are shown in Fig. 2.7, where $\gamma_0 = 0.7$, $\gamma_1 = 0.5$, $\omega = 0.125$, and $k_\mu = 120$. Figures 2.7a and 2.7c show that only the periodic motion exists in the one-term and three-term Galerkin truncations. Contrarily, the periodic motion occurs for small values of the string stiffness in the two-term and four-term Galerkin truncations (Fig. 2.7b and 2.7d), and the complicated motions are shown in these cases. However, in the two-term Galerkin truncation, the period window within the complicated motion exists more frequently.

The bifurcation diagrams for the dynamic viscosity k_μ are depicted in Fig. 2.8, wherein $\gamma_0 = 0.5$, $\gamma_1 = 0.8$, $\omega = 0.125$, and $k_E = 400$. In the one-term Galerkin truncation case, only the periodic motion occurs (Fig. 2.8a). Figures 2.8b, 2.8c, and 2.8d show the two-term, three-term, and four-term Galerkin truncations. As shown in these figures, the complicated motion occurs when the dynamic viscosity is small and is replaced by the periodic motion when the dynamic viscosity increases. In Figs. 2.9 and 2.10, the chaotic motion and regular motion are demonstrated by phase trajectories and Poincaré maps for two-term and four-term truncations, respectively.

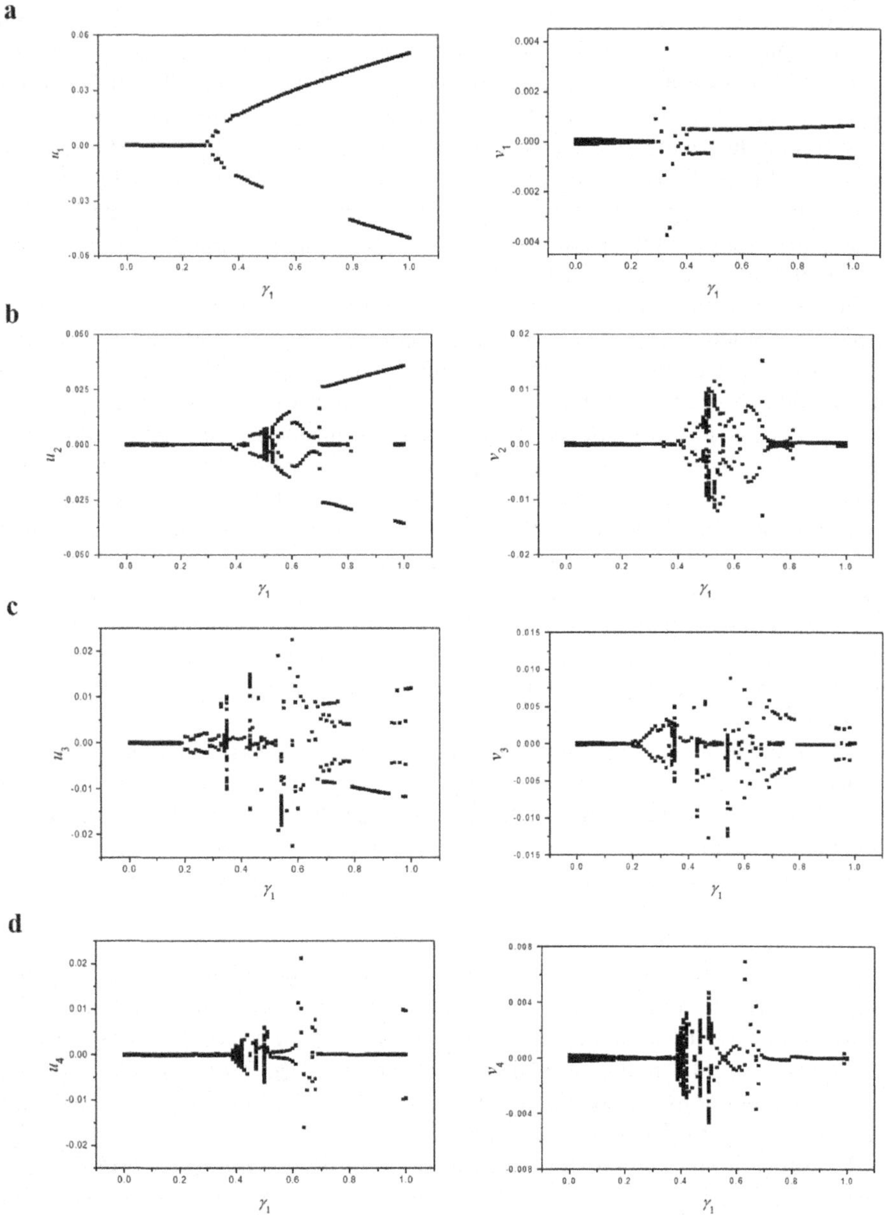

Fig. 2.5 Bifurcation versus the amplitude of axial speed fluctuation: **a** one-term Galerkin truncation, **b** two-term Galerkin truncation, **c** three-term Galerkin truncation, **d** four-term Galerkin truncation

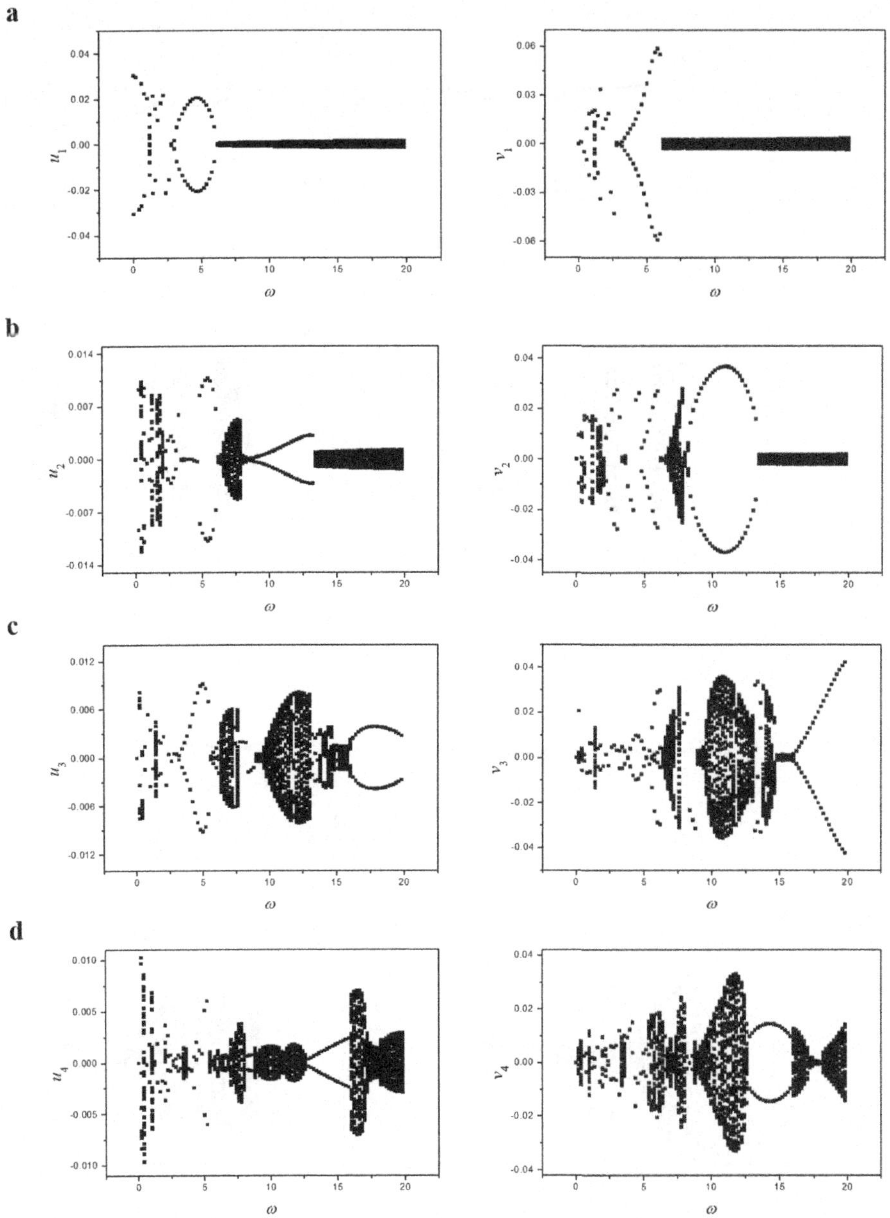

Fig. 2.6 Bifurcation versus the frequency of axial speed fluctuation: **a** one-term Galerkin truncation, **b** two-term Galerkin truncation, **c** three-term Galerkin truncation, **d** four-term Galerkin truncation

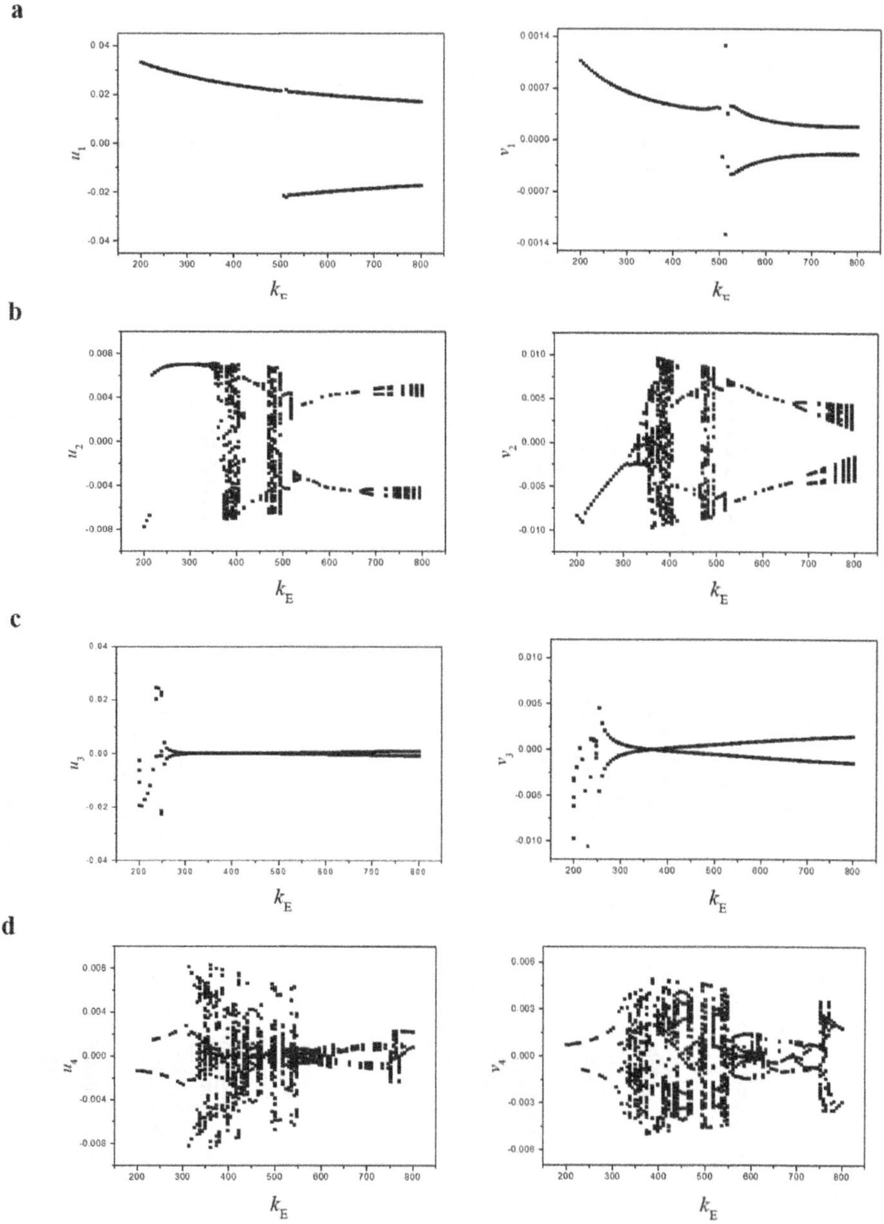

Fig. 2.7 Bifurcation versus the string stiffness: **a** one-term Galerkin truncation, **b** two-term Galerkin truncation, **c** three-term Galerkin truncation, **d** four-term Galerkin truncation

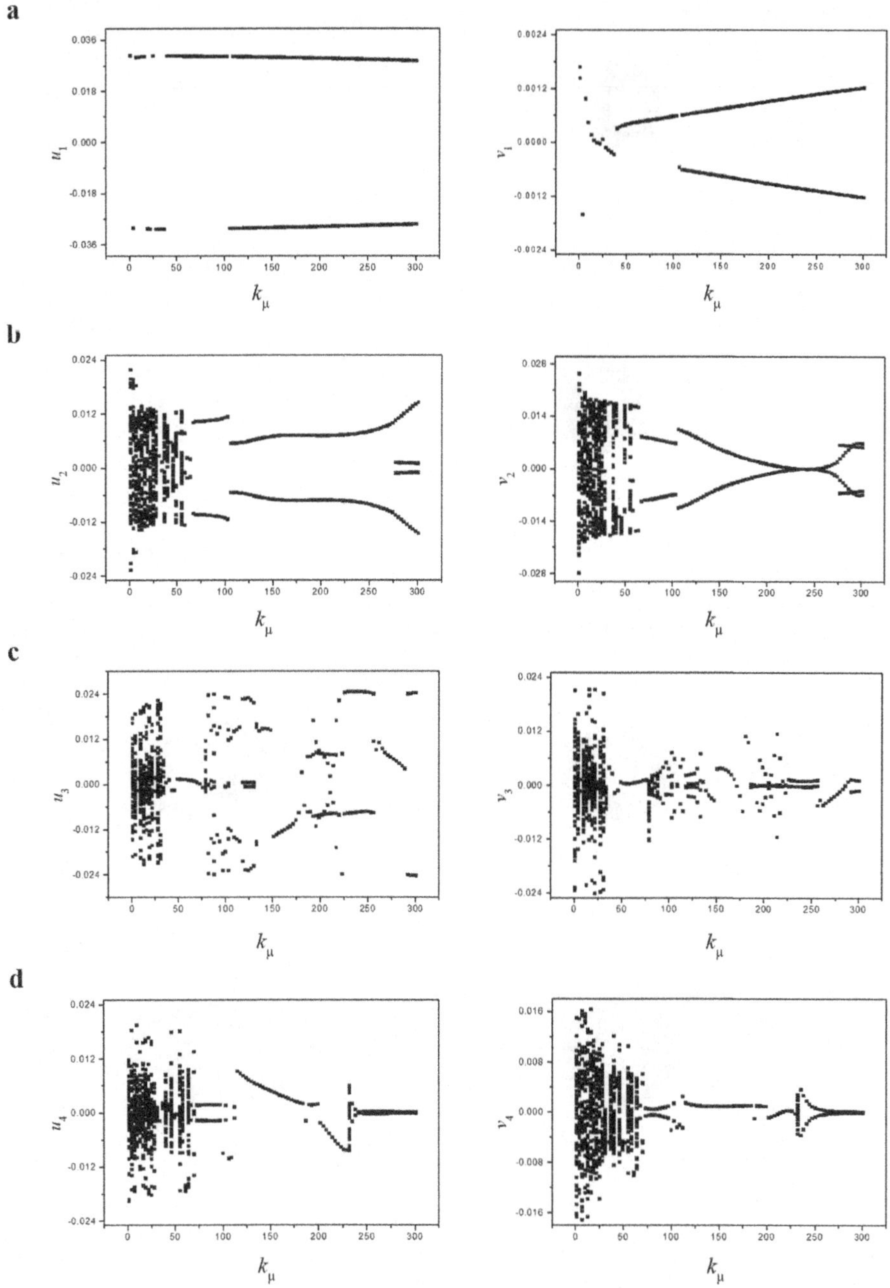

Fig. 2.8 Bifurcation versus the string dynamic viscosity: **a** one-term Galerkin truncation, **b** two-term Galerkin truncation, **c** three-term Galerkin truncation, **d** four-term Galerkin truncation

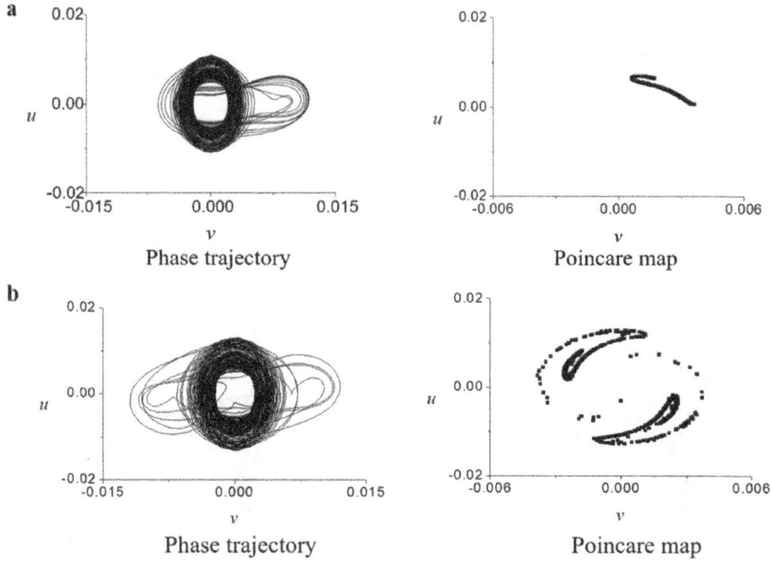

Fig. 2.9 Regular motion for $\gamma_1 = 0.9$, $\gamma_0 = 0.5$, $\omega = 0.125$, $k_E = 400$, and $k_\mu = 60$: **a** two-term Galerkin truncation and **b** four-term Galerkin truncation

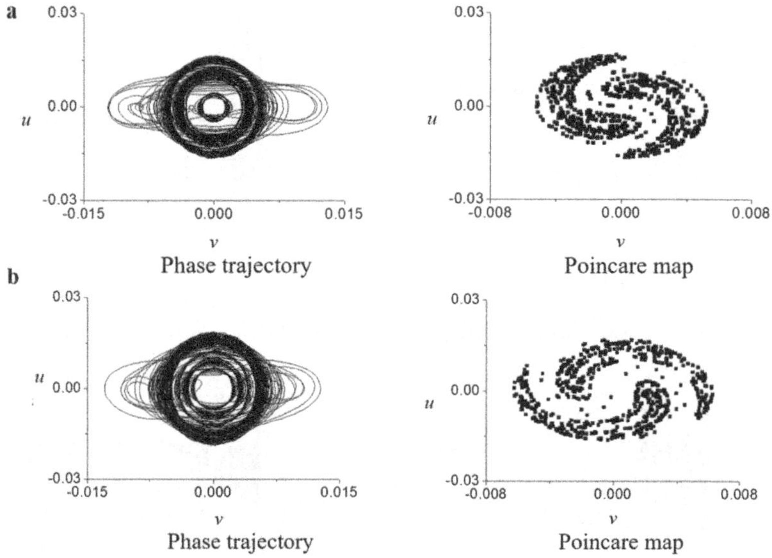

Fig. 2.10 Chaotic motion for $\gamma_1 = 0.9$, $\gamma_0 = 0.5$, $\omega = 0.125$, $k_E = 400$, and $k_\mu = 10$: **a** two-term Galerkin truncation and **b** four-term Galerkin truncation

On the whole, bifurcation diagrams corresponding to the two-term and four-term truncations are the same qualitatively. Contrarily, these bifurcation diagrams are fundamentally different from those corresponding to the two-term and three-term truncations. Since the gyroscopic coupling is considered only in the even-term truncations, the more even-term truncation can lead to more precise results. Therefore, the four-term Galerkin truncation produced the best result in the present work. However, convergence was not achieved quantitatively. This is possibly due to the poor convergence properties of the eigenfunctions of stationary strings.

References

Bapat VA, Srinivasan P (1967) Nonlinear transverse oscillations in traveling strings by the method of harmonic balance. J Appl Mech 34(3):775–777

Bapat VA, Srinivasan P (1971) Nonlinear transverse oscillations on traveling strings by a direct linearization method. J Indian Indust Sci 53:120–125

Chen LQ (2005) Principal parametric resonance of axially accelerating viscoelastic strings constituted by the Boltzmann superposition principle. P Roy Soc Lond A Mat 461:2701–2720

Chen LQ, Zhao WJ (2004) A computationmethod for nonlinear vibration of axially accelerating viscoelastic strings. Appl Math Comput 162(1):305–310

Chen LQ, Zhao WJ (2008) Iterative algorithm for axially accelerating strings with integral constitutive law. Acta Mech Solida Sin 21(5):449–456

Chen LQ, Zhang NH (2005) Nonlinear dynamics of axially moving viscoelastic strings based on translating eigenfunctions. Chaos Solitons Fractals 24(4):1065–1074

Chen LQ, Zhang NH, Zu JW (2003a) The regular and chaotic vibrations of an axially moving viscoelastic string based on 4-order Galerkin truncation. J Sound Vib 261(4):764–773

Chen LQ, Zu JW, Wu J (2003b) Steady-state response of the parametrically excited axially moving string constituted by the Boltzmann superposition principle. Acta Mech 162(1):143–155

Chen LQ, Wu J, Zu JW (2004a) Asymptotic nonlinear behaviors in transverse vibration of an axially accelerating viscoelastic string. Nonlinear Dyn 35(4):347–360

Chen LQ, Wu J, Zu JW (2004b) The chaotic response of the viscoelastic traveling string: an integral constitutive law. Chaos Solitons Fractals 21(2):349–357

Chen LQ, Zhao WJ, Zu JW (2004c) Transient responses of an axially accelerating viscoelastic string constituted by a fractional differentiation law. J Sound Vib 278(4–5):861–871

Chen LQ, Zhao WJ, Zu JW (2005) Simulations of transverse vibrations of an axially moving string: a modified difference approach. Appl Math Comput 166(3):596–607

Chen LQ, Zhao WJ, Ding H (2009) On Galerkin discretization of axially moving nonlinear strings. Acta Mech Solida Sin 22(4):369–376

Christensen RM (1982) Theory of viscoelasticity. Academic, New York

Drozdov A, Kalamkarov AL (1996) Constitutive model for nonlinear viscoelastic behavior of polymer. Polym Eng Sci 36(14):1907–1919

Fung RF, Huang JS, Chen YC (1997) The transient amplitude of the viscoelastic travelling string: an integral constitutive law. J Sound Vib 201(2):153–167

Fung RF, Huang JS, Chen YC, Yao CM (1998a) Nonlinear dynamic analysis of the viscoelastic string with a harmonically varying transport speed. Comput Struct 66(6):777–784

Fung RF, Wang PH, Lee MJ (1998b) Nonlinear vibration analysis of a traveling string with time-dependent length by finite element method. J Chin Inst Eng 21(1):109–117

Ghangrekar SG (1975) On nonlinear oscillation of moving string. Labdev J Sci Tech Phys Sci A 13:255–257

Ghayesh MH (2008) Nonlinear transversal vibration and stability of an axially moving viscoelastic string supported by a partial viscoelastic guide. J Sound Vib 314(3–5):757–774
Ghayesh MH, Balar S (2008) Non-linear parametric vibration and stability of axially moving viscoelastic Rayleigh beams. Int J Solids Struct 45(25–26):6451–6467
Ghayesh MH, Moradian N (2011) Nonlinear dynamic response of axially moving, stretched viscoelastic strings. Arch Appl Mech 81(6):781–799
Grimshaw R (1993) Nonlinear ordinary differential equations. CRC Press, United States of America
Hou Z, Zu JW (2001) Parametric vibration of viscoelastic moving belts using standard linear solid model. In: International design engineering technical conferences and computers and information in engineering conference. American Society of Mechanical Engineers; Pittsburgh, vol 80289, pp 1117–1124
Hou Z, Zu JW (2002) Non-linear free oscillations of moving belts with standard viscoelastic model. Mech Mach Theory 37(9):925–940
Jha RK, Parker RG (2000) Spatial discretization of axially moving media vibration problems. J Vib Acoust 122(3):290–294
Kartik V, Wickert JA (2006) Vibration and guiding of moving media with edge weave imperfections. J Sound Vib 291(1–2):419–436
Korde KR (1985) On nonlinear oscillation of moving string. J Appl Mech 52(2):493–494
Lakshmikumaran AV, Wickert JA (1996) On the vibration of coupled traveling string and air bearing systems. J Vib Acoust-Trans ASME 118(3):398–405
Li YH, Gao Q, Jian KL, Yin XG (2003) Dynamic responses of viscoelastic axially moving belt. Appl Math Mech-Engl 24(11):1348–1354
Liang F, Yang XD, Qian YJ, Zhang W (2017) Free vibration analysis of pipes conveying fluid based on linear and nonlinear complex mode approach. Int J Appl Mech 9(8):1750112
Liang F, Yang XD, Qian YJ, Zhang W (2018) Forced response analysis of pipes conveying fluid by nonlinear normal modes method and iterative approach. J Comput Nonlinear Dyn 13(1):014502
Mahalingam S (1957) Transverse vibrations of power transmission chains. Br J Appl Phys 8:145–148
Malookani RA, van Horssen WT (2015) On resonances and the applicability of Galerkin's truncation method for an axially moving string with time-varying velocity. J Sound Vib 344:1–17
Mote CD (1966) On the nonlinear oscillation of an axially moving string. J Appl Mech 33(2):463–464
Mote CD, Thurman AL (1971) Oscillation modes of an axially moving material. J Appl Mech 38(1):279–280
Moustafa MA, Salman FK (1976) Dynamic properties of a moving thread line. J Eng Ind 98(3):868–875
Nayfeh AH, Nayfeh JF, Mook DT (1992) On methods for continuous systems with quadratic and cubic nonlinearities. Nonlinear Dyn 3(2):145–162
Pakdemirli M (1994) A comparison of two perturbation methods for vibrations of systems with quadratic and cubic nonlinearities. Mech Res Commun 21(2):203–208
Pakdemirli M, Boyaci H (1995) Comparison of direct-perturbation methods with discretization-perturbation methods for nonlinear vibrations. J Sound Vib 186(5):837–845
Pakdemirli M, Boyaci H (2002) Effect of non-ideal boundary conditions on the vibrations of continuous systems. J Sound Vib 249(4):815–823
Pakdemirli M, Ulsoy AG, Ceranoglu A (1994) transverse vibration of an axially accelerating string. J Sound Vib 169(2):179–196
Pakdemirli M, Nayfeh SA, Nayfeh AH (1995) Analysis of one-to-one autoparametric resonances in cables: discretization versus direct treatment. Nonlinear Dyn 8(1):65–83
Parker RG (1999) Supercritical speed stability of the trivial equilibrium of an axially-moving string on an elastic foundation. J Sound Vib 221(2):205–219
Pellicano F, Catellani G, Fregolent A (2004) Parametric instability of belt: theory and experiments. Comput Struct 82(1):81–91
Sack RA (1954) Transverse oscillations in travelling strings. Br J Appl Phys 5(6):224

Sandilo SH, van Horssen WT (2014) On variable length induced vibrations of a vertical string. J Sound Vib 333(11):2432–2449
van Horssen WT (2003) On the influence of lateral vibrations of supports for an axially moving string. J Sound Vib 268(2):323–330
Wickert JA (1993) Analysis of self-excited longitudinal vibration of a moving tape. J Sound Vib 160(3):455–463
Wickert JA (1994) Response solutions for the vibration of a traveling string on an elastic-foundation. J Vib Acoust-Trans ASME 116(1):137–139
Wickert JA, Mote CD (1988) Linear transverse vibration of an axially moving string-particle system. J Acoust Soc Am 84(3):963–969
Wickert JA, Mote CD (1990) Classical vibration analysis of axially moving continua. J Appl Mech 57(3):738–744
Wickert JA, Mote CD (1991) Response and discretization method for axially moving materials. Appl Mech Rev 44(11):279–284
Yang XD, Wu H, Qian YJ, Zhang W, Lim CW (2017) Nonlinear vibration analysis of axially moving string based on gyroscopic modes decoupling. J Sound Vib 393:308–320
Yurddas A, Ozkaya E, Boyaci H (2013) Nonlinear vibrations of axially moving multi-supported strings having non-ideal support conditions. Nonlinear Dyn 73(3):1223–1244
Zhang NH (2008) Dynamic analysis of an axially moving viscoelastic string by the Galerkin method using translating string eigenfunctions. Chaos Solitons Fractals 35(2):291–302
Zhang L, Zu JW (1999) Nonlinear vibration of parametrically excited moving belts, part 1: dynamic response. J Appl Mech 66(2):396–402
Zhang L, Zu JW, Zhong Z (2002) Transient response for viscoelastic moving belts using block-by-block method. Int J Struct Stab Dyn 2:265–280
Zhang NH, Wang JJ, Cheng CJ (2007) Complex-mode Galerkin approach in transverse vibration of an axially accelerating viscoelastic string. Appl Math Mech-Engl 28(1):1–9
Zhu WD, Ni J (2000) Energetics and stability Aof translating media with an arbitrarily varying length. J Vib Acoust 122(3):295–304
Zhu WD, Ni J, Huang J (2001) Active control of translating media with arbitrarily varying length. J Vib Acoust 123(3):347–358
Zu JW, Hou Z (2000) Comparison of different viscoelastic models for nonlinear free vibrations of moving belts. Vib Cont Continuous Syst 107:55–61

Chapter 3
Beam Model

Abstract This chapter provides thorough analyses of the dynamics of axially moving beams. The equations of motion are derived using diverse beam theories: the Euler–Bernoulli, Rayleigh, and Timoshenko beam theories. Other aspects of axially moving beams such as the cantilever beam, viscoelastic beam, fluid-conveying pipe, nanoscale beam, and composite beam are also discussed. One approximate model based on the Galerkin method is presented.

Keywords Axially moving beams · Lateral vibration · Euler–Bernoulli beam · Rayleigh beam · Timoshenko beam · Laminate composite material · Fluid-conveying pipe · Nonhomogenous boundaries · Galerkin method · Stability · Bifurcations · Chaos

In Chap. 2, we discussed the string model, which is the most fundamental model of axially moving systems by disregarding the bending stiffness. When the material's bending stiffness is quite large, we need a more adequate model to predict the dynamic behaviors of the AMS. This chapter is dedicated to the beam model, which describes AMSs with significant bending stiffness.

3.1 Introduction

As previously mentioned, the beam model describes an AMS where the bending stiffness is significant (i.e., the bending stiffness is also known as the flexural rigidity of a beam). The dynamic model of a longitudinally moving beam is developed based on beam theories, such as the Euler–Bernoulli, Rayleigh, and Timoshenko beam theories. The Euler–Bernoulli beam addresses tension and bending stiffness, whereas the Rayleigh beam further examines the rotary inertia effect. Mathematically, the dynamics of the Euler–Bernoulli and Rayleigh beams are described by one fourth-order PDE. Contrary to the Euler–Bernoulli and Rayleigh beams, the Timoshenko beam, which further considers shear deformation, is characterized by two second-order PDEs. Numerous studies on axially moving beams using these three theories, including elastic and viscoelastic beams, have been published in the last 55 years

© The Author(s), under exclusive license to Springer Nature Singapore Pte Ltd. 2022
K.-S. Hong et al., *Control of Axially Moving Systems*,
https://doi.org/10.1007/978-981-16-2915-0_3

(Mote 1965; Öz and Pakdemirli 1999; Chen and Yang 2005; Ghayesh and Balar 2008; Li et al. 2018). In recent years, due to the development of various materials, laminated composite beams comprising two or more layers of orthotropic materials with different properties and nanoscale beams have been widely employed for AMSs in the industry. Their dynamics have been investigated in Ghayesh (2011), Marynowski (2012), Li et al. (2018), Wang and Shen (2019), Sarparast et al. (2020), Alshaqaq and Hawwa (2020), Wang and Sun (2020), and Wang et al. (2020).

Since the beam model addresses bending stiffness, it is appropriate for the dynamic analysis of thick axially moving materials. In mechanical engineering, saws of metal-cutting band-saw machines, steel rods in the continuous manufacturing process, and steel wires in automatic wire-winding machines are typical examples of axially moving beams. Furthermore, translating cantilevers, such as flexible robotic end-effectors, drill strings (Sahebkar et al. 2011), aerospace structures (Wang et al. 2009), and hosting cables (Zhu and Ni 2000), can be modeled as axially moving cantilever beams.

3.2 Dynamic Models of Axially Moving Beams

3.2.1 Euler–Bernoulli Beam

The classical Euler–Bernoulli beam theory is the most straightforward beam theory wherein the beam cross section is always perpendicular to the neutral axis during deformation. In the Euler–Bernoulli beam theory, the deflection of an axially moving beam accounts for only the tension and bending moment. This theory is appreciated for investigating slender beams. Due to its mathematical simplicity, the Euler–Bernoulli beam theory has been widely utilized to develop dynamic models of axially moving beams (Mote 1965; Wickert and Mote 1989; Öz and Pakdemirli 1999).

Accounting to the Euler–Bernoulli beam theory, we consider the lateral vibration $w(x, t)$ of a uniform moving beam of length l and cross-sectional area A, as shown in Fig. 3.1. The beam material is elastic and homogeneous. The dynamic model of the beam is established based on the extended Hamilton principle. The kinetic energy K of the considered beam is similar to the string case [i.e., (2.3)]. In contrast, the potential energy U of the beam is more complicated due to the bending stiffness, that is

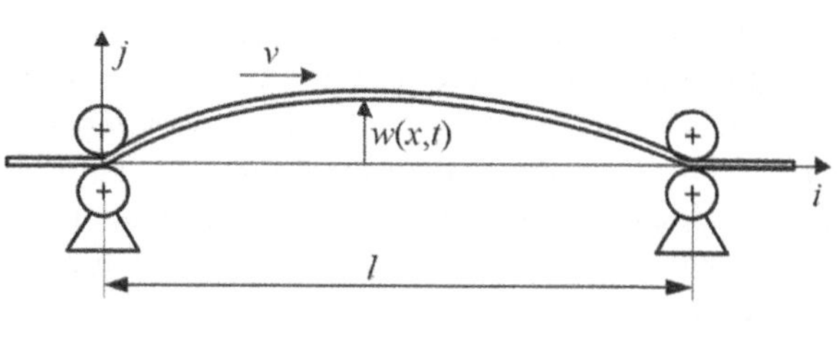

Fig. 3.1 Axially moving beam

$$U = \frac{1}{2}\int\limits_0^l T(x,t)w_x^2\mathrm{d}x + \frac{1}{2}\int\limits_0^l M(x,t)w_{xx}\mathrm{d}x, \tag{3.1}$$

where $T(x, t)$ is the axial tension and $M(x, t)$ is the bending moment. The second term in (3.1) is the potential energy owing to the bending moment. Using the extended Hamilton principle results in

$$\begin{aligned} 0 &= \frac{1}{2}\int\limits_{t_1}^{t_2}\delta\left(\int\limits_0^l \left[\rho A v_0^2 + \rho A(w_t + v_0 w_x)^2 - T(x,t)w_x^2 - M(x,t)w_{xx}\right]\mathrm{d}x\right)\mathrm{d}t \\ &= \frac{1}{2}\int\limits_{t_1}^{t_2}\int\limits_0^l \delta L_\mathrm{b}\mathrm{d}x\mathrm{d}t - \frac{1}{2}\int\limits_{t_1}^{t_2}\int\limits_0^l \delta(M(x,t)w_{xx})\mathrm{d}x\mathrm{d}t, \end{aligned} \tag{3.2}$$

where $L_\mathrm{b} = \rho A v_0^2 + \rho A(w_t + v_0 w_x)^2 - T(x,t)w_x^2$. By applying the same procedure of integration by parts applied to the string in Sect. 2.2.1, (3.2) can be expressed as follows:

$$\begin{aligned} 0 &= \frac{1}{2}\int\limits_{t_1}^{t_2}\int\limits_0^l \left[\frac{\partial L_\mathrm{b}}{\partial w} - \frac{\partial}{\partial x}\frac{\partial L_\mathrm{b}}{\partial w_x} - \frac{\partial}{\partial t}\frac{\partial L_\mathrm{b}}{\partial w_t}\right]\delta w\mathrm{d}x\mathrm{d}t \\ &\quad + \frac{1}{2}\int\limits_{t_1}^{t_2}\frac{\partial L_\mathrm{b}}{\partial w_x}\delta w\Bigg|_{x=0}^{x=l}\mathrm{d}t + \frac{1}{2}\int\limits_0^l \frac{\partial L_\mathrm{b}}{\partial w_t}\delta w\Bigg|_{t=t_1}^{t=t_2}\mathrm{d}x \\ &\quad - \int\limits_{t_1}^{t_2}\int\limits_0^l (M(x,t))_{xx}\delta w\mathrm{d}x\mathrm{d}t - \int\limits_{t_1}^{t_2} M(x,t)\delta w_x\Bigg|_{x=0}^{x=l}\mathrm{d}t + \int\limits_{t_1}^{t_2}(M(x,t))_x\delta w\Bigg|_{x=0}^{x=l}\mathrm{d}t \\ &= \int\limits_{t_1}^{t_2}\int\limits_0^l \left[-\rho A\left(w_{tt} + 2v_0 w_{xt} + v_0^2 w_{xx}\right) + (T(x,t)w_x)_x - (M(x,t))_{xx}\right]\delta w\mathrm{d}x\mathrm{d}t \\ &\quad - \int\limits_{t_1}^{t_2} M(x,t)\delta w_x\Bigg|_{x=0}^{x=l}\mathrm{d}t + \int\limits_{t_1}^{t_2}\left[\frac{1}{2}\frac{\partial L_\mathrm{b}}{\partial w_x} + (M(x,t))_x\right]\delta w\Bigg|_{x=0}^{x=l}\mathrm{d}t \\ &\quad + \int\limits_0^l \frac{1}{2}\frac{\partial L_\mathrm{b}}{\partial w_t}\delta w\Bigg|_{t=t_1}^{t=t_2}\mathrm{d}x. \end{aligned} \tag{3.3}$$

Since $\delta w(x, t_1) = \delta w(x, t_2) = 0$, the last term vanishes. Additionally, because the beam is simply supported at both ends by rollers, δw vanishes at $x = 0$ and $x = l$. Therefore, the mathematical model of the axially moving Euler–Bernoulli beam is given as follows:

$$\rho A\left(w_{tt}+2v_0w_{xt}+v_0^2w_{xx}\right)-(T(x,t)w_x)_x+(M(x,t))_{xx}=0, \tag{3.4}$$

where v_0 is the axial velocity. The boundary conditions are

$$w(0,t)=0,\ w(l,t)=0,\ w_{xx}(0,t)=0,\ w_{xx}(l,t)=0. \tag{3.5}$$

The dynamic model shown in (3.4) is a fourth-order PDE, wherein the last term on the left side corresponds to the bending moment. According to the Euler–Bernoulli beam theory, the bending moment is proportional to the small slope curvature w_{xx}, namely

$$M(x,t)=EIw_{xx}, \tag{3.6}$$

where I is the area moment of inertia, and EI is called the flexural rigidity. In addition, if the tension approximates the initial tension T_0, (3.4) can be simplified to

$$\rho A\left(w_{tt}+2v_0w_{xt}+v_0^2w_{xx}\right)-T_0w_{xx}+EIw_{xxxx}=0, \tag{3.7}$$

which is the linear model that describes the lateral vibration of an axially moving beam.

Furthermore, the dynamics of axially moving beams with large vibrational amplitude and time-varying velocity have been investigated. The equation of motion of an axially moving beam with varying velocity is given as follows.

$$\rho A\left(w_{tt}+\dot{v}(t)w_x+2v(t)w_{xt}+v(t)^2w_{xx}\right)-T_0w_{xx}+EIw_{xxxx}=0. \tag{3.8}$$

On the other hand, the dynamic model considering large vibration amplitude in the beam's potential energy, i.e.,

$$U=\int_0^l\left(T_0+\frac{1}{2}A\sigma(x,t)\right)\varepsilon(x,t)\mathrm{d}x+\frac{1}{2}\int_0^l EIw_{xx}^2\mathrm{d}x, \tag{3.9}$$

is described by the following nonlinear equation:

$$\rho A\left(w_{tt}+2v_0w_{xt}+v_0^2w_{xx}\right)-T_0w_{xx}-\frac{3EA}{2}w_x^3w_{xx}+EIw_{xxxx}=0. \tag{3.10}$$

In most of the dynamic models considered in the literature, the equation of motion of the beam is often described by a partial differential equation. In Wickert (1992) and Chen and Yang (2005), another form of the governing equation of a moving beam was also established based on the quasi-static stretch assumption. With this assumption, the strain is assumed to be a function of time only, and the spatiotemporal strain is replaced by an average value. Accordingly, the mathematical model of an axially moving beam is derived, namely

$$\rho A\left(w_{tt}+2v_0 w_{xt}+v_0^2 w_{xx}\right)-T_0 w_{xx}-w_{xx}\int_0^l \frac{EA}{2l} w_x^2 \mathrm{d}x+EIw_{xxxx}=0. \tag{3.11}$$

Equation (3.11) shows that the dynamics of the beam can be described by an integro-partial differential equation (IPDE). The entire derivation of this equation and the quasi-static stretch assumption are discussed in Chap. 5. Chen and Yang (2005) performed a study comparing the dynamics of axially moving beams based on the PDE model and IPDE model.

3.2.2 Varying-Length Beam

In this section, the mathematical model of axially moving beams with varying length $l(t)$ is discussed. We further investigate the cantilever beam case, as shown in Fig. 3.2. This model exists in many systems, such as spacecraft antennas, flexible robotic end-effectors, and hoisting cables.

In Fig. 3.2, one end of the beam is simply supported, whereas another end is attached by an end mass m_e. The axial force along the beam, $P(x, t)$, and the kinetic energy of the considered beam can be given as (2.23) and (2.24), respectively. The

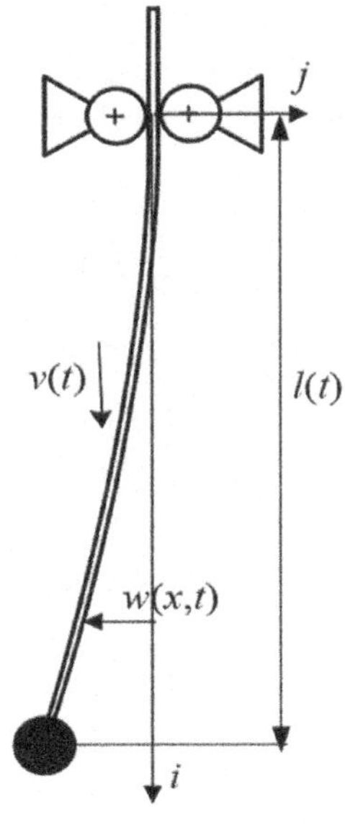

Fig. 3.2 Axially moving beam with an end mass

potential energy, including the energy owing to the bending moment, is given as $U = 1/2\int_0^{l(t)} \left(P(x,t)w_x^2 + EIw_{xx}^2\right)\mathrm{d}x$. The domains of integration for spatial variable x in the kinetic and potential energies also depend on time. The dynamic model of the moving cantilever beam can be developed by an approach similar to that discussed in Sect. 2.2.4. Using the extended Hamilton principle yields

$$
\begin{aligned}
0 &= \int_{t_1}^{t_2}\left(\delta\int_0^{l(t)} L_\mathrm{b}(x,t)dx + \delta L_\mathrm{e}(t)\right)\mathrm{d}t - \frac{1}{2}\int_{t_1}^{t_2}\left(\delta\int_0^{l(t)} EIw_{xx}^2\mathrm{d}x\right)\mathrm{d}t \\
&= \int_{t_1}^{t_2}\left[L_\mathrm{b}(l(t),t) + EIw_{xx}^2(l(t),t)\right]\delta l\mathrm{d}t \\
&+ \int_{t_1}^{t_2}\int_0^{l(t)}\left[\frac{\partial L_\mathrm{b}}{\partial w}\delta w + \frac{\partial L_\mathrm{b}}{\partial w_x}\delta w_x + \frac{\partial L_\mathrm{b}}{\partial w_t}\delta w_t\right]\mathrm{d}x\mathrm{d}t \\
&+ \int_{t_1}^{t_2}\delta L_\mathrm{e}(t)\mathrm{d}t - \frac{1}{2}\int_{t_1}^{t_2}\int_0^{l(t)}\delta EIw_{xx}^2\mathrm{d}x\mathrm{d}t
\end{aligned}
\tag{3.12}
$$

where

$$
L_\mathrm{b} = \frac{1}{2}\rho A\left[\dot{l}(t)^2 + w_t^2 + 2\dot{l}(t)w_t w_x + \dot{l}(t)^2 w_x^2\right] - \frac{1}{2}P(x,t)w_x^2, \tag{3.13}
$$

$$
L_\mathrm{e} = \frac{1}{2}m_\mathrm{e}\left[\dot{l}(t)^2 + \left(w_t(l(t),t) + \dot{l}(t)w_x(l(t),t)\right)^2\right]. \tag{3.14}
$$

Following the Leibniz integral rule and procedure for integration by parts, (3.12) can be rewritten as follows:

$$
\begin{aligned}
0 = &-\int_{t_1}^{t_2}\int_0^{l(t)}\left[\rho A\left(w_{tt} + \ddot{l}(t)w_x + 2\dot{l}(t)w_{xt} + \dot{l}(t)^2 w_{xx}\right)\right. \\
&\left.-(P(x,t)w_x)_x + EIw_{xxxx}\right]\delta w\mathrm{d}x\mathrm{d}t \\
&-\int_{t_1}^{t_2}\left[\left[m_\mathrm{e}\left(w_{tt} + \ddot{l}(t)w_x + 2\dot{l}(t)w_{xt} + \dot{l}(t)^2 w_{xx}\right)\right.\right. \\
&\left.\left.+P(x,t)w_x - EIw_{xxx}\right]\delta w\right]_{x=l(t)}\mathrm{d}t \\
&+\int_{t_1}^{t_2}\left[\rho A\dot{l}(t)\left(w_t + \dot{l}(t)w_x\right) - P(x,t)w_x + EIw_{xxx}\right]\delta w\Big|_{x=0}\mathrm{d}t
\end{aligned}
$$

$$+\int_{t_1}^{t_2} EIw_{xx}\delta w_x|_{x=0}^{x=l(t)}\mathrm{d}t + \int_{t_1}^{t_2} \left[L_b(l(t),t) + EIw_{xx}^2(l(t),t)\right]\delta l\mathrm{d}t. \quad (3.15)$$

The resulting governing equations for the cantilever beam and the end mass, respectively, are

$$\rho A\left(w_{tt} + \ddot{l}(t)w_x + 2\dot{l}(t)w_{xt} + \dot{l}(t)^2 w_{xx}\right) - (P(x,t)w_x)_x + EIw_{xxxx} = 0, \quad (3.16)$$

$$m_e\left[w_{tt}(l(t),t) + \ddot{l}(t)w_x(l(t),t) + 2\dot{l}(t)w_{xt}(l(t),t) + \dot{l}(t)^2 w_{xx}(l(t),t)\right] + P(l(t),t)w_x(l(t),t) - EIw_{xxx}(l(t),t) = 0, \quad (3.17)$$

and the boundary conditions are $w(0,t) = w_x(0,t) = 0$ and $w_{xx}(l(t),t) = 0$.

3.2.3 *Viscoelastic Beams*

In the study of axially moving strings, the viscoelastic property of the material can be considered. The influence of material damping on the dynamics can be modeled as some types of viscoelastic strain–stress constitutive relations such as the Kelvin viscoelastic model (Ghayesh 2008; Zhang et al. 2007), fractional differentiation model (Yang et al. 2013), Zener model (Zhao et al. 2006), Burgers model (Zhao and Chen 2002), and Boltzmann's superposition principle (Zhao and Chen 2008). The material's viscoelastic models can be directly applied to the study of axially moving beams.

Although the elastic model can describe most dynamics of the axially moving material, the material's damping can reveal further phenomena since new plastics and composite materials have been used in the industry. The general one-dimensional constitutive equation for differential-type materials obeys the following relationship (Marynowski and Kapitaniak 2002)

$$P_V\sigma(x,t) = Q_V\varepsilon(x,t) \quad (3.18)$$

where P_V and Q_V are differential operators defined as

$$P_V = \sum_{j=0}^{n} a_j\frac{\partial^j}{\partial t^j}, \quad Q_V = \sum_{j=0}^{m} b_j\frac{\partial^j}{\partial t^j}. \quad (3.19)$$

Different parameters a_j and b_j yield a series of viscoelastic models. The most fundamental and simplest two-parameter viscoelastic mode of material is the Kelvin viscoelastic model. The general model (3.18) reduces to

$$a_0\sigma(x,t) = b_0\varepsilon(x,t) + b_1\varepsilon_t(x,t) \tag{3.20}$$

where

$$a_0 = 1,\ b_0 = E,\ b_1 = \mu. \tag{3.21}$$

The two viscoelastic parameters (elastic modulus E and viscosity μ) determine the material properties. The bending moment M applied to the beam model is then

$$M = EIw_{xx} + \mu I w_{xxt}. \tag{3.22}$$

The nonlinear governing equation can be obtained, from (3.4), as follows.

$$\rho A\left(w_{tt} + 2v_0 w_{xt} + v_0^2 w_{xx}\right) - (Tw_x)_x + M_{xx} = 0 \tag{3.23}$$

where the perturbed axial stress is the superposition of an initial tension force T_0 producing the required initial stress and the nonlinear perturbation force with the strain equation given by

$$\varepsilon(x,t) = \frac{1}{2}w_x^2(x,t). \tag{3.24}$$

Substituting (3.20), (3.22), and (3.24) into (3.23) yields the nonlinear governing equation of an axially moving beam with the Kelvin viscoelastic material as follows.

$$\begin{aligned}&\rho A\left(w_{tt} + 2v_0 w_{xt} + v_0^2 w_{xx}\right) + EIw_{xxxx} + \mu I w_{xxxxt} - T_0 w_{xx} - \frac{3}{2}EAw_x^2 w_{xx}\\&- 2\mu A\left(w_x w_{xt} w_{xx} + v_0 w_x w_{xx}^2\right) - \mu A\left(w_x^2 w_{xxt} + v_0 w_x^2 w_{xxxx}\right) = 0.\end{aligned} \tag{3.25}$$

The other two-parameter viscoelastic model is the Maxwell model with the general model (3.18) recovering to

$$a_1\sigma_t + a_0\sigma = b_1\varepsilon_t \tag{3.26}$$

where

$$b_1 = 1,\ a_0 = \frac{1}{\mu},\ b_0 = \frac{1}{E}. \tag{3.27}$$

By integration manipulation, the Maxwell constitutive equation becomes

$$\sigma(x,t) = \mathrm{e}^{-\frac{E}{\mu}t}\left(\sigma_0 + E\int_0^t \frac{\partial\varepsilon(x,\tau)}{\partial\tau}\mathrm{e}^{\frac{E}{\mu}\tau}\mathrm{d}\tau\right). \tag{3.28}$$

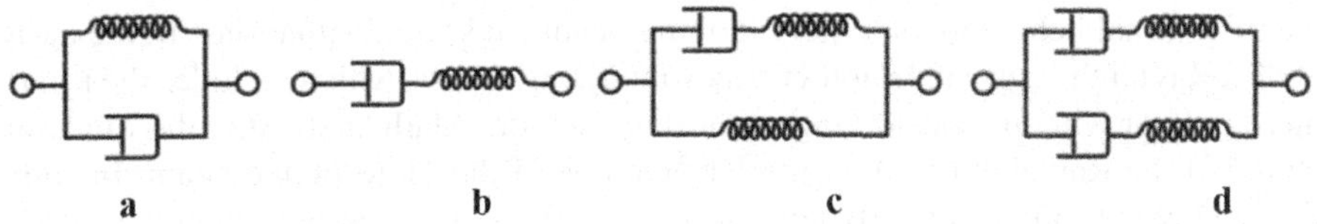

Fig. 3.3 Diagrams of viscoelastic models: **a** Kelvin model, **b** Maxwell model, **c** standard linear solid model, **d** Burgers model

It is clear that the Maxwell constitutive equation is not suitable to the solid materials, since the strain will be infinite if a finite static stress is applied. The most common three-parameter viscoelastic model is the standard linear solid model (or the Zener model). In this case, the general model (3.18) recovers to

$$a_1\sigma_t(x,t) + a_0\sigma(x,t) = b_1\varepsilon_t(x,t) + b_0\varepsilon(x,t) \tag{3.29}$$

where

$$a_1 = \mu,\ a_0 = E_2,\ b_1 = (E_1 + E_2)\mu,\ b_0 = E_1E_2. \tag{3.30}$$

The viscoelastic modes can be illustrated by the spring-damper combinations as presented in Fig. 3.3. Further, a more complex model (four-parameter Burgers model) can be considered. Then, the general model (3.18) becomes

$$a_2\sigma_{tt}(x,t) + a_1\sigma_t(x,t) + a_0\sigma(x,t) = b_2\varepsilon_{tt}(x,t) + b_1\varepsilon_t(x,t) \tag{3.31}$$

where

$$a_2 = \mu_1\mu_2,\ a_1 = (E_1 + E_2)\mu_1 + E_1\mu_2,\ a_0 = E_1E_2,\ b_2 = E_1\mu_1\mu_2,\ b_1 = E_1E_2\mu_1. \tag{3.32}$$

The application of viscoelastic models makes the governing equation more complicated with multiple linear and nonlinear terms as presented in (3.25). The detailed governing equations can be found in Marynowski and Kapitaniak (2002). It has been concluded that the simple Kelvin viscoelastic model and the complicated Burgers model give similar results for small values of internal damping coefficients (smaller than 3×10^{-5}). For the materials with a large damping coefficient, a model with more parameters gives more reliable results, with the price of more computation.

3.2.4 Rayleigh Beam

The Euler–Bernoulli beam theory is a simple beam-modeling theory. However, since this theory disregards the rotary inertia and shear effects, it cannot accurately predict the vibrational behaviors of axially moving beams in several situations. In the early 1890s, Rayleigh suggested another beam-modeling theory, which includes the rotary inertia effect without considering shear deformation. Mathematically, the dynamic model of the Rayleigh beam is simpler than that of the Timoshenko beam. In addition, this theory models the dynamics of the beam more accurately than the Euler–Bernoulli beam theory. The Rayleigh beam theory was utilized to establish the governing equations of axially moving beams in Ghayesh and Balar (2008) and Rezaee and Lotfan (2015).

We now develop the equation of motion of the axially moving beam demonstrated in Fig. 3.1 by using the Rayleigh beam theory. Considering the rotary inertia effect, the kinetic energy of the Rayleigh beam includes the term corresponding to this effect, namely

$$K_I = \frac{1}{2}\rho I \int_0^l (w_{xt} + v_0 w_{xx})^2 \mathrm{d}x \tag{3.33}$$

where ρI is the mass moment of inertia per unit length about the neutral axis of the beam. The kinetic energy of the beam is given as follows:

$$K = \frac{1}{2}\rho A \int_0^l \left[v_0^2 + (w_t + v_0 w_x)^2\right]\mathrm{d}x + \frac{1}{2}\int_0^l \rho I (w_{xt} + v_0 w_{xx})^2 \mathrm{d}x. \tag{3.34}$$

Using the extended Hamilton principle yields the following equation:

$$\begin{aligned} 0 = {} & \frac{1}{2}\int_{t_1}^{t_2}\int_0^l \delta L_b \mathrm{d}x\mathrm{d}t - \frac{1}{2}\int_{t_1}^{t_2}\int_0^l \delta(M(x,t)w_{xx})\mathrm{d}x\mathrm{d}t \\ & + \frac{1}{2}\int_{t_1}^{t_2}\int_0^l \delta\left[\rho I (w_{xt} + v_0 w_{xx})^2\right]\mathrm{d}x\mathrm{d}t. \end{aligned} \tag{3.35}$$

An interesting observation is that the difference between (3.35) and (3.2) is the presence of the last term on the right-hand side of (3.35). We express this term as follows:

$$\begin{aligned} \frac{\rho I}{2}\int_{t_1}^{t_2}\int_0^l \delta(w_{xt} + v_0 w_{xx})^2 \mathrm{d}x\mathrm{d}t = {} & \frac{1}{2}\rho I \int_{t_1}^{t_2}\int_0^l \left[\frac{\partial(w_{xt} + v_0 w_{xx})^2}{\partial w_{xx}}\delta w_{xx}\right. \\ & \left. + \frac{\partial(w_{xt} + v_0 w_{xx})^2}{\partial w_{xt}}\delta w_{xt}\right]\mathrm{d}x\mathrm{d}t \end{aligned}$$

$$
\begin{aligned}
&= \rho I \int_{t_1}^{t_2} \int_0^l [v_0(w_{xt} + v_0 w_{xx})\delta w_{xx} \\
&\quad + (w_{xt} + v_0 w_{xx})\delta w_{xt}]\mathrm{d}x\mathrm{d}t \\
&= \rho I \int_{t_1}^{t_2} \int_0^l (v_0 w_{xxxt} \\
&\quad + v_0^2 w_{xxxx} + w_{xxtt} + v_0 w_{xxxt})\delta w \mathrm{d}x\mathrm{d}t \\
&\quad + \rho I v_0 \int_{t_1}^{t_2} [(w_{xt} + v_0 w_{xx})\delta w_x]|_{x=0}^{x=l}\mathrm{d}t \\
&\quad - \rho I v_0 \int_{t_1}^{t_2} [(w_{xxt} + v_0 w_{xxx})\delta w]|_{x=0}^{x=l}\mathrm{d}t \\
&\quad + \rho I \int_{t_1}^{t_2} [(w_{xt} + v_0 w_{xx})\delta w_t]|_{x=0}^{x=l}\mathrm{d}t \\
&\quad - \rho I \int_0^l [(w_{xxt} + v_0 w_{xxx})\delta w]|_{t=t_1}^{t=t_2}\mathrm{d}x.
\end{aligned}
\tag{3.36}
$$

Substituting (3.36) into (3.35) and applying the same procedure that was applied in Sect. 3.2.1, (3.35) can be rewritten as follows:

$$
\begin{aligned}
0 &= \int_{t_1}^{t_2} \int_0^l \left[-\rho A\left(w_{tt} + 2v_0 w_{xt} + v_0^2 w_{xx}\right) + (T(x,t)w_x)_x - (M(x,t))_{xx}\right. \\
&\quad \left. + \rho I\left(2v_0 w_{xxxt} + v_0^2 w_{xxxx} + w_{xxtt}\right)\right]\delta w \mathrm{d}x\mathrm{d}t \\
&\quad - \int_{t_1}^{t_2} [M(x,t) - \rho I v_0(w_{xt} + v_0 w_{xx})]\delta w_x|_{x=0}^{x=l}\mathrm{d}t \\
&\quad + \int_0^l \left[\frac{1}{2}\frac{\partial L_\mathrm{b}}{\partial w_t} - \rho I(w_{xxt} + v_0 w_{xxx})\right]\delta w\Bigg|_{t=t_1}^{t=t_2} \mathrm{d}x \\
&\quad + \rho I \int_{t_1}^{t_2} [(w_{xt} + v_0 w_{xx})\delta w_t]|_{x=0}^{x=l}\mathrm{d}t \\
&\quad + \int_{t_1}^{t_2} \left[\frac{1}{2}\frac{\partial L_\mathrm{b}}{\partial w_x} + (M(x,t))_x - \rho I v_0(w_{xxt} + v_0 w_{xxx})\right]\delta w\Bigg|_{x=0}^{x=l} \mathrm{d}t.
\end{aligned}
\tag{3.37}
$$

According to (3.37), we obtain the dynamic model of an axially moving Rayleigh beam as follows.

$$\rho A\left(w_{tt} + 2v_0 w_{xt} + v_0^2 w_{xx}\right) - (T(x,t)w_x)_x + (M(x,t))_{xx} \\ - \rho I\left(2v_0 w_{xxxt} + v_0^2 w_{xxxx} + w_{xxtt}\right) = 0, \tag{3.38}$$

and the boundary conditions are

$$w(0,t) = 0,\ w(l,t) = 0,\ w_{xx}(0,t) = 0,\ w_{xx}(l,t) = 0. \tag{3.39}$$

3.2.5 Timoshenko Beam

The above analysis of axially moving beam is based on the Euler–Bernoulli or Rayleigh beam models. Actually, the Euler–Bernoulli beam model yields increasing inaccuracy with the increasing slenderness ratio defined by the ratio of the beam length to the radius of gyration of the cross section because the model neglects the effects of shear deformation and cross-sectional rotation. The Rayleigh beam model takes into account the rotary effect. In the early 1920s, a beam theory was proposed by Stephen Timoshenko (1991, 1992), wherein the shear effect and the effect of rotation of the cross section are added to the Euler–Bernoulli beam. This theory is known as the Timoshenko beam theory. The Timoshenko model is a significant improvement for thick and short beams, and high-frequency responses since the influences of shear or rotary of the cross section are considerable in these cases (Han et al. 1999).

The Timoshenko model should be used to study axially moving thick beams to account for the nonslenderness of a beam. A simply-supported uniform Timoshenko beam, with length l, mass density ρ, initial tension T, cross-sectional area A, and moment of inertia I, axially moves at the time-dependent moving velocity $v(t)$. The bending vibration is characterized by two variables dependent on time t and axial coordinate x, namely lateral vibration $w(x,t)$ and the rotation angle $\theta(x,t)$ of the cross section by cause of the bending moment. According to balance equations of both lateral and rotation directions, the coupled governing equations are obtained as

$$\rho A \frac{\mathrm{D}^2 w}{\mathrm{D}t^2} - T_0 w_{xx} - \kappa AG(w_{xx} - \theta_x) = 0, \tag{3.40}$$

$$\rho I \frac{\mathrm{D}^2 \theta}{\mathrm{D}t^2} - EI\theta_{xx} - \kappa AG(w_x - \theta) = 0, \tag{3.41}$$

where κ is the shape factor, E is the modulus of elasticity, and G is the shearing modulus of the beam. Taking the second derivatives of (3.40) with respect to x and t, respectively, and substituting the two resulting equations into the derivative of (3.41) with respect to x yields

$$EI\frac{\partial^4 w}{\partial x^2\partial t^2}-\rho I\frac{\partial^4 w}{\partial x^2\partial t^2}+\left(\frac{\rho I}{\kappa GA}\frac{\partial^2}{\partial t^2}-\frac{EI}{\kappa GA}\frac{\partial^2}{\partial x^2}+1\right)\left(\rho A\frac{\mathrm{D}^2 w}{\mathrm{D}t^2}-T_0\frac{\partial^2 w}{\partial x^2}\right)=0, \tag{3.42}$$

where the material derivative for varying axial velocity is

$$\frac{\mathrm{D}^2 w}{\mathrm{D}t^2}=\frac{\partial^2 w}{\partial t^2}+2v\frac{\partial^2 w}{\partial x\partial t}+v^2\frac{\partial^2 w}{\partial x^2}+\dot{v}\frac{\partial w}{\partial x}. \tag{3.43}$$

The equation of motion of the longitudinally translating Timoshenko beam reduces to that of the Euler-Bernoulli beam after ignoring the second term and letting $\kappa \to \infty$.

Introduce the following dimensionless variables and parameters

$$\overline{w}=\frac{w}{l},\ \overline{x}=\frac{x}{l},\ \overline{t}=t\sqrt{\frac{T_0}{\rho AL^2}},\ \overline{v}_0=v_0\sqrt{\frac{\rho A}{T_0}},$$
$$k_0=\frac{\kappa\rho T_0}{GA^2l^2},\ k_1=\frac{\kappa IE}{GAl^2},\ k_2=\frac{EA}{T_0},\ k_3=\frac{I}{Al^2},\ k_4=\frac{EI}{T_0l^2}. \tag{3.44}$$

The dimensionless governing equation of the longitudinally translating Timoshenko beam with constant velocity v_0 is then

$$\begin{aligned}&\overline{w}_{\overline{t}\overline{t}}+2\overline{v}_0\overline{w}_{\overline{x}\overline{t}}+(\overline{v}_0^2-1)\overline{w}_{\overline{x}\overline{x}}-(k_0+k_1+k_3-k_0\overline{v}_0^2)\overline{w}_{\overline{x}\overline{x}\overline{t}\overline{t}}\\&+(k_1+k_4-k_1\overline{v}_0^2)\overline{w}_{\overline{x}\overline{x}\overline{x}\overline{x}}+k_0(\overline{w}_{\overline{t}\overline{t}\overline{t}\overline{t}}+2\overline{v}_0\overline{w}_{\overline{x}\overline{t}\overline{t}\overline{t}})\\&-2k_1\overline{v}_0\overline{w}_{\overline{x}\overline{x}\overline{x}\overline{t}}=0.\end{aligned} \tag{3.45}$$

In the subsequent analysis, both hinged ends are assumed. Then, the boundary conditions are

$$\overline{w}(0,t)=\overline{w}(1,t)=0,\ \overline{w}_{\overline{x}\overline{x}}(0,\overline{t})=\overline{w}_{\overline{x}\overline{x}}(1,\overline{t})=0. \tag{3.46}$$

Substitution of the solution

$$\overline{w}(\overline{x},\overline{t})=\varphi_n(\overline{x})\mathrm{e}^{\mathrm{i}\omega_n\overline{t}}+\varphi_n^*(\overline{x})\mathrm{e}^{-\mathrm{i}\omega_n\overline{t}} \tag{3.47}$$

into (3.45) and (3.46) yields

$$\begin{aligned}&k_0\varphi_n\omega_n^4-2\mathrm{i}k_0\overline{v}_0\varphi_n'\omega_n^3+[(k_0+k_1+k_3-k_0\overline{v}_0^2)\varphi_n''-\varphi_n]\omega_n^2\\&+2\mathrm{i}\overline{v}_0(\varphi_n'-k_1\varphi_n''')\omega_n+(\overline{v}_0^2-1)\varphi_n''+(k_1+k_4-k_1\overline{v}_0^2)\varphi_n''''=0,\end{aligned} \tag{3.48}$$

$$\varphi_n(0) = 0, \varphi_n(1) = 0, \varphi_n''(0) = 0, \varphi_n''(1) = 0. \tag{3.49}$$

The complex mode functions take the form

$$\varphi_n(\overline{x}) = C_{1n}\left(e^{i\beta_{1n}\overline{x}} + C_{2n}e^{i\beta_{2n}\overline{x}} + C_{3n}e^{i\beta_{3n}\overline{x}} + C_{4n}e^{i\beta_{4n}\overline{x}}\right). \tag{3.50}$$

The boundary conditions require the following condition.

$$\begin{pmatrix} 1 & 1 & 1 & 1 \\ \beta_{1n}^2 & \beta_{2n}^2 & \beta_{3n}^2 & \beta_{4n}^2 \\ e^{i\beta_{1n}} & e^{i\beta_{2n}} & e^{i\beta_{3n}} & e^{i\beta_{4n}} \\ \beta_{1n}^2 e^{i\beta_{1n}} & \beta_{2n}^2 e^{i\beta_{2n}} & \beta_{3n}^2 e^{i\beta_{3n}} & \beta_{4n}^2 e^{i\beta_{4n}} \end{pmatrix} \begin{pmatrix} 1 \\ C_{2n} \\ C_{3n} \\ C_{4n} \end{pmatrix} C_{1n} = 0. \tag{3.51}$$

Further, substituting (3.50) back to (3.48) yields

$$\begin{aligned} &(k_1 + k_4 - k_1\overline{v}_0^2)\beta_{in}^4 - 2\overline{v}_0 k_1 \omega_n \beta_{in}^3 - [(k_0 + k_1 + k_3 - k_0\overline{v}_0^2)\omega_n^2 \\ &+ (1 - \overline{v}_0^2)]\beta_{in}^2 + (2k_0\overline{v}_0\omega_n^3 - 2\overline{v}_0\omega_n)\beta_{in} + k_0\omega_n^4 - \omega_n^2 = 0. \end{aligned} \tag{3.52}$$

From the above two Eqs. (3.51) and (3.52), the complex mode functions and the corresponding natural frequencies can be obtained. The nth complex mode functions are expressed as follows.

$$\begin{aligned} \varphi_n(x) = {} & e^{i\beta_{1n}x} - \frac{\left(\beta_{4n}^2 - \beta_{1n}^2\right)\left(e^{i\beta_{3n}} - e^{i\beta_{1n}}\right)}{\left(\beta_{4n}^2 - \beta_{2n}^2\right)\left(e^{i\beta_{3n}} - e^{i\beta_{2n}}\right)} e^{i\beta_{2n}x} - \frac{\left(\beta_{4n}^2 - \beta_{1n}^2\right)\left(e^{i\beta_{2n}} - e^{i\beta_{1n}}\right)}{\left(\beta_{4n}^2 - \beta_{3n}^2\right)\left(e^{i\beta_{2n}} - e^{i\beta_{3n}}\right)} e^{i\beta_{3}x} \\ & + \left(-1 + \frac{\left(\beta_{4n}^2 - \beta_{1n}^2\right)\left(e^{i\beta_{3n}} - e^{i\beta_{1n}}\right)}{\left(\beta_{4n}^2 - \beta_{2n}^2\right)\left(e^{i\beta_{3n}} - e^{i\beta_{2n}}\right)} + \frac{\left(\beta_{4n}^2 - \beta_{1n}^2\right)\left(e^{i\beta_{2n}} - e^{i\beta_{1n}}\right)}{\left(\beta_{4n}^2 - \beta_{3n}^2\right)\left(e^{i\beta_{2n}} - e^{i\beta_{3n}}\right)}\right) e^{i\beta_{4n}x}. \end{aligned} \tag{3.53}$$

By a given set of parameters, the varying natural frequencies of the first two orders are plotted in Fig. 3.4. If the parameters $\overline{v}_0 = 1$ and $k_4 = 1$ are taken, the

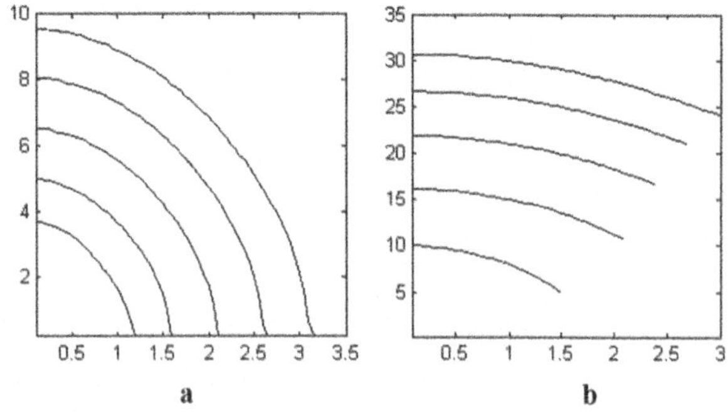

Fig. 3.4 Varying natural frequencies with axial moving velocity: **a** the first-order, **b** the second-order

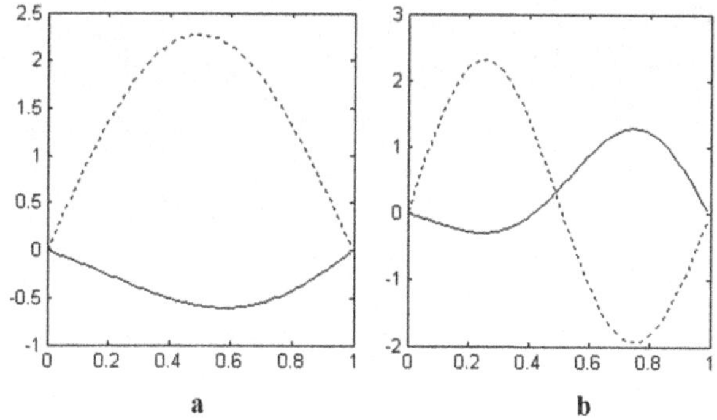

Fig. 3.5 Complex mode functions: **a** the first-order, **b** the second-order

complex mode functions of the first two orders are shown in Fig. 3.5. The mentioned viscoelastic models can be used for the dynamics investigation of axially moving Timoshenko beams. The axially moving Timoshenko beam with Kelvin viscoelastic model can be obtained by substituting modulus E by the viscoelastic operator $E+\mu\frac{\partial}{\partial t}$ in (3.22).

3.2.6 Beams with Nonhomogenous Boundaries

The nonhomogenous boundary problems of axially moving beams are treated in this subsection. In details, two dynamic models of a belt–pulley system with support wheels and a one-way clutch are derived from the Euler–Bernoulli beam theory.

3.2.6.1 Beams with Support Wheels

In practice, axially moving systems is usually supported or restrained by the wheels. Therefore, the boundary of the traveling system is not perfectly simply supported or fixed. Under the effects of the supports or constraints of the wheels, the boundary conditions of the translating system become nonhomogenous (Wang and Mote 1986; Kong and Parker 2005; Zhu and Parker 2008; Ding and Zu 2014). Nonhomogeneity is an inevitable property when considering the boundary of the belt-wheel contact.

A typical belt-drive device consists of a traveling beam supported by rotating wheels. Figure 3.6 demonstrates the schema of an axially moving beam and two wheels. In this figure, r_1 and r_2 denote the radii of the left and right wheels, respectively. Suppose that the support shafts of the wheels undergo a harmonic oscillation $A_e\cos(\Omega t)$ in the vertical direction, where A_e denotes the vibration magnitude, and Ω is the angular frequency. Thus, the entire dynamical system is subject to harmonic excitation. Using Newton's second law, the lateral and longitudinal vibrations' governing equations are derived for the traveling beam as follows.

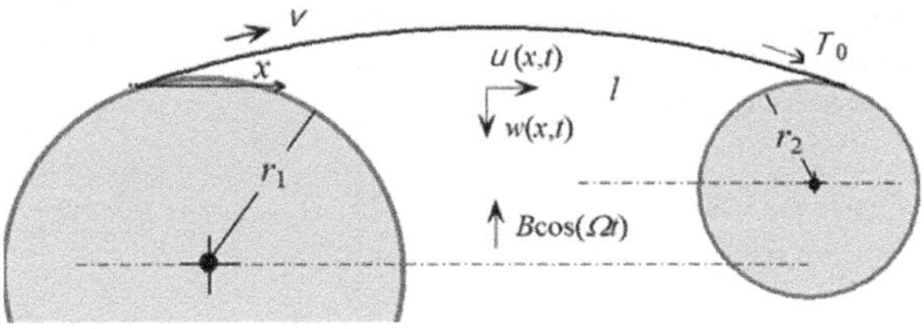

Fig. 3.6 Schema of a typical belt-drive device

$$\begin{aligned}&\rho A\left(w_{tt}+2v_0w_{xt}+v_0^2w_{xx}\right)+M_{xx}(x,t)-(Tw_x)_x=A_e\rho A\Omega^2\cos(\Omega t),\\&\rho A\left(u_{tt}+2v_0u_{xt}+v_0^2u_{xx}\right)-(T_0+T_d)_x=0\end{aligned}\tag{3.54}$$

where $M(x, t)$ denotes the bending moment. Since the longitudinal displacement is insignificant, the beam's longitudinal vibration can be neglected. The total tension T, including the static tension T_0 and the dynamic tension T_d, is given as follows.

$$T=T_0+T_\mathrm{d}=T_0+\frac{1}{2}EAw_x^2\tag{3.55}$$

Under the quasi-static condition, T_d can be expressed in an integral form as follows.

$$T_\mathrm{d}=\frac{1}{2}EAw_x^2\approx\frac{EA}{2l}\int_0^l w_x^2\mathrm{d}x.\tag{3.56}$$

The Kelvin model is employed to describes the viscoelastic property of the axially moving material with the following constitutional relation.

$$M(x,t)=\left(E+\mu\frac{\partial}{\partial t}\right)Iw_{xx}.\tag{3.57}$$

Therefore, the dynamic model for the lateral vibration of the system can be obtained as follows.

$$\begin{aligned}&\rho A\left(w_{tt}+2v_0w_{xt}+v_0^2w_{xx}\right)-T_0w_{xx}+\left(E+\mu\frac{\partial}{\partial t}\right)Iw_{xxxx}-\frac{EA}{2l}w_{xx}\int_0^l w_x^2dx\\&=A_e\rho A\Omega^2\cos(\Omega t).\end{aligned}\tag{3.58}$$

The nonhomogeneous boundary conditions are derived as follows.

$$\begin{aligned}&w(0,t)=0,\ w(l,t)=0,\\&EIw_{xx}(0,t)=EI/r_1,\ EIw_{xx}(l,t)=EI/r_2.\end{aligned}\tag{3.59}$$

The entire derivation of the mentioned boundary conditions can be found in Wang and Mote (1986). Nonhomogeneous boundary conditions depend on the wheels' radii and the bending stiffness. The nonhomogeneous boundary conditions will become simply supported boundary conditions if a wheel's radius tends to infinity. Additionally, the axially translating beam model will become a moving string model if the bending stiffness is zero. In this case, the second pair of boundary conditions in (3.59) does not exist.

Now, introduce the following nondimensional parameters.

$$k_1 = \frac{EA}{T_0}, \alpha = \frac{Ic}{l^3\sqrt{\rho A T_0}}, k_f = \sqrt{\frac{EI}{T_0 l^2}}, \overline{a}_e = \frac{A_e}{l}, \overline{\omega} = \Omega\sqrt{\frac{\rho A l^2}{T_0}} \tag{3.60}$$

Then, the dimensionless equation of motion and boundary conditions is derived as follows.

$$\begin{aligned} &\overline{w}_{\overline{t}\overline{t}} + 2\overline{v}_0\overline{w}_{\overline{x}\overline{t}} + \left(\overline{v}_0^2 - 1\right)\overline{w}_{\overline{x}\overline{x}} + k_f^2\overline{w}_{\overline{x}\overline{x}\overline{x}\overline{x}} + \alpha\overline{w}_{\overline{x}\overline{x}\overline{x}\overline{x}\overline{t}} \\ &= \frac{k_1}{2}\overline{w}_{\overline{x}\overline{x}}\int_0^1 \overline{w}_{\overline{x}}^2 \mathrm{d}\overline{x} + \overline{a}_e\overline{\omega}^2\cos\left(\overline{\omega}\overline{t}\right), \end{aligned} \tag{3.61}$$

$$\begin{aligned} &\overline{w}\left(0, \overline{t}\right) = 0, \overline{w}\left(1, \overline{t}\right) = 0, \\ &\overline{w}_{\overline{x}\overline{x}}\left(0, \overline{t}\right) = l/r_1, \overline{w}_{\overline{x}\overline{x}}\left(1, \overline{t}\right) = l/r_2. \end{aligned} \tag{3.62}$$

The equilibrium deformation, $w_\infty(x)$, of (3.61) satisfies

$$\left(\overline{v}_0^2 - 1\right)w_\infty'' + k_f^2 w_\infty'''' = \frac{k_1}{2}w_\infty''\int_0^1 {w_\infty'}^2 \mathrm{d}x. \tag{3.63}$$

Correspondingly, the equilibrium boundary conditions satisfy

$$\begin{aligned} &w_\infty(0) = 0, w_\infty(1) = 0, \\ &w_\infty''(0) = l/r_1, w_\infty''(1) = l/r_2. \end{aligned} \tag{3.64}$$

The numerical methods, such as the quadrature method (Ding et al. 2018), can be used to derive the equilibrium configurations. The differential and integral quadrature methods (DIQMs) transform (3.63) and (3.64) into a set of algebraic equations as follows

$$\sum_{k=1}^{N}\left[\left(\overline{v}_0^2 - 1\right)A_{jk}^{(2)} + k_f^2 A_{jk}^{(4)}\right.$$

$$
\begin{aligned}
&-\frac{k_1}{2}\sum_{g=1}^{N}\left\{I_g\left[\sum_{k=1}^{N}A_{gk}^{(1)}w_{\infty k}\right]^2\right\}A_{jk}^{(2)}\Bigg]w_{\infty k}=0,\quad (j=3,4,\ldots,N-2)\\
&\sum_{k=1}^{N}A_{2k}^{(2)}w_{\infty k}=\frac{l}{r_1},\\
&\sum_{k=1}^{N}A_{(N-1)k}^{(2)}w_{\infty k}=\frac{l}{r_2},\\
&w_{\infty 1}=w_{\infty N}=0.
\end{aligned}
\tag{3.65}
$$

It is noted that the integral weighting coefficients I_g ($g = 1, 2, \ldots, N$) in (3.65) are calculated as follows.

$$
\begin{pmatrix} I_1\\ I_2\\ \vdots\\ I_{N-1}\\ I_N \end{pmatrix}
=
\begin{pmatrix} 1\\ 1/2\\ \vdots\\ 1/(N-1)\\ 1/N \end{pmatrix}
\begin{pmatrix}
1 & 1 & \cdots & 1 & 1\\
\overline{x}_1 & \overline{x}_2 & \cdots & \overline{x}_{N-1} & \overline{x}_N\\
\vdots & \vdots & & \vdots & \vdots\\
\overline{x}_1^{N-2} & \overline{x}_2^{N-2} & \cdots & \overline{x}_{(N-1)}^{N-2} & \overline{x}_N^{N-2}\\
\overline{x}_1^{N-1} & \overline{x}_2^{N-1} & \cdots & \overline{x}_{(N-1)}^{N-1} & \overline{x}_N^{N-1}
\end{pmatrix}^{-1}
\tag{3.66}
$$

where $\overline{x}_i$ ($i = 1, 2, \ldots, N$) denote the sampling points. The solution of the derivative weighting coefficients in (3.65) for discretizing the first-order derivative in the neutral axis coordinate $\overline{x}$ is shown as follows.

$$
A_{jk}^{(1)}=\begin{cases}
\dfrac{\prod\limits_{m=1,m\neq j}^{N}(\overline{x}_j-\overline{x}_m)}{(\overline{x}_j-\overline{x}_k)\prod\limits_{m=1,m\neq k}^{N}(\overline{x}_k-\overline{x}_m)} & \text{for } k\neq j,\ \text{with } j,k=1,2,\ldots,N,\\
\sum\limits_{m=1,m\neq j}^{N}\dfrac{1}{\overline{x}_j-\overline{x}_m} & \text{for } k=j.
\end{cases}
\tag{3.67}
$$

The higher-order weighting coefficients are derived via a recurrence relationship, namely

$$
A_{jk}^{(r)}=\sum_{m=1}^{N}A_{jm}^{(1)}A_{mk}^{(r-1)}\quad \text{for } j,k=1,2,\ldots,N,\ \text{and } r=2,3,\ldots
\tag{3.68}
$$

Then, we can utilize an iterative procedure to solve the nonlinear algebraic equations in (3.65). The iterative scheme is shown as follows.

$$
\begin{aligned}
&w_{\infty 1}=0,\\
&w_{\infty 2}=\frac{\frac{l}{r_1}-\sum_{k=1,k\neq 2}^{N}A_{2k}^{(2)}w_{\infty k}}{A_{22}^{(2)}},
\end{aligned}
$$

$$
w_{\infty j} = \frac{-\sum_{k=1,k\neq j}^{N}\left\{\left\{\overline{v}_0^2 - 1 - \frac{k_1}{2}\sum_{g=1}^{N}\left\{I_g\left[\sum_{k=1}^{N}A_{gk}^{(1)}w_{\infty k}\right]^2\right\}\right\}A_{jk}^{(2)} + k_f^2 A_{jk}^{(4)}\right\}w_{\infty k}}{\left(\overline{v}_0^2 - 1 - \frac{k_1}{2}\sum_{g=1}^{N}\left\{I_g\left[\sum_{k=1}^{N}A_{gk}^{(1)}w_{\infty k}\right]^2\right\}\right)A_{jj}^{(2)} + k_f^2 A_{jj}^{(4)}},
$$

$$(j = 3, 4, \ldots, N-2),$$

$$
w_{\infty(N-1)} = \frac{\frac{l}{r_2} - \sum_{k=1,k\neq N-1}^{N} A_{(N-1)k}^{(2)} w_{\infty k}}{A_{(N-1)(N-1)}^{(2)}},
$$

$$
w_{\infty N} = 0. \tag{3.69}
$$

The values for the first step iteration are set as follows.

$$
w_{\infty j} = 0.0001\sin(\pi\overline{x}_j), \quad (j = 1, 2, \ldots, N). \tag{3.70}
$$

Table 3.1 shows the geometrical and physical parameters of a traveling beam. Plots in Fig. 3.7 demonstrate the effects of system parameters, including the support wheel's size, the translating speed, the cross section's height, and the static tension, on the beam's equilibrium deformation with $N = 27$. When the wheel's size increases, the equilibrium curvature decreases. It is observed that the equilibrium deformation increases with the increase of the traveling speed, especially for a relatively high speed. This observation points out that the dynamical characteristics of axial moving systems can be changed by the traveling speed. In addition, the lateral equilibrium deformation of the system can be affected by the axial traveling speed. As shown in Fig. 3.7c, d, when the static tension and the cross section's height of the moving beam increase, the equilibrium curvature increases.

The lateral vibration can be analyzed based on the nontrivial equilibrium solution. The total displacement $\overline{w}(\overline{x}, \overline{t})$ is then replaced by $\tilde{w}(\overline{x}, \overline{t}) + w_\infty(\overline{x})$, where $\tilde{w}(\overline{x}, \overline{t})$ is the beam's vibration displacement measured from the curved equilibrium configuration. Thus, the following equation of motion and boundary conditions of

Table 3.1 System parameters of the considered system

Symbol	Parameters	Value
h	Height	0.005 m
b	Width	0.02 m
l	Span length	0.3 m
ρ	Density	1200 kg/m^3
E	Young's modulus	2×10^8 N/m^2
T_0	Static tension	300 N
r_1	Left wheel's radius	0.05 m
r_2	Right wheel's radius	0.02 m
c	Viscous damping	5×10^5 N s/m^2
A_e	Excitation amplitude	0.001 m

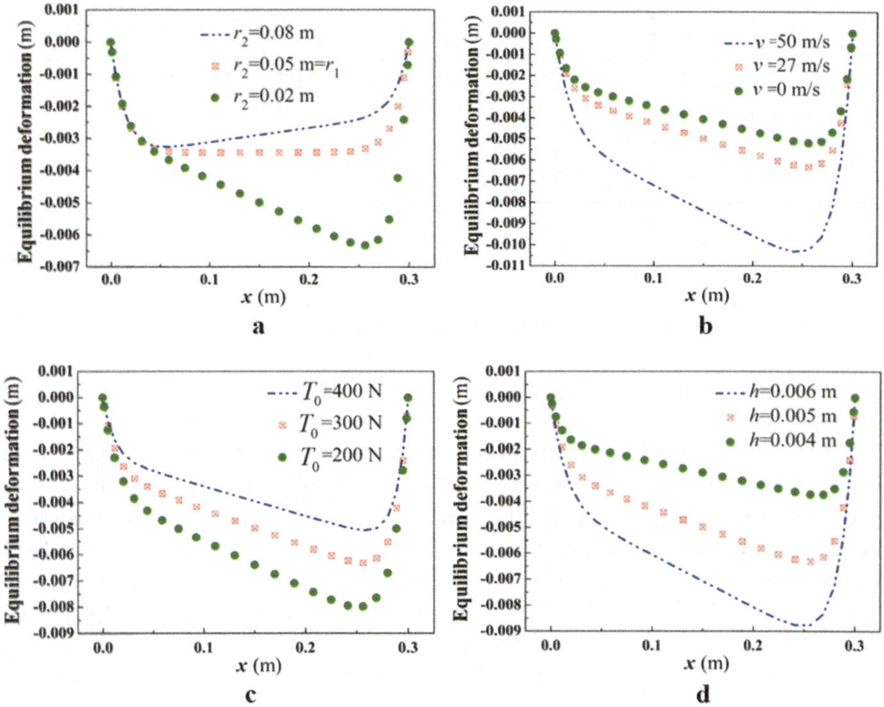

Fig. 3.7 Equilibrium deformation: **a** effect of the right wheel's radius, **b** effect of the axial traveling speed, **c** effect of the initial tension, **d** effect of the cross section's height

the vibrations around the equilibrium is obtained.

$$\tilde{w}_{\bar{t}\bar{t}} + 2\bar{v}_0\tilde{w}_{\bar{x}\bar{t}} + k_{\mathrm{f}}^2\tilde{w}_{\bar{x}\bar{x}\bar{x}\bar{x}} + \alpha\tilde{w}_{\bar{x}\bar{x}\bar{x}\bar{x}\bar{t}} + \left(\bar{v}_0^2 - 1\right)\tilde{w}_{\bar{x}\bar{x}} - k_1 w'' \int_0^1 w'_\infty \tilde{w}_x \mathrm{d}\bar{x} - \frac{k_1}{2}\tilde{w}_{\bar{x}\bar{x}} \int_0^1 {w'_\infty}^2 \mathrm{d}\bar{x}$$
$$= \frac{k_1}{2} w''_\infty \int_0^1 \tilde{w}_{\bar{x}}^2 \mathrm{d}\bar{x} + k_1\tilde{w}_{\bar{x}\bar{x}} \int_0^1 w'_\infty \tilde{w}_{\bar{x}} \mathrm{d}\bar{x} + \frac{k_1}{2}\tilde{w}_{\bar{x}\bar{x}} \int_0^1 \tilde{w}_{\bar{x}}^2 \mathrm{d}\bar{x} + \bar{a}_{\mathrm{e}}\bar{\omega}^2\cos\left(\bar{\omega}\bar{t}\right) \quad (3.71)$$

and

$$\tilde{w}\left(0, \bar{t}\right) = 0,\ \tilde{w}_{\bar{x}\bar{x}}\left(0, \bar{t}\right) = 0,\ \tilde{w}\left(1, \bar{t}\right) = 0,\ \tilde{w}_{\bar{x}\bar{x}}\left(1, \bar{t}\right) = 0. \quad (3.72)$$

To determine the characteristics of the beam's natural vibration, we ignore the excitation term, the nonlinear terms, and the damping term. Accordingly, the equation of motion for the free vibration of the axially moving beam is derived as follows.

$$\tilde{w}_{\overline{tt}} + 2\overline{v}_0\tilde{w}_{\overline{xt}} + k_f^2\tilde{w}_{\overline{xxxx}} + \left(\overline{v}_0^2 - 1\right)\tilde{w}_{\overline{xx}} = k_1 w_\infty'' \int_0^1 w_\infty' \tilde{w}_x \mathrm{d}\overline{x}$$
$$+ \frac{k_1}{2}\tilde{w}_{\overline{xx}} \int_0^1 w_\infty'^{\,2} \mathrm{d}\overline{x}. \tag{3.73}$$

Equation (3.73) can be solved via the Galerkin truncation method based on the numerical solutions of the equilibrium configuration. The lateral vibration is assumed as follows.

$$\tilde{w}\left(\overline{x}, \overline{t}\right) = \sum_{m=1}^{M} \varphi_m(\overline{x}) q_m\left(\overline{t}\right), \varphi_m = \sin(m\pi\overline{x}), \quad m = 1, 2, \ldots, M \tag{3.74}$$

where M denotes the truncation order. The IPDE in (3.73) is converted to a set of ordinary differential equations, namely

$$\sum_{m=1}^{M} \ddot{q}_m\varphi_m + 2\overline{v}_0 \sum_{m=1}^{M} \dot{q}_m\varphi_m' + \left(\overline{v}_0^2 - 1\right) \sum_{m=1}^{M} q_m\varphi_m'' + k_f^2 \sum_{m=1}^{M} q_m\varphi_m''''$$
$$= \sum_{m=1}^{M} \left[\left(k_1 w_\infty'' \int_0^1 w_\infty'\varphi_m' \mathrm{d}\overline{x}\right) q_m\right] + \frac{1}{2}k_1^2 \left[\int_0^1 w_\infty'^{\,2} \mathrm{d}\overline{x}\right] \sum_{m=1}^{M} q_m\varphi_m''. \tag{3.75}$$

The weight functions in the Galerkin method are the following.

$$\varphi_m = \sin(m\pi\overline{x}), \quad m = 1, 2, \ldots, M. \tag{3.76}$$

The truncation procedure leads to a set of ordinary differential equations as follows.

$$\sum_{m=1}^{M} M_{nm}\ddot{q}_m + \sum_{m=1}^{M} G_{nm}\dot{q}_m + \sum_{m=1}^{M} K_{nm}q_m = 0 \tag{3.77}$$

where

$$M_{mn} = \int_0^1 \sin(m\pi\overline{x}) \sin(n\pi\overline{x}) \mathrm{d}\overline{x} = \frac{1}{2}\delta_{nm},$$

$$G_{mn} = 2\overline{v}_0 \int_0^1 m\pi \cos(m\pi\overline{x}) \sin(n\pi\overline{x}) \mathrm{d}\overline{x} = \begin{cases} 0, & m = n, \\ \frac{4mn\overline{v}_0}{m^2 - n^2}, & n \neq m, \end{cases}$$

$$K_{mn} = \left(\overline{v}_0^2 - 1 - \frac{1}{2} k_1^2 \int_0^1 w_\infty'^2 \mathrm{d}\overline{x} \right) \int_0^1 (-m^2 \pi^2) \sin(m\pi\overline{x}) \sin(n\pi\overline{x}) \mathrm{d}\overline{x}$$
$$+ k_f^2 \int_0^1 (m^4 \pi^4) \sin(m\pi\overline{x}) \sin(n\pi\overline{x}) \mathrm{d}\overline{x}$$
$$- k_1 \int_0^1 [w_\infty'' \sin(n\pi\overline{x})] \mathrm{d}\overline{x} \int_0^1 [w_\infty' m\pi \cos(m\pi\overline{x})] \mathrm{d}\overline{x}. \tag{3.78}$$

Generalized coordinates are substituted by

$$q_j(\overline{t}) = Q_j e^{i\omega\overline{t}}, \quad j = 1, 2, \ldots, M \tag{3.79}$$

where ω denotes the natural frequency of the lateral vibration around the beam's equilibrium configuration. Equation (3.77) can be written in the matrix–vector form.

$$[-\omega^2 \mathbf{M} + i\omega \mathbf{G} + \mathbf{K}] \mathbf{Q} = 0. \tag{3.80}$$

Matrices M_{nm}, G_{nm}, and K_{nm} (n, m = 1, 2, ..., M) are obtained via the Galerkin procedure. The nontrivial solution exists when the determinant of the coefficient matrix in (3.80) must be zero. Accordingly, we can derive the natural frequencies ω by the following equation.

$$|-\omega^2 \mathbf{M} + i\omega \mathbf{G} + \mathbf{K}| = 0. \tag{3.81}$$

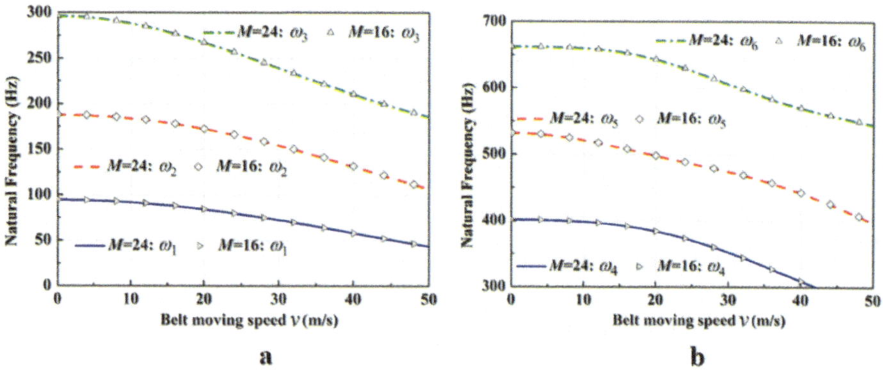

Fig. 3.8 Natural frequencies of the traveling belt with $M = 16$ and 24: **a** the first, second, and third orders, **b** the fourth, fifth, and sixth orders

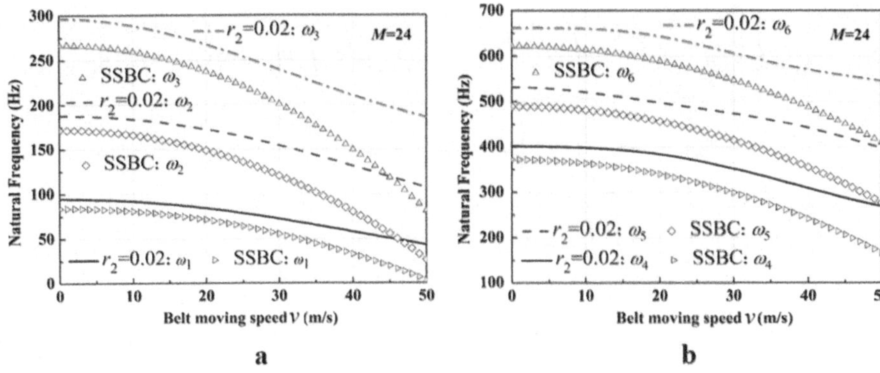

Fig. 3.9 Effects of the support wheel's radius on natural frequencies of the traveling belt: **a** the first, second, and third orders, **b** the fourth, fifth, and sixth orders

Figure 3.8 demonstrates the natural frequencies of a translating beam (i.e., the system parameters are shown in Table 3.1). The truncation method combined with the differential and integral quadrature methods is used to determine the natural frequencies, in which the number of the quadrature sampling points is $N = 27$.

Figure 3.8 shows the first eight natural frequencies of the axially moving beam with the truncation order $M = 16$ and $M = 24$. As shown in Fig. 3.8, when the axial speed decreases, the natural frequencies of the beam's lateral vibration increase. It is also observed that natural frequencies in the case $M = 16$ show the converging result to the exact solution. The effects of the nonhomogeneous boundaries on the beam's natural frequencies are demonstrated in Fig. 3.9. In this figure, "SSBC" denotes the simply-supported boundary conditions. According to the numerical results, we draw two key conclusions: (i) The impacts of the wheel radius on natural frequencies increase when the beam's axial speed increases and (ii) the smaller radius of the support wheel corresponds to the larger natural frequencies. The exploration by Ding et al. (2018) on the effects of the nonhomogeneous boundary condition on the vibration properties of the traveling beam contribute fundamentally and significantly to the understanding of the dynamics of axially moving systems.

If damping and excitation are considered, the steady-state responses of the nonhomogeneous system can be calculated by the Galerkin truncation method or the DIQMs. Based on (3.71), the set of the second-order ordinary differential equations is then

$$\ddot{q}_i = \overline{a}_e \overline{\omega}^2 \cos(\overline{\omega}\overline{t}) \int_0^1 \sin(i\pi\overline{x}) d\overline{x} - 2\sum_{m=1}^{M} G_{im}\dot{q}_m - \alpha(i\pi)^4 \dot{q}_m - k_f^2 (i\pi)^4 q_i$$

$$+\left\{\left[\overline{v}_0^2 - 1 - \frac{k_1}{2}\int_0^1 w_\infty'^{\,2}\mathrm{d}\overline{x} - k_1\sum_{m=1}^{M} m\pi \int_0^1 w_\infty' \cos(m\pi\overline{x})\mathrm{d}\overline{x}q_m - \frac{k_1\pi^2}{4}\left(q_1^2 + 4q_2^2\right)\right]i^2\pi^2\right\}q_i$$
$$+2\left\{k_1\sum_{m=1}^{M} m\pi\left[\int_0^1 w_\infty' \cos(m\pi\overline{x})\mathrm{d}\overline{x}\right]q_m + \frac{k_1\pi^2}{4}\left(q_1^2 + 4q_2^2\right)\right\}\int_0^1 \sum_{m=1}^{M} w_\infty'' \sin(i\pi\overline{x})\mathrm{d}\overline{x}, \quad i = 1, 2, \ldots, M. \tag{3.82}$$

The stable steady-state periodic response of the generalized coordinates can be determined via numerical approaches. The local maximum of the midpoint of the axially moving material at the steady-state phase can be abstracted as the magnitude of the lateral vibration.

Now, we investigate the influence of axial speed and wheel size on the resonance of an axially moving beam. The foundation's frequency is set as $\Omega = 500$ rad/s to investigate the forced vibration near the first primary resonance. Figure 3.10a shows that the axial speed significantly impacts the resonant intensity and resonant region of an axially moving beam. As shown in this figure, the amplitude and frequency of the first-order primary resonance reduce with the increase of the axial speed. Besides that, Fig. 3.10a also shows that the second-order resonance gradually disappears, and the third-order resonance amplitude gradually increases when the axial speed increases. Figure 3.10b shows steady-state amplitude–frequency responses with different boundary conditions. It is observed that the influence of the wheel size on the system resonance is significant. Thus, ignoring the wheel sizes at boundaries leads to a misunderstanding of the traveling beam's forced vibrations. In other

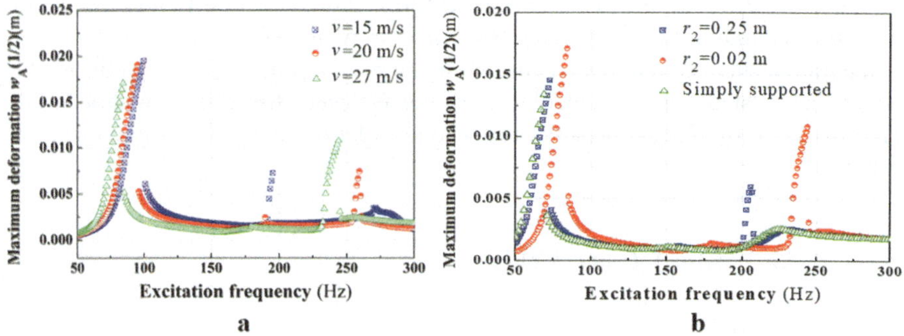

Fig. 3.10 Amplitude–frequency relationship: **a** effect of the axial velocity, **b** effect of the right wheel's radius

words, nonhomogeneous boundary conditions of axially moving materials cannot be neglected.

3.2.6.2 Beam with One-Way Clutch

Belt-drive systems in practice are usually connected with heavy accessories, such as pumps, alternators, fans, and compressors. The effects of resonances of the heavy accessories on the systems can be eliminated by using one-way clutches to remove the opposite direction torque transmission. In belt-driver systems, the driven pulley is connected with a heavy accessory via a one-way clutch.

Figure 3.11 demonstrates the schema of a belt-drive system with a one-way clutch. We assume that the power transmission belt in the considered system is modeled by a uniform axially moving beam of length l. In this schema, K_{d} denotes the stiffness of the wrap spring; v_0 and T_0 are the axial speed and static tension of the axially moving beam, respectively; M_1 indicates the preload between the accessory and the driven pulley; J_{a} denotes the rotational inertia of the accessory. The driving pulley and driven pulley are assumed to be of the same size. The radius and rotational inertia of these pulleys are denoted by r and J, respectively. x_1 and $w_1(x_1, t)$ are the neutral axis coordinate and the lateral displacements of the top belt span, respectively, and x_2 and $w_2(x_1, t)$ are that of the bottom belt span. Additionally, the angular displacements of the driven pulley, driving pulley, and accessory at time t are defined by $\theta_i(t)$ $(i = 1, 2, 3)$, respectively.

The one-way clutch is modeled based on the relative angular displacement of the driven pulley and the accessory. The accessory is treated as a load part, and the driven pulley connects with it via a wrap spring. When the driven pulley and the shaft are disengaged, the spring is disconnected (Xu and Lowen 1994; Ding and Zu 2014). The one-way clutch disengages when the angular displacement of the accessory is larger than that of the driven pulley. In other words, there is no mechanical link between the dynamic system and its accessory.

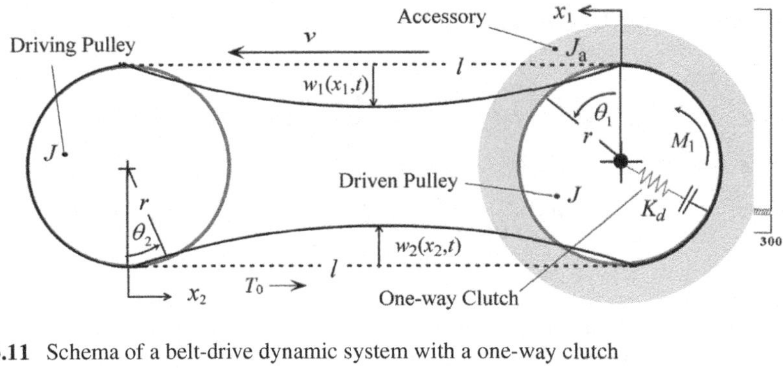

Fig. 3.11 Schema of a belt-drive dynamic system with a one-way clutch

According to Newton's second law, the following nonlinear PDEs describing the lateral vibrations of an axially moving viscoelastic beam are derived (Ding and Li 2014; Ding 2015).

$$\rho A\left(w_{i\,tt} + 2v_0 w_{ix_it} + v_0^2 w_{i\,x_ix_i}\right) - (T_0 + T_i)w_{i\,x_ix_i} + EIw_{i\,x_ix_ix_ix_i} + \mu I w_{i\,x_ix_ix_ix_it} = 0 \tag{3.83}$$

with the boundary conditions as

$$w_i(0,t) = 0,\ w_i(l,t) = 0,\ w_{ix_ix_i}(0,t) = \frac{1}{r},\ w_{ix_ix_i}(l,t) = \frac{1}{r} \tag{3.84}$$

where $T_i(t)$ ($i = 1, 2$) are the dynamic tensions, and the subscripts of the spatial variables x_i and the temporal variable t indicate partial differentiation with respect to x_i and t, respectively. The rotational motions for the pulleys and the accessory are derived as follows.

$$\begin{aligned} J\ddot{\theta}_2 + c_b\dot{\theta}_2 &= (T_2 - T_1)r, \\ J\ddot{\theta}_1 + c_b\dot{\theta}_1 &= (T_1 - T_2)r - f(\delta\theta)\left[K_d(\delta\theta) + c_d\left(\delta\dot{\theta}\right)\right] + M_1, \\ J_a\ddot{\theta}_a + c_a\dot{\theta}_3 &= f(\delta\theta)\left[K_d(\delta\theta) + c_d\left(\delta\dot{\theta}\right)\right] - M_1 \end{aligned} \tag{3.85}$$

where c_a, c_b, and c_d are the damping coefficients of the rotating accessory, the friction between belt and pulley, and the wrap spring, respectively. $\delta\theta = \theta_1 - \theta_3$ denotes the relative angular displacement between the driven pulley and the accessory shaft. The function $f(\delta\theta)$ in (3.85) is given as follows.

$$f(\delta\theta) = \begin{cases} 1, & \delta\theta > 0, \\ 0, & \delta\theta \le 0. \end{cases} \tag{3.86}$$

Under the quasi-static stretch assumption (Wickert 1992), we assume that the dynamic tension in each span is uniform throughout the span. Accordingly, the dynamic tensions can be defined as follows.

$$\begin{aligned} T_1 &= \frac{EA}{l}r(\theta_2 - \theta_1) + \frac{EA}{2l}\int_0^l (\overline{w}_{1\overline{x}_1})^2 \mathrm{d}\overline{x}_1, \\ T_2 &= \frac{EA}{l}r(\theta_1 - \theta_2) + \frac{EA}{2l}\int_0^l (\overline{w}_{2\overline{x}_2})^2 \mathrm{d}\overline{x}_2. \end{aligned} \tag{3.87}$$

The equations of motion for the lateral vibrations of an axially moving viscoelastic beam coupled with rotational vibrations of the pulleys and the accessory are converted into the following dimensionless normalized form.

$$
\begin{aligned}
&\overline{w}_{1\bar{t}\bar{t}} + 2\overline{v}_0\overline{w}_{1\bar{x}_1\bar{t}} + \left(\overline{v}_0^2 - 1\right)\overline{w}_{1\bar{x}_1\bar{x}_1} + k_{\rm f}^2\overline{w}_{1\bar{x}_1\bar{x}_1\bar{x}_1\bar{x}_1} + \alpha\overline{w}_{1\bar{x}_1\bar{x}_1\bar{x}_1\bar{x}_1\bar{t}} = T_1\overline{w}_{1\bar{x}_1\bar{x}_1},\\
&\overline{w}_{2\bar{t}\bar{t}} + 2\overline{v}_0\overline{w}_{2\bar{x}_2\bar{t}} + \left(\overline{v}_0^2 - 1\right)\overline{w}_{2\bar{x}_2\bar{x}_2} + k_{\rm f}^2\overline{w}_{2\bar{x}_2\bar{x}_2\bar{x}_2\bar{x}_2} + \alpha\overline{w}_{2\bar{x}_2\bar{x}_2\bar{x}_2\bar{x}_2\bar{t}} = T_2\overline{w}_{2\bar{x}_2\bar{x}_2},\\
&\overline{J}\ddot{\theta}_2 = -\alpha_{\rm b}\dot{\theta}_2 + T_2 - T_1,\\
&\overline{J}\ddot{\theta}_1 = -\alpha_{\rm b}\dot{\theta}_1 + T_1 - T_2 + \overline{M}_1 - \frac{r_{\rm a}}{r_1} f(\delta\theta)\left[k_{\rm d}(\theta_1 - \theta_3) + \alpha_{\rm d}\left(\dot{\theta}_1 - \dot{\theta}_3\right)\right],\\
&\overline{J}_{\rm a}\ddot{\theta}_3 = -\alpha_{\rm a}\dot{\theta}_3 - \frac{r\overline{M}_1}{r_{\rm a}} + f(\delta\theta)\left[k_{\rm d}(\theta_1 - \theta_3) + \alpha_{\rm d}\left(\dot{\theta}_1 - \dot{\theta}_3\right)\right],\\
&T_1 = \frac{k_1 r}{l}(\theta_2 - \theta_1) + \frac{k_1}{2}\int_0^1 (\overline{w}_{1\bar{x}_1})^2 {\rm d}\overline{x}_1,\\
&T_2 = \frac{k_1 r}{l}(\theta_1 - \theta_2) + \frac{k_1}{2}\int_0^1 (\overline{w}_{2\bar{x}_2})^2 {\rm d}\overline{x}_2
\end{aligned}
\tag{3.88}
$$

with the dimensionless boundary conditions as

$$
\overline{w}_i\left(0, \bar{t}\right) = 0, \overline{w}_i\left(1, \bar{t}\right) = 0, \overline{w}_{i\bar{x}_i\bar{x}_i}\left(0, \bar{t}\right) = \frac{l}{r}, \overline{w}_{i\bar{x}_i\bar{x}_i}\left(1, \bar{t}\right) = \frac{l}{r} \tag{3.89}
$$

where the dimensionless variables and parameters are defined by

$$
\begin{aligned}
&\overline{x}_i = \frac{x_i}{l}, \overline{w}_i = \frac{w_i}{l}, \bar{t} = t\sqrt{\frac{T_0}{\rho A l^2}}, \overline{v}_0 = v_0\sqrt{\frac{\rho A}{T_0}}, \overline{M}_1 = \frac{M_1}{T_0 r}, \alpha = \frac{I\mu}{l^3\sqrt{\rho A T_0}},\\
&k_1 = \frac{EA}{T_0}, k_{\rm f} = \sqrt{\frac{EI}{T_0 l^2}}, k_{\rm d} = \frac{K_{\rm d}}{T_0 r_{\rm a}}, \alpha_{\rm a} = \frac{c_{\rm a}}{l r_{\rm a}}\sqrt{\frac{1}{\rho A T_0}}, \alpha_{\rm b} = \frac{c_{\rm b}}{l r}\sqrt{\frac{1}{\rho A T_0}},\\
&\alpha_{\rm d} = \frac{c_{\rm d}}{l r_{\rm a}}\sqrt{\frac{1}{\rho A T_0}}, \overline{J}_{\rm a} = \frac{J_{\rm a}}{\rho A r_{\rm a} l^2}, \overline{J} = \frac{J}{\rho A r l^2},
\end{aligned}
\tag{3.90}
$$

where k_1 is the nonlinear coefficient representing the effect of nonlinearity of the axially moving beam, whereas $k_{\rm f}{}^2$ corresponds to the beam's flexural rigidity.

Equilibrium solutions $w_{\infty 1}(\overline{x}_1)$ and $w_{\infty 2}(\overline{x}_2)$ of (3.88) without the one-way clutch satisfy

$$
\begin{aligned}
&\left(\overline{v}_0^2 - 1 - \widehat{T}\right)w''_{\infty i} + k_{\rm f}^2 w''''_{\infty i} = 0, \quad i = 1, 2,\\
&\widehat{T} = \frac{k_1}{4}\left(\int_0^1 w'_{\infty 1}{}^2 {\rm d}\overline{x}_1 + \int_0^1 w'_{\infty 2}{}^2 {\rm d}\overline{x}_2\right),\\
&\widehat{\theta}_1 - \widehat{\theta}_3 = \frac{r\overline{M}_1}{r_{\rm a} k_{\rm d}}
\end{aligned}
\tag{3.91}
$$

where the subscript $_\infty$ indicates the sense of the equilibrium displacement, and the prime indicates differentiation with respect to x_i, and $\widehat{T} = (T_2 + T_1)/2$. The boundary conditions of the axially moving beam in (3.91) are as follows

$$\hat{w}_i(0) = 0,\ \hat{w}_i(1) = 0,\ w''_{\infty i}(0) = \frac{l}{r},\ w''_{\infty i}(1) = \frac{l}{r},\quad i = 1, 2. \tag{3.92}$$

The equilibrium (3.91) is similar to the wheel supported case (3.63) with equal diameters. The nontrivial equilibrium solutions to (3.91) can be determined by numerical methods.

Substituting the nontrivial static and dynamic solution into (3.88) and using the Galerkin truncation method, a set of second-order ordinary differential equations are obtained as follows

$$\begin{aligned}
\ddot{q}_i &= -4\overline{v}_0 \sum_{k=1}^{N_1} G_{ik}\dot{q}_k - \alpha(i\pi)^4\dot{q}_k \\
&+ \left[\left(\overline{v}_0^2 - 1 - \hat{T} - k_1 \sum_{k=1}^{N_1} k\pi I_{1k}q_k - T_1\right)i^2\pi^2 - k_f^2(i\pi)^4\right]q_i \\
&+ 2\left(k_1 \sum_{k=1}^{N_1} k\pi I_{1k}q_k + T_1\right)H_{1i},\quad i = 1, 2, \ldots, N_1, \\
\ddot{q}_{N_1+m} &= -4\overline{v}_0 \sum_{k=N_1+1}^{N_1+N_2} G_{m(k-N_1)}\dot{q}_k - \alpha(m\pi)^4\dot{q}_{N_1+m} \\
&+ \left\{\left[\overline{v}_0^2 - 1 - \widehat{T} - I_2k_1 \sum_{k=N_1+1}^{N_1+N_2} (k - N_1)\pi I_{2k}q_k - T_2\right]m^2\pi^2\right. \\
&\left. -k_f^2(m\pi)^4\right\}q_{N_1+m} \\
&+ 2\left[k_1 \sum_{k=N_1+1}^{N_1+N_2} (k - N_1)\pi I_{2k}q_k + T_2\right]H_{2m},\quad m = 1, 2, \ldots, N_2, \\
\ddot{\vartheta}_1 &= -\frac{c_b}{J}\dot{\vartheta}_1 + \frac{T_1}{J} - \frac{T_2}{J} + \frac{k_1\pi}{J}\left[\sum_{k=1}^{N_1} kI_{1k}q_k - \sum_{k=N_1+1}^{N_1+N_2} (k - N_1)I_{2k}q_k\right] \\
&- f\left(\vartheta_1 - \vartheta_3 + \frac{r\overline{M}_1}{r_ak_d}\right)\left[k_d\left(\vartheta_1 - \vartheta_3 + \frac{r\overline{M}_1}{r_ak_d}\right) + \alpha_d\left(\dot{\vartheta}_1 - \dot{\vartheta}_3\right)\right]\frac{r_a}{Jr} + \frac{\overline{M}_1}{J}, \\
\ddot{\vartheta}_3 &= -\frac{c_a}{J_a}\dot{\vartheta}_3 + \frac{1}{J_a}f\left(\vartheta_1 - \vartheta_3 + \frac{r\overline{M}_1}{r_ak_d}\right)\left[k_d\left(\vartheta_1 - \vartheta_3 + \frac{r\overline{M}_1}{r_ak_d}\right) + \alpha_d\left(\dot{\vartheta}_1 - \dot{\vartheta}_3\right)\right] \\
&- \frac{r\overline{M}_1}{J_ar_a}, \\
\vartheta_2 &= A_m \cos\omega t
\end{aligned} \tag{3.93}$$

where

$$I_{1k} = \int_0^1 {w'_{\infty 1}}^2 \cos(k\pi \overline{x}_1)\mathrm{d}\overline{x}_1, \quad k = 1, 2, \ldots, N_1,$$

$$I_{2k} = \int_0^1 {w'_{\infty 2}}^2 \cos[(k - N_1)\pi \overline{x}_2]\mathrm{d}\overline{x}_2, \quad k = N_1 + 1, N_1 + 2, \ldots, N_1 + N_2,$$

$$H_{1i} = \int_0^1 \sum_{k=1}^{N_1} w''_{\infty 1} \sin(i\pi \overline{x}_1)\mathrm{d}\overline{x}_1, \quad i = 1, 2, \ldots, N_1,$$

$$H_{2m} = \int_0^1 w''_{\infty 2} \sin(m\pi \overline{x}_2)\mathrm{d}\overline{x}_2, \quad m = 1, 2, \ldots, N_2,$$

$$\widehat{T} = \frac{k_1}{4}\left(\int_0^1 (w'_{\infty 1})^2 \mathrm{d}\overline{x}_1 + \int_0^1 (w'_{\infty 2})^2 \mathrm{d}\overline{x}_2\right),$$

$$T_1 = \frac{k_1 r}{l}(\vartheta_2 - \vartheta_1) + \frac{k_1 \pi^2}{4}(q_1^2 + 4q_2^2),$$

$$T_2 = \frac{k_1 r}{l}(\vartheta_1 - \vartheta_2) + \frac{k_1 \pi^2}{4}(q_3^2 + 4q_4^2),$$

$$G_{ik} = \int_0^1 k\pi \cos(k\pi \overline{x}) \sin(i\pi \overline{x})\mathrm{d}\overline{x} = \begin{cases} \frac{2ik}{i^2 - k^2}, & i \neq k \text{ and} (i + k) \text{ is even}, \\ 0, & \text{otherwise}. \end{cases} \tag{3.94}$$

For the given parameters in Table 3.1, q_j ($j = 1, 2, \ldots, N_1 + N_2$), and ϑ_1 and ϑ_3 are numerically solved from (3.93) by utilizing the fourth-order Runge–Kutta method.

We assume that the excitation magnitude of the driving pulley is $A_{\mathrm{m}} = 0.0015$. Based on a sixteen-term Galerkin truncation ($N_1 = 16$ and $N_2 = 16$), the numerical solutions of the nonlinear vibrations of the spans are computed (i.e., Figs. 3.12 and 3.13). The steady-state responses of the rotation angles of the driven pulley and the accessory are shown in Fig. 3.12a. It is observed that the resonance exists in three frequency regions: between 300 and 500 rad/s, between 600 and 700 rad/s, and between 1100 and 1200 rad/s. Figure 3.12a reveals that while the excitation frequency is out of the resonance regions, the driven pulley's response coincides with the accessory's response. Nevertheless, the accessory's vibration is more significant than the driven pulley's vibration. The magnitudes of the lateral vibrations of twos spans with a one-way clutch are shown in Fig. 3.12b. It is observed that there has no significant difference between the vibration magnitude of the two spans. In Fig. 3.13, we compare the steady-state responses of the system in two cases: with a one-way clutch and without any one-way clutch (i.e., engaged state). Figure 3.13a shows the

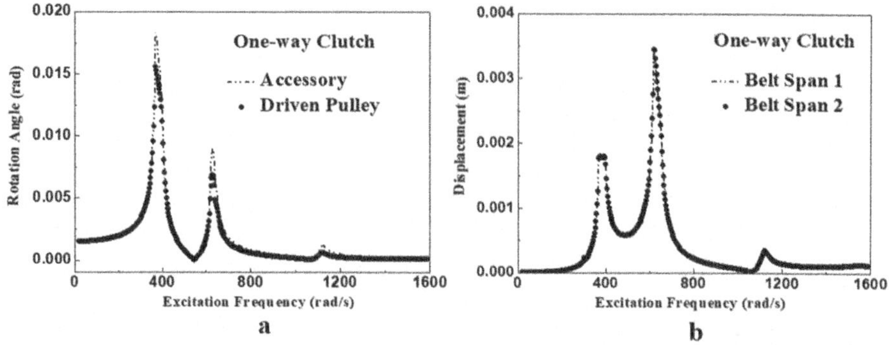

Fig. 3.12 Steady-state response of the pulley–belt system with a one-way clutch via sweep frequency: **a** driven pulley and accessory, **b** lateral vibrations of the belt spans

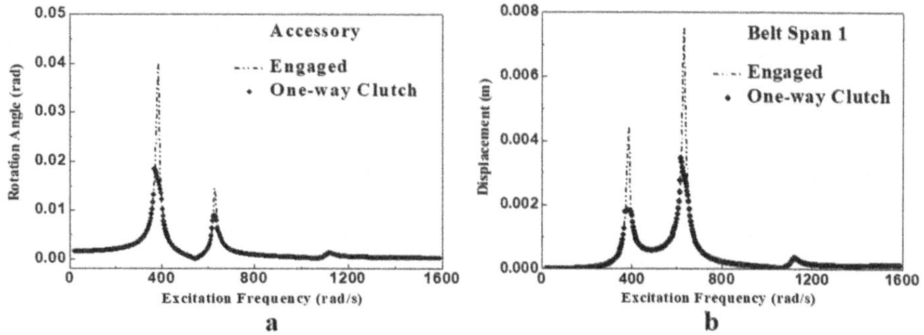

Fig. 3.13 Steady-state response of the pulley–belt system with and without a one-way clutch via sweep frequency: **a** driven pulley and accessory, **b** lateral vibrations of the belt spans

response of the accessory, whereas Fig. 3.13b shows that of span 1. As shown in Fig. 3.13, the magnitude of the system without the one-way clutch (i.e., engaged state) is much larger than the system with the one-way clutch.

Additionally, as shown in Figs. 3.12 and 3.13, jumping, a typical nonlinear phenomenon, exists in the resonance response of the system with the one-way clutch. This phenomenon presents softening nonlinear characteristics. Thus, the softening nonlinearities are consequences of the dynamic system with the one-way clutch. These nonlinear vibration characteristics revealed by Ding and Zu (2013), Ding et al. (2014), and Ding (2015) deepen the understanding of vibration suppression by one-way clutches and provide essential and important information for the optimal design of engineering systems.

3.2.7 Fluid-Conveying Pipe

Fluid-conveying pipes are regarded as a particular type of AMSs. In this section, we investigate the dynamic of fluid-conveying pipes. The dynamic model of a vertically fluid-conveying pipe, treated as a uniform tubular beam as presented in Fig. 3.14, is established based on the Euler–Bernoulli beam theory.

The force balances of an arbitrary fluid segment and the pipe segment are considered as shown in Figs. 3.15 and 3.16, respectively. The x- and y-directional-force balances of the fluid segment yield the two equations

$$F_{\mathrm{f}}\frac{\partial w}{\partial x} - m_{\mathrm{f}}\dot{v}_{\mathrm{f}} - A_{\mathrm{f}}\frac{\partial P_{\mathrm{f}}}{\partial x} - \hat{q}S_{\mathrm{c}} = 0, \tag{3.95}$$

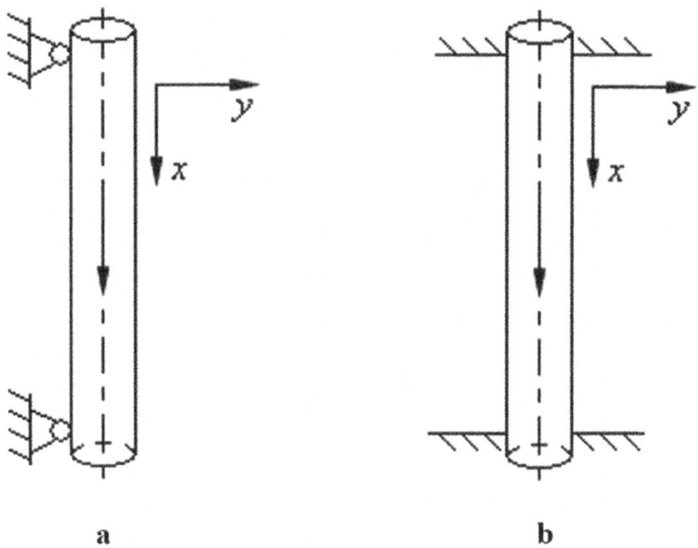

Fig. 3.14 Two models of pipe conveying fluid: **a** pined-pined, **b** fixed–fixed

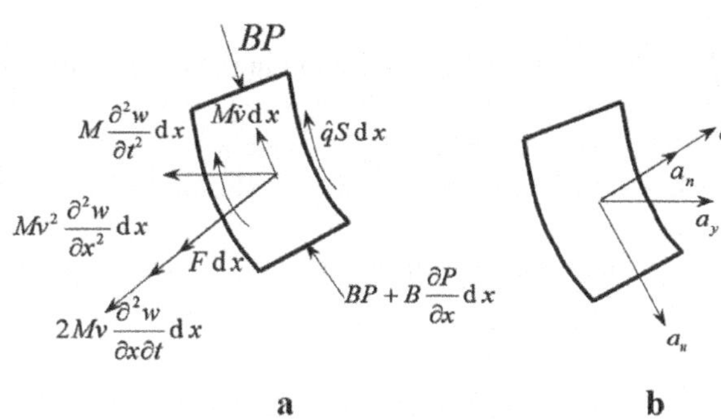

Fig. 3.15 Analysis of a fluid element: **a** force, **b** acceleration

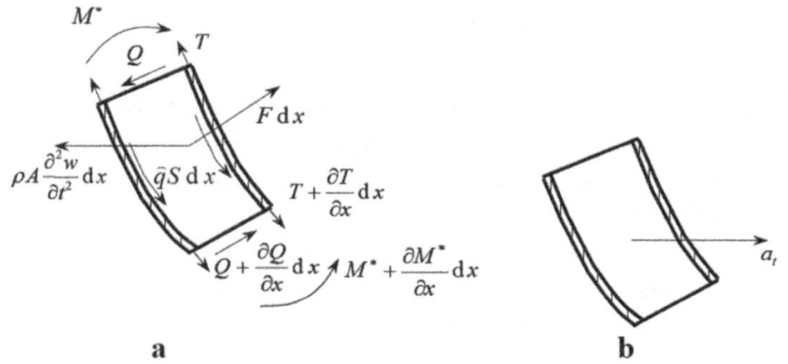

Fig. 3.16 Analysis of a pipe element: **a** force, **b** acceleration

$$m_{\mathrm{f}}\left(\frac{\partial}{\partial t}+v_{\mathrm{f}}\frac{\partial}{\partial x}\right)^2 w+\hat{q}S_{\mathrm{c}}\frac{\partial w}{\partial x}+m_{\mathrm{f}}\dot{v}_{\mathrm{f}}\frac{\partial w}{\partial x}+A_{\mathrm{f}}P_{\mathrm{f}}\frac{\partial^2 w}{\partial x^2}+A_{\mathrm{f}}\frac{\partial P_{\mathrm{f}}}{\partial x}\cdot\frac{\partial w}{\partial x}+F_{\mathrm{f}}=0 \tag{3.96}$$

where m_{f} is the linear density of the fluid, v_{f} is the fluid velocity, P_{f} is the fluid pressure, S_{c} is the circumference of the cross section, F_{f} is the pressure between the fluid and the structure, A_{f} is the cross-sectional area of the fluid, and $\hat{q}$ is the shear force between the fluid and the pipe wall.

The force balances in x- and y-directions of the pipe segment also give two equations

$$\frac{\partial T}{\partial x}-F_{\mathrm{f}}\frac{\partial w}{\partial x}+\hat{q}S_{\mathrm{c}}=0, \tag{3.97}$$

$$F+\frac{\partial Q}{\partial x}+T\frac{\partial^2 w}{\partial x^2}+\frac{\partial T}{\partial x}\cdot\frac{\partial w}{\partial x}+\hat{q}S_{\mathrm{c}}\frac{\partial w}{\partial x}-\rho A\frac{\partial^2 w}{\partial t^2}=0, \tag{3.98}$$

and

$$Q=-\frac{\partial M^*}{\partial x} \tag{3.99}$$

where M^* is the bending moment applied to the pipe segment, Q is the shear force, T is the axial tension, A is the cross-sectional area of the pipe, and ρ is the density of the pipe material. Manipulating the above equations leads to the governing equation of the fluid-conveying pipe lateral vibration

$$\begin{aligned}&EIw_{xxxx}+(m_{\mathrm{f}}+\rho A)w_{tt}+2m_{\mathrm{f}}v_{\mathrm{f}}w_{xt}+(m_{\mathrm{f}}v_{\mathrm{f}}^2-T+A_{\mathrm{f}}P_{\mathrm{f}})w_{xx}\\&+A_{\mathrm{f}}P_{\mathrm{f}\,x}w_x-T_xw_x+m_{\mathrm{f}}\dot{v}_{\mathrm{f}}w_x=0.\end{aligned} \tag{3.100}$$

The governing equation becomes the axially moving Euler–Bernoulli beam model, if we let the pipe density ρ and pressure P_{f} to be zero.

3.2.8 Nanoscale Beams

Thus far, we have considered only axially moving macroscale beams using local elasticity theory (Mase and Mase 1992), in which the stress at a point in the beam is dependent uniquely on the strain at the same point (i.e., the stress–strain relationship is characterized by algebraic equations). This theory disregards the size-dependent effects, and the beam model based on this theory produces a satisfactory result in numerous cases. In nanoscale beams, the size-dependent effects become important; therefore, the local elasticity theory is not appropriate. The nonlocal elasticity theory that is presented by Eringen and Edelen (1972) is utilized to consider the small-scale effect of nanoscale beams. This theory is based on the atomic theory of lattice dynamics and experimental observations on phonon dispersion. According to Eringen's theory, the stress at any point on the beam depends on the strain at all points of the beam. This theory leads to the following nonlocal constitutive relation for a nanobeam:

$$\sigma_{\mathrm{nl}}(x,t) - (e_0 a_{\mathrm{nl}})^2 (\sigma_{\mathrm{nl}}(x,t))_{xx} = E\varepsilon(x,t) \tag{3.101}$$

where $\sigma_{\mathrm{nl}}(x,t)$ denotes the nonlocal stress, e_0 is a constant that is appropriate to each material determined from the atomic dispersion relation or experimentally, and a_{nl} is a characteristic length, such as the lattice parameter, the grain size, or the granular displacement. The nonlocal bending moment and axial strain are given as follows:

$$M_{\mathrm{nl}}(x,t) = -\int_A y\sigma_{\mathrm{nl}}(x,t)\mathrm{d}A, \tag{3.102}$$

$$\varepsilon(x,t) = -yw_{xx}(x,t), \tag{3.103}$$

where A is the cross-sectional area of the beam and y is the coordinate measured along the j-axis from the middle plane of the beam. Substituting (3.103) into (3.101), multiplying both sides of (3.101) by y, and then integrating over the cross-sectional area yields

$$-\int_A y\sigma_{\mathrm{nl}}(x,t)\mathrm{d}A - (e_0 a_{\mathrm{nl}})^2 \left(-\int_A y\sigma_{\mathrm{nl}}(x,t)\mathrm{d}A\right)_{xx} = Ew_{xx}(x,t)\int_A y^2\mathrm{d}A. \tag{3.104}$$

By using (3.102), (3.104) can be rewritten as

$$M_{\mathrm{nl}}(x,t) - (e_0 a_{\mathrm{nl}})^2 (M_{\mathrm{nl}}(x,t))_{xx} = EIw_{xx}(x,t). \tag{3.105}$$

The nonlocal bending moment can be determined by (3.105). Based on the nonlocal bending moment, we can develop a dynamic model of an axially moving nanoscale beam. The general form of the equation of motion of a moving Euler–Bernoulli beam was given in Sect. 3.2.1, namely

$$\rho A\big(w_{tt} + 2v_0 w_{xt} + v_0^2 w_{xx}\big) - (T(x,t)w_x)_x + (M_{\text{nl}}(x,t))_{xx} = 0. \tag{3.106}$$

Differentiating both sides of (3.105) and (3.106) with respect to x yields:

$$\rho A\big(w_{xxtt} + 2v_0 w_{xxxt} + v_0^2 w_{xxxx}\big) - (T(x,t)w_x)_{xxx} + (M_{\text{nl}}(x,t))_{xxxx} = 0, \tag{3.107}$$

$$(M_{\text{nl}}(x,t))_{xx} - (e_0 a_{\text{nl}})^2 (M_{\text{nl}}(x,t))_{xxxx} = EI w_{xxxx}(x,t). \tag{3.108}$$

Multiplying both sides of (3.107) by $(e_0 a_{\text{nl}})^2$ and using (3.106) and (3.108), the governing equation describing the dynamics of an axially moving nanoscale beam is derived as follows.

$$\begin{aligned}&\rho A\big(w_{tt} + 2v_0 w_{xt} + v_0^2 w_{xx}\big) - (T(x,t)w_x)_x + EI w_{xxxx}(x,t)\\ &\quad - (e_0 a_{\text{nl}})^2\big[\rho A\big(w_{xxtt} + 2v_0 w_{xxxt} + v_0^2 w_{xxxx}\big) - (T(x,t)w_x)_{xxx}\big] = 0,\end{aligned} \tag{3.109}$$

where the last term of (3.109) is associated with the size-dependent effects of the nanoscale material. The dynamic model given by (3.109) was developed based on the Euler–Bernoulli beam theory. Furthermore, the equation of motion of a moving nanoscale viscoelastic beam based on the Rayleigh beam theory can be also established by a similar approach (Rezaee and Lotfan 2015). Concerning this case, the differential equation that describes the nonlocal bending moment can be expressed as

$$M_{\text{nl}}(x,t) - (e_0 a_{\text{nl}})^2 (M_{\text{nl}}(x,t))_{xxxx} = EI w_{xx}(x,t) + \mu I (w_{xxt}(x,t) + v_0 w_{xxx}(x,t)). \tag{3.110}$$

The dynamic model of the moving viscoelastic Rayleigh nanobeam is derived as follows:

$$\begin{aligned}&\rho A\big(w_{tt} + 2v_0 w_{xt} + v_0^2 w_{xx}\big) - (T(x,t)w_x)_x - \rho I\big(2v_0 w_{xxxt} + v_0^2 w_{xxxx} + w_{xxtt}\big)\\ &\quad + \mu I (w_{xxxxt}(x,t) + v_0 w_{xxxxx}(x,t))\\ &\quad - (e_0 a_{\text{nl}})^2\big[\rho A\big(w_{xxtt} + 2v_0 w_{xxxt} + v_0^2 w_{xxxx}\big)\\ &\quad + EI w_{xxxx}(x,t) - (T(x,t)w_x)_{xxx} - \rho I\big(2v_0 w_{xxxxxt} + v_0^2 w_{xxxxxx} + w_{xxxxtt}\big)\big]\\ &= 0.\end{aligned} \tag{3.111}$$

3.2.9 Laminated Composite Beams

All formulations of the AMSs presented in the previous sections are exclusively applicable to one-layered materials. In recent years, in addition to these materials, laminated composite materials that comprise two or more layers have also been employed in AMSs, such as band saws, fluid pipes, aerospace structures, and paper sheets. A difference between the conventional materials and the laminated composite materials exists in the stress–strain relation. In this section, we investigate the dynamic model of a laminated composite material that moves axially.

A laminated composite structure (i.e., refer to Fig. 3.17) is characterized by stacking many laminae in the j-axis (Kaw 1997). Before deriving the stress–strain relationship of laminated composite materials, we discuss the material properties of laminae. A lamina is a thin layer of a composite material constructed by fibers bonded by a metal or polymer matrix (Fig. 3.17). The lamina is an orthotropic material, and its stress–strain relationship depends on the direction of the fibers, that is

$$\begin{bmatrix} \sigma_X \\ \sigma_Z \\ \sigma_{XZ} \end{bmatrix} = \mathbf{T}_L^{-1} \begin{bmatrix} Q_{11} & Q_{12} & Q_{16} \\ Q_{12} & Q_{22} & Q_{26} \\ Q_{16} & Q_{26} & Q_{66} \end{bmatrix} \mathbf{R}\mathbf{T}_L\mathbf{R}^{-1} \begin{bmatrix} \varepsilon_X \\ \varepsilon_Z \\ \varepsilon_{XZ} \end{bmatrix} \tag{3.112}$$

where ε_X, ε_Z, and ε_{XZ} denote the strain components in the x- and z-coordinates; σ_X, σ_Z, and σ_{XZ} indicate the stress components in the x- and z-coordinates; Q_{ij} $(i, j = 1, 2, 6)$ are the stiffness elements of the material; and $\mathbf{T}_L$ and $\mathbf{R}$ are referred to as the transformation matrix and the Reuter matrix, respectively, which are defined by

$$\mathbf{T}_L = \begin{bmatrix} \cos^2\theta_a & \sin^2\theta_a & -2\cos\theta_a \sin\theta_a \\ \sin^2\theta_a & \cos^2\theta_a & 2\cos\theta_a \sin\theta_a \\ \cos\theta_a \sin\theta_a & -\cos\theta_a \sin\theta_a & \cos^2\theta_a - \sin^2\theta_a \end{bmatrix}; \mathbf{R} = \begin{bmatrix} 1 & 0 & 0 \\ 0 & 1 & 0 \\ 0 & 0 & 2 \end{bmatrix} \tag{3.113}$$

where θ_a represents the angle from the i-axis to the normal direction of the fiber (Fig. 3.17). In most laminated composite materials, laminae with differential fiber

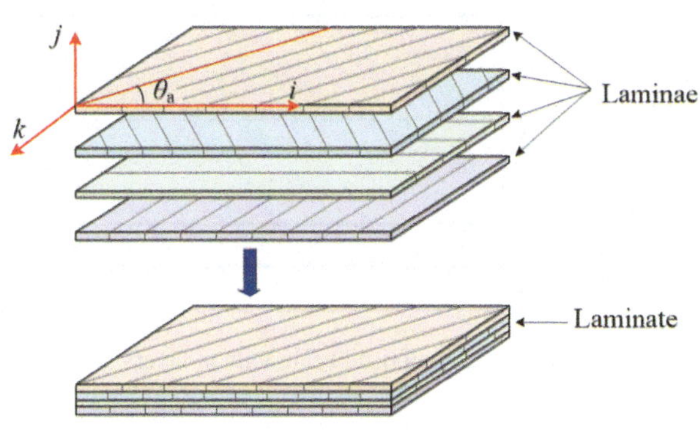

Fig. 3.17 Laminated composite material

orientations are staked to improve the stiffness of the material. The stress–strain relation of laminated composite beams is derived based on the classical laminate theory, which is an extension of the classical plate theory (Shokrieh 2014). By applying this theory to a composite beam with m layers, as shown in Fig. 3.18, the stress–strain relation of the nth layer is given as follows:

$$\sigma^n = \overline{Q}_{11}^n \varepsilon,$$
$$\overline{Q}_{11}^n = Q_{11}^n \cos^4\theta_\text{a}^n + Q_{11}^n \sin^4\theta_\text{a}^n + 2\left(Q_{11}^n + 2Q_{11}^n \sin^2\theta_\text{a}^n \cos^2\theta_\text{a}^n\right), \tag{3.114}$$

where θ_a^n represents the angle θ_a in the n-th layer, and Q_{ij}^n represents the stiffness elements of the material in the n-th layer. For an orthotropic material, the stiffness elements are given as follows:

$$Q_{11}^n = \frac{E_\text{X}^n}{1 - \upsilon_\text{XY}^n \upsilon_\text{YX}^n},\ Q_{22}^n = \frac{E_\text{Y}^n}{1 - \upsilon_\text{XY}^n \upsilon_\text{YX}^n},\ Q_{12}^n = \frac{E_\text{X}^n \upsilon_\text{XY}^n}{1 - \upsilon_\text{XY}^n \upsilon_\text{YX}^n},\ Q_{66}^n = G^n, \tag{3.115}$$

where E_X^n, E_Y^n, υ_XY^n, and υ_YX^n are Young's moduli and Poisson's numbers in the principal directions, whereas G^n is the shear modulus. The strain is given by (3.103), and the bending moment is derived as follows.

$$\begin{aligned} M(x,t) &= -\int_A y\sigma(x,t)\mathrm{d}A = -\int_A y\overline{Q}_{11}\varepsilon(x,t)\mathrm{d}A \\ &= b\int_{-h/2}^{h/2} \overline{Q}_{11} y^2 \mathrm{d}y w_{xx}(x,t) = bD_{11} w_{xx}(x,t) \end{aligned} \tag{3.116}$$

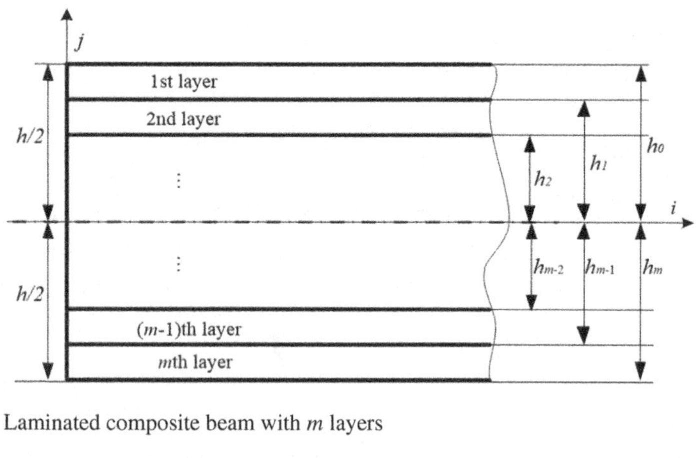

Fig. 3.18 Laminated composite beam with m layers

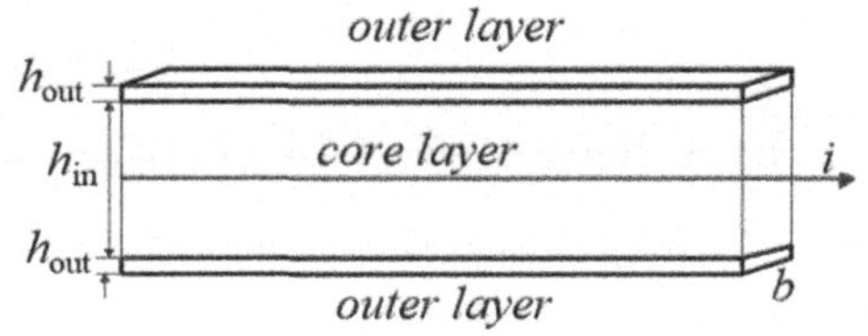

Fig. 3.19 Structure of a sandwich beam

where h and b are the height and the width of the beam, respectively, and D_{11} can be computed as follows.

$$D_{11} = \int_{-h/2}^{h/2} \overline{Q}_{11} y^2 \mathrm{d}y = \sum_{n=1}^{m} \int_{h_{n-1}}^{h_n} \overline{Q}_{11}^{n} y^2 \mathrm{d}y = \frac{1}{3} \sum_{n=1}^{m} \overline{Q}_{11}^{n} \left(h_n^3 - h_{n-1}^3\right), \quad (3.117)$$

where h_n denotes the distance from the external boundary of the n-th layer to the i-axis. According to the classical laminate theory, the governing equation of an axially moving laminated composite beam is expressed as follows.

$$\rho A\left(w_{tt} + 2v_0 w_{xt} + v_0^2 w_{xx}\right) + (T(x,t)w_x)_x + bD_{11}w_{xxxx} = 0. \quad (3.118)$$

One of the typical composite beams is the beam with a sandwich structure (i.e., Fig. 3.19), wherein the material properties of the core and outer layers are significantly different. For example, the core is composed of a viscoelastic material, whereas the outer layers are composed of elastic materials (Marynowski 2012). A dynamic model of axially moving beams with this structure was developed by Marynowski (2012) and Lv et al. (2014). Another example of axially moving beams with sandwich structures was presented by Wei et al. (2017). A mathematical model of a moving beam with a magnetorheological fluid core and aluminum outer layers with a magnetic field was established in this work.

3.3 Approximate Model

The governing equations of axially moving beams are nonlinear partial differential equations, which are difficult to analyze. In some circumstances, approximate models based on the truncation via the Galerkin method are widely used in the dynamic analysis of axially moving systems. In this section, an Euler–Bernoulli beam is taken as an example to show the Galerkin truncation procedure. The Galerkin method can also be applied to more complicated beam models.

The PDE of an axially moving Euler-Bernoulli beam in Eq. (3.25)

$$\overline{w}_{\overline{t}\overline{t}} + 2\overline{v}\overline{w}_{\overline{t}\overline{x}} + \dot{\overline{v}}\overline{w}_{\overline{x}} + \left(\overline{v}^2 - 1\right)\overline{w}_{\overline{x}\overline{x}} + k_f^2\overline{w}_{\overline{x}\overline{x}\overline{x}\overline{x}} + \alpha\overline{w}_{\overline{x}\overline{x}\overline{x}\overline{x}\overline{t}} = \frac{3}{2}k_1^2\overline{w}_{\overline{x}\overline{x}}(u_{\overline{x}})^2$$
$$+ \alpha k_2\left[2\overline{w}_{\overline{x}}\overline{w}_{\overline{t}\overline{x}}\overline{w}_{\overline{x}\overline{x}} + (\overline{w}_{\overline{x}})^2\overline{w}_{\overline{x}\overline{x}\overline{t}}\right] \tag{3.119}$$

where

$$\overline{w} = \frac{w}{l}, \overline{x} = \frac{x}{l}, \overline{t} = t\sqrt{\frac{T_0}{\rho A l^2}}, \overline{v} = v\sqrt{\frac{\rho A}{T_0}}, k_f^2 = \frac{EI}{T_0 l^2}$$
$$\alpha = \frac{I\mu}{l^3\sqrt{\rho A T_0}}, k_1 = \sqrt{\frac{EA}{T_0}}, k_2 = \frac{A\mu}{l\sqrt{\rho A T_0}}. \tag{3.120}$$

In (3.119), the variation of the axial speed has been considered and some high-order nonlinear terms dependent on the axial speed have been neglected. The beam with simple supports at both ends gives the following boundary conditions in the dimensionless form.

$$\overline{w}(0, t) = \overline{w}\left(1, \overline{t}\right) = 0, \ \overline{w}_{\overline{x}\overline{x}}(0, t) = \overline{w}_{\overline{x}\overline{x}}(1, t) = 0. \tag{3.121}$$

The Galerkin method will be used to (3.119) under the boundary condition (3.121). Suppose that the solution to (3.119) takes the form

$$\overline{w}(\overline{x}, \overline{t}) = \sum_{i=1}^{\infty} q_i(\overline{t})\varphi_i(\overline{x}) \tag{3.122}$$

where $q_i(\overline{t})$ and $\phi_i(\overline{x})$ denote the set of generalized displacements of the beam the set of the trial functions satisfying the boundary conditions (3.121), respectively. $\psi_i(\overline{x})$ is the set of the weight functions. Then, (3.119) is discretized into a set of infinite ordinary differential equations

$$\Bigg\langle \sum_{i=1}^{\infty} \ddot{q}_i(\overline{t})\varphi_i(\overline{x}) + 2\overline{v}\sum_{i=1}^{\infty} \dot{q}_i(\overline{t})\varphi_i'(\overline{x}) + \dot{\overline{v}}\sum_{i=1}^{\infty} q_i(\overline{t})\varphi_i'(\overline{x}) + \left(\overline{v}^2 - 1\right)\sum_{i=1}^{\infty} q_i(\overline{t})\varphi_i''(\overline{x})$$
$$+ k_{\mathrm{f}}^2\sum_{i=1}^{\infty} q_i(\overline{t})\varphi_i''''(\overline{x}) + \alpha\sum_{i=1}^{\infty} \dot{q}_i(\overline{t})\varphi_i''''(\overline{x}) - \frac{3}{2}k_1^2\sum_{i=1}^{\infty} q_i(\overline{t})\varphi_i''(\overline{x})\left[\sum_{i=1}^{\infty} q_i(\overline{t})\varphi_i'(\overline{x})\right]^2$$
$$- \alpha k_2\Bigg\{2\sum_{i=1}^{\infty} q_i(\overline{t})\varphi_i'(\overline{x})\sum_{i=1}^{\infty} \dot{q}_i(\overline{t})\varphi_i'(\overline{x})\sum_{i=1}^{\infty} q_i(\overline{t})\varphi_i'(\overline{x})$$
$$+\left[\sum_{i=1}^{\infty} q_i(\overline{t})\varphi_i'(\overline{x})\right]^2\sum_{i=1}^{\infty} \dot{q}_i(\overline{t})\varphi_i''(\overline{x})\Bigg\}, \psi_n(\overline{x})\Bigg\rangle = 0, \quad n = 1, 2, \ldots, \tag{3.123}$$

where the inner production of two functions f and g on [0, 1] is defined as

$$\langle f, g\rangle = \int_0^1 f(\overline{x})g(\overline{x})\mathrm{d}\overline{x}. \tag{3.124}$$

Application of the distributive law of the inner product to (3.123) leads to

$$\begin{aligned}
&\sum_{i=1}^{\infty}\ddot{q}_i(\overline{t})\langle\varphi_i, \psi_n\rangle + 2\overline{v}\sum_{i=1}^{\infty}\dot{q}_i(\overline{t})\langle\varphi_i', \psi_n\rangle + (\overline{v}^2 - 1)\sum_{i=1}^{\infty}q_i(\overline{t})\langle\varphi_i'', \psi_n\rangle \\
&\quad + k_f^2\sum_{i=1}^{\infty}q_i(\overline{t})\langle\varphi_i'''', \psi_n\rangle + \alpha\sum_{i=1}^{\infty}\dot{q}_i(\overline{t})\langle\varphi_i'''', \psi_n\rangle \\
&\quad - \frac{3}{2}k_1^2\sum_{i=1}^{\infty}\sum_{j=1}^{\infty}\sum_{k=1}^{\infty}q_i(\overline{t})q_j(\overline{t})q_k(\overline{t})\langle\varphi_i'\varphi_j'\varphi_k'', \psi_n\rangle \\
&\quad - \alpha k_2\sum_{i=1}^{\infty}\sum_{i=1}^{\infty}\sum_{i=1}^{\infty}q_i(\overline{t})q_j(\overline{t})\dot{q}_k(\overline{t})\big(\langle 2\varphi_i'\varphi_k'\varphi_j'', \psi_n\rangle \\
&\quad + \langle\varphi_i'\varphi_j'\varphi_k'', \psi_n\rangle\big) = 0, \quad n = 1, 2, \ldots
\end{aligned} \tag{3.125}$$

Truncating (3.125) by retaining first m terms yields

$$\begin{aligned}
&\sum_{i=1}^{m}\ddot{q}_i(\overline{t})\langle\varphi_i, \psi_n\rangle + 2\overline{v}\sum_{i=1}^{m}\dot{q}_i(\overline{t})\langle\varphi_i', \psi_n\rangle + (\overline{v}^2 - 1)\sum_{i=1}^{m}q_i(\overline{t})\langle\varphi_i'', \psi_n\rangle \\
&\quad + k_f^2\sum_{i=1}^{m}q_i(\overline{t})\langle\varphi_i'''', \psi_n\rangle + \alpha\sum_{i=1}^{m}\dot{q}_i(\overline{t})\langle\varphi_i'''', \psi_n\rangle \\
&\quad - \frac{3}{2}k_1^2\sum_{i=1}^{m}\sum_{j=1}^{m}\sum_{k=1}^{m}q_i(\overline{t})q_j(\overline{t})q_k(\overline{t})\langle\varphi_i'\varphi_j'\varphi_k'', \psi_n\rangle \\
&\quad - \alpha k_2\sum_{i=1}^{m}\sum_{i=1}^{m}\sum_{i=1}^{m}q_i(\overline{t})q_j(\overline{t})\dot{q}_k(\overline{t})\big(\langle 2\varphi_i'\varphi_k'\varphi_j'', \psi_n\rangle \\
&\quad + \langle\varphi_i'\varphi_j'\varphi_k'', \psi_n\rangle\big) = 0, \quad n = 1, 2, \ldots
\end{aligned} \tag{3.126}$$

which is a set of m ordinary differential equations.

In most cases, both the weight and trial functions are selected as the eigenfunctions of the linear stationary beam with the boundary condition in (3.121), namely

$$\psi_n(\overline{x}) = \sin(n\pi\overline{x}), \varphi_i(\overline{x}) = \sin(i\pi\overline{x}) \quad (n, i = 1, 2, \ldots, m). \tag{3.127}$$

Inserting (3.127) into (3.126), determining the corresponding inner products, regrouping the nonlinear terms to combine the same terms, and then eliminating all null terms in the resulting equation, we can obtain the following equation.

$$
\begin{aligned}
&\ddot{q}_n(\bar{t}) + 4 \sum_{i+n \text{ is odd}}^{m} \frac{ni}{n^2 - i^2}\left[2\bar{v}\dot{q}_i(\bar{t}) + \dot{\bar{v}}q_i(\bar{t})\right] \\
&\quad + \left(\bar{v}^2 - 1\right)n^2\pi^2 q_n(\bar{t}) + k_f^2 n^4\pi^4 q_n + \alpha n^4\pi^4 \dot{q}_n \\
&= \frac{7n\pi^4}{16}\left\{\sum_{k=n+1}^{m+n}\left\{(k-n)q_{k-n}\sum_{i=\max\{1,k-m\}}^{\min\{k-1,m\}}\left[i(k-i)q_i\left(k_1^2 q_{k-i} + \alpha k_2 \dot{q}_{k-i}\right)\right]\right\}\right. \\
&\quad \left. + \sum_{k=2}^{m-n}\left\{(n+k)q_{n+k}\sum_{i=1}^{k-1}\left[i(k-i)q_i\left(k_1^2 q_{k-i} + \alpha k_2 \dot{q}_{k-i}\right)\right]\right\}\right\} \\
&\quad - \frac{5n\pi^4}{16}\sum_{k=2}^{n-1}\left\{(n-k)q_{n-k}\sum_{i=1}^{k-1}\left[i(k-i)q_i\left(k_1^2 q_{k-i} + \alpha k_2 \dot{q}_{k-i}\right)\right]\right\} \\
&\quad - \frac{n\alpha k_2\pi^4}{8}\left\{\sum_{k=n+1}^{m+n}\left[(k-n)\dot{q}_{k-n}\sum_{i=\max\{1,k-m\}}^{\min\{k-1,m\}} i(k-i)q_i q_{k-i}\right]\right. \\
&\quad + \sum_{k=2}^{m-n}\left[(n+k)\dot{q}_{n+k}\sum_{i=1}^{k-1} i(k-i)q_i q_{k-i}\right] \\
&\quad \left. + \sum_{k=2}^{n-1}\left[(n-k)\dot{q}_{n-k}\sum_{i=1}^{k-1} i(k-i)q_i q_{k-i}\right]\right\}
\end{aligned}
\tag{3.128}
$$

where the sum is zero if its upper limit is smaller than its lower limit. Although (3.128) seems rather complicated, it is very efficient if used for computer implementation since most nonlinear terms which are repeated were put together, and the terms with zero coefficients were eliminated.

3.4 Dynamic Analysis of Beam Models

3.4.1 Equilibrium Solutions

Equilibrium configurations of axially moving beams (i.e., both Euler–Bernoulli and Timoshenko beams) and the critical speeds are derived in this section. Based on the linear part of (3.128), the eigenvalues can be determined numerically and the dynamic characteristics can be concluded.

Figure 3.20 shows the effect of the beam's moving speed on the eigenvalues' real parts and imaginary parts. It is noted that the imaginary parts denote the beam's natural frequencies, whereas the real parts denote the corresponding damping-like quantity. According to stability analysis, the following three cases are implied:

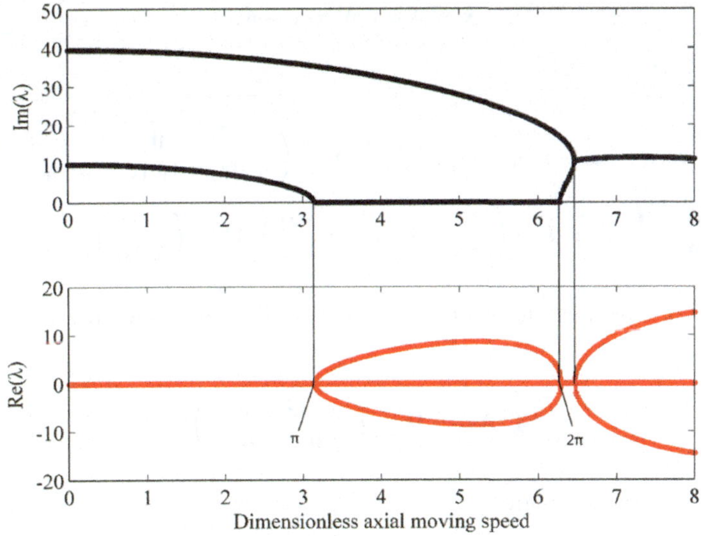

Fig. 3.20 Bifurcation phenomena of the eigenvalues

Table 3.2 Bifurcation series for axial translating beam systems

$0 \leq v < \pi$	$\pi < v < 2\pi$	$2\pi < v < 6.4614$	$6.4614 < v$
Stable	Divergence	Stable	Flutter (one pair repeated frequencies)

- Case 1: The system is stable if the real parts of all the eigenvalues are negative or zero.
- Case 2: The system undergoes divergence and becomes unstable if one or more eigenvalues have positive real parts and the imaginary parts of all the eigenvalues are zero.
- Case 3:The flutter occurs and leads the system instability if one or more eigenvalues have both the positive real part and nonzero imaginary part.

In Table 3.2, the bifurcation series of the dynamic behaviors is listed for different axial moving velocities.

3.4.2 *Stability Analysis*

The parametric and external excitations to the axially moving beam can be investigated by perturbation methods either to the PDE or to the truncated ODEs. For example, the averaging method can be used for the truncated ODEs. The two-term Galerkin truncation of the axially moving beam with a pulsating speed can be written as follows.

$$\dot{\mathbf{x}} = \mathbf{A}\mathbf{x} + \mathbf{A}_1\mathbf{x}\cos\omega\bar{t}, \tag{3.129}$$

where

$$\mathbf{x} = (x_1, x_2, x_3, x_4)^{\mathrm{T}} = (q_1, q_2, \dot{q}_1, \dot{q}_2)^{\mathrm{T}}, \mathbf{A} = \begin{pmatrix} \mathbf{0} & \mathbf{I} \\ \pi^2(\bar{v}_0^2 - 1)\mathbf{G} - \pi^4\mathrm{H} & -\frac{16}{3}\bar{v}_0\mathbf{J} \end{pmatrix},$$
$$\mathbf{A}_1 = \begin{pmatrix} \mathbf{0} & \mathbf{0} \\ \mathbf{0} & -\pi^2\delta\mathbf{G} \end{pmatrix}, \mathbf{J} = \begin{pmatrix} 0 & 1 \\ -1 & 0 \end{pmatrix}, \mathbf{G} = \begin{pmatrix} 1 & 0 \\ 0 & 4 \end{pmatrix}, \mathbf{H} = \begin{pmatrix} 1 & 0 \\ 0 & 16 \end{pmatrix}. \tag{3.130}$$

There exists canonical transformation matrix $\mathbf{T}$, which satisfies the following condition

$$\mathbf{T}^{\mathrm{T}}\mathbf{A}\mathbf{T} = \boldsymbol{\Omega} = \begin{pmatrix} -\omega_1\mathbf{J} & \mathbf{0} \\ \mathbf{0} & -\omega_2\mathbf{J} \end{pmatrix}. \tag{3.131}$$

Considering a transformation

$$\mathbf{y} = \mathbf{T}\mathbf{x}, \tag{3.132}$$

one can rewrite (3.129) in terms of $\mathbf{y}$. A new time variable $\tau = \omega\bar{t}$ is introduced. Assume that instability occurs at a small vicinity of ω_0. Setting $\omega = \omega_0(1 + \Delta)$, where Δ is a small variance, one obtains the following equation.

$$\dot{\mathbf{y}} = \boldsymbol{\Omega}_0\mathbf{y} - \Delta\boldsymbol{\Omega}_0\mathbf{y} + \frac{\mathbf{D}}{\omega_0}\mathbf{y}\cos\tau, \tag{3.133}$$

where

$$\boldsymbol{\Omega}_0 = \frac{\boldsymbol{\Omega}}{\omega_0} = \begin{pmatrix} -k_1\mathbf{J} & \mathbf{0} \\ \mathbf{0} & -k_2\mathbf{J} \end{pmatrix}, k_1 = \frac{\omega_1}{\omega_0}, k_2 = \frac{\omega_2}{\omega_0}, \mathbf{D} = \mathbf{T}^{\mathrm{T}}\mathbf{A}_1\mathbf{T}. \tag{3.134}$$

Consider $\mathbf{y}$ in an amplitude-phase form

$$y_1 = a_1\cos\varphi_1, y_2 = a_1\sin\varphi_1, y_3 = a_2\cos\varphi_2, y_4 = a_2\sin\varphi_2, \tag{3.135}$$

where

$$\varphi_1 = k_1\tau + \theta_1, \varphi_2 = k_2\tau + \theta_2, \tag{3.136}$$

and a_1, a_2, θ_1, and θ_2 are the variants about time τ. Using (3.135) and their derivatives, a set of variant equations are obtained as follows.

$$\begin{aligned} \dot{a}_1 &= g_1\cos\varphi_1 + g_2\sin\varphi_1, a_1\dot{\theta}_1 = g_2\cos\varphi_1 - g_1\sin\varphi_1, \\ \dot{a}_2 &= g_3\cos\varphi_2 + g_4\sin\varphi_2, a_2\dot{\theta}_2 = g_4\cos\varphi_2 - g_3\sin\varphi_2. \end{aligned} \tag{3.137}$$

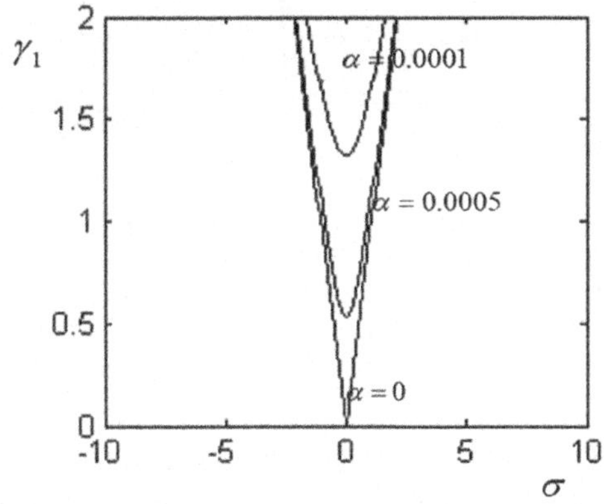

Fig. 3.21 Stability boundaries for the summation resonance of beams with simple supports

Let $k_1 = 1/2$. Applying the averaging operator $\lim_{t\to\infty} \frac{1}{t}\int_0^t (\cdot)\mathrm{d}\tau$ to (3.137) yields the equations of motion upon the first approximation.

$$\begin{aligned} \dot{a}_1 &= (U_1 h \cos 2\theta_1 + V_1 h \sin 2\theta_1 + M_1\alpha)a_1, \\ a_1\dot{\theta}_1 &= \left(V_1 h \cos 2\theta_1 - U_1 h \sin 2\theta_1 + N_1\alpha - \frac{1}{2}\Delta\right)a_1. \end{aligned} \tag{3.138}$$

The trial solution conditions give the boundary conditions of stability. Figure 3.21 shows the stability boundaries, where the structural damping has been supplied. In the enclosed areas, the system loses stability due to subharmonic resonances.

The perturbation method can be directly applied to PDEs. The linear counterpart of (3.119) is

$$\overline{w}_{\overline{t}\overline{t}} + 2\overline{v}\overline{w}_{\overline{t}\overline{x}} + \dot{\overline{v}}\overline{w}_{\overline{x}} + \left(\overline{v}^2 - 1\right)\overline{w}_{\overline{x}\overline{x}} + k_{\mathrm{f}}^2\overline{w}_{\overline{x}\overline{x}\overline{x}\overline{x}} + \alpha\overline{w}_{\overline{x}\overline{x}\overline{x}\overline{x}\overline{t}} = 0. \tag{3.139}$$

We assume that the axial speed of the system is a small simple harmonic variation about the constant mean speed γ_0, namely

$$\overline{v}(\overline{t}) = \gamma_0 + \varepsilon\gamma_1 \sin\omega\overline{t} \tag{3.140}$$

where $\varepsilon\gamma_1$ and ω are the magnitude and frequency of the harmonic variation, respectively. ε is the bookkeeping device utilized to indicate that the fluctuation amplitude is small, with the same order as the dimensionless viscosity coefficient. Although the connection between the amplitude of the variation and the dimensionless viscosity coefficient through ε is apparent, these terms are actually independent since each of them includes, respectively, an arbitrary parameter α or γ_1 of order one. Substituting (3.140) into (3.139) and neglecting higher order ε terms results in the following equation.

$$M\frac{\partial^2 \overline{w}}{\partial \overline{t}^2} + G\frac{\partial \overline{w}}{\partial \overline{t}} + K\overline{w} = -2\varepsilon\gamma_1 \sin \omega \overline{t}\frac{\partial^2 \overline{w}}{\partial \overline{x}\partial \overline{t}} - 2\varepsilon\gamma_0\gamma_1 \sin \omega \overline{t}\frac{\partial^2 \overline{w}}{\partial \overline{x}^2} - \varepsilon\omega\gamma_1 \cos \omega \overline{t}\frac{\partial \overline{w}}{\partial \overline{x}} - \varepsilon\alpha\frac{\partial^5 \overline{w}}{\partial \overline{x}^4 \partial \overline{t}} \tag{3.141}$$

where the mass, the gyroscopic, and the linear stiffness operators, respectively, are defined as

$$\widetilde{M} = I\,,\ \widetilde{G} = 2\gamma_0\frac{\partial}{\partial \overline{x}}\,,\ \widetilde{K} = \left(\gamma_0^2 - 1\right)\frac{\partial^2}{\partial \overline{x}^2} + k_f^2\frac{\partial^4}{\partial \overline{x}^4}. \tag{3.142}$$

The multiple scales method will be employed to solve (3.141) directly. A first-order uniform approximation is sought in the form

$$\overline{w}\left(\overline{x}, \overline{t}; \varepsilon\right) = \overline{w}_0(\overline{x}, \tau_0, \tau_1) + \varepsilon\overline{w}_1(\overline{x}, \tau_0, \tau_1) + \cdots \tag{3.143}$$

where $\tau_0 = \overline{t}$ and $\tau_1 = \varepsilon\overline{t}$ are the fast scale and slow scale time variables, respectively. τ_0 characterizes the motions occurring at ω_k, i.e., one of the natural frequencies of the corresponding unperturbed linear system. τ_1 characterizes the modulation of the phases and amplitudes by cause of viscoelasticity and possible resonance. Substituting (3.143) into (3.141) using the following relationship

$$\frac{\partial}{\partial \overline{t}} = \frac{\partial}{\partial \tau_0} + \varepsilon\frac{\partial}{\partial \tau_1} + \cdots,\ \frac{\partial^2}{\partial \overline{t}^2} = \frac{\partial^2}{\partial \tau_0^2} + 2\varepsilon\frac{\partial^2}{\partial \tau_0 \partial \tau_1} + \cdots \tag{3.144}$$

and checking the coefficients of ε^0 and ε^1 lead to

$$\widetilde{M}\frac{\partial^2 \overline{w}_0}{\partial \tau_0^2} + \widetilde{G}\frac{\partial \overline{w}_0}{\partial \tau_0} + \widetilde{K}\overline{w}_0 = 0 \tag{3.145}$$

and

$$\widetilde{M}\frac{\partial^2 \overline{w}_1}{\partial \tau_0^2} + \widetilde{G}\frac{\partial \overline{w}_1}{\partial \tau_0} + \widetilde{K}\overline{w}_1 = -2\frac{\partial^2 \overline{w}_0}{\partial \tau_0 \partial \tau_1} - 2\gamma_0\frac{\partial^2 \overline{w}_0}{\partial \overline{x}\partial \tau_1} - 2\gamma_1 \sin \omega T_0\left(\frac{\partial^2 \overline{w}_0}{\partial \overline{x}\partial \tau_0} + \gamma_0\frac{\partial^2 \overline{w}_0}{\partial \overline{x}^2}\right) - \gamma_1\omega \cos \omega\tau_0\frac{\partial \overline{w}_0}{\partial \overline{x}} - \alpha\frac{\partial^5 \overline{w}_0}{\partial \overline{x}^4 \partial \tau_0}. \tag{3.146}$$

The solution to (3.145) is

$$\overline{w}_0(\overline{x}, \tau_0, \tau_1) = \sum_{k=0,1,\ldots}\left[\varphi_k(\overline{x})A_k(\tau_1)\mathrm{e}^{\mathrm{i}\omega_k\tau_0} + \overline{\varphi}_k(\overline{x})\overline{A}_k(\tau_1)\mathrm{e}^{-\mathrm{i}\omega_k\tau_0}\right]. \tag{3.147}$$

It is noted that the over bar in (3.147) denotes the complex conjugation. The kth complex eigenfunction and the kth natural frequency can be derived by using boundary conditions.

The summation parametric resonance may occur if the variational frequency ω approaches the sum of any two natural frequencies of (3.145). In order to quantify the deviation of ω from $\omega_n + \omega_m$, we introduce the detuning parameter σ. Accordingly, ω can be described as follows.

$$\omega = \omega_n + \omega_m + \varepsilon\sigma \tag{3.148}$$

where ω_m and ω_n are the mth and nth natural frequencies of (3.145), respectively. To investigate the summation parametric response, (3.147) can be expressed as

$$\overline{w}_0(\overline{x}, \tau_0, \tau_1) = \varphi_n(\overline{x})A_n(\tau_1)e^{i\omega_n\tau_0} + \varphi_m(\overline{x})A_m(\tau_1)e^{i\omega_m\tau_0} + cc \tag{3.149}$$

where cc denotes the complex conjugate of all preceding terms on the right-hand side of the equation. Substituting (3.148) and (3.149) into (3.146) and expressing the trigonometric functions in exponential forms yield

$$\begin{aligned}
\widetilde{M}\frac{\partial^2 w_1}{\partial T_0^2} + \widetilde{G}\frac{\partial w_1}{\partial T_0} + \widetilde{K}w_1 = \{&-2\dot{A}_n\left(i\omega_n\varphi_n + \gamma_0\varphi_n'\right) \\
&+ \gamma_1\left[\frac{1}{2}(\omega_m - \omega_n)\overline{\varphi}_m' + i\gamma_0\overline{\varphi}_m''\right]\overline{\varphi}_m e^{i\sigma\tau_1} \\
&-i\alpha\omega_n A_n\varphi_n''''\}e^{i\omega_n\tau_0} \\
&+\left\{-2\dot{A}_m\left(i\omega_m\varphi_m + \gamma_0\varphi_m'\right) + \gamma_1\left[\frac{1}{2}(\omega_n - \omega_m)\overline{\varphi}_n'\right.\right. \\
&\left.+i\gamma_0\overline{\varphi}_n''\right]\overline{\varphi}_n e^{i\sigma\tau_1} - i\alpha\omega_m A_m\varphi_m''''\}e^{i\omega_m\tau_0} + cc + NST
\end{aligned} \tag{3.150}$$

where *NST* stands for the terms that will not bring secular terms into the solution. The prime and dot denote differentiations with respect to the dimensionless spatial variable x and the slow time variable τ_1, respectively. Equation (3.150) has a bounded solution only if a solvability condition holds. The solvability condition demands the orthogonal relationships.

$$\begin{aligned}
&\left\langle -2\dot{A}_n\left(i\omega_n\varphi_n + \gamma_0\varphi_n'\right) + \gamma_1\left[\frac{1}{2}(\omega_m - \omega_n)\overline{\varphi}_m' + i\gamma_0\overline{\varphi}_m''\right]\right. \\
&\left.\overline{\varphi}_m e^{i\sigma\tau_1} - i\alpha\omega_n A_n\varphi_n'''', \varphi_n\right\rangle = 0, \\
&\left\langle -2\dot{A}_m\left(i\omega_m\varphi_m + \gamma_0\varphi_m'\right) + \gamma_1\left[\frac{1}{2}(\omega_n - \omega_m)\overline{\varphi}_n' + i\gamma_0\overline{\varphi}_n''\right]\right. \\
&\left.\overline{\varphi}_n e^{i\sigma\tau_1} - i\alpha\omega_m A_m\varphi_m'''', \varphi_m\right\rangle = 0
\end{aligned} \tag{3.151}$$

where the inner product is defined for complex functions on [0, 1] as follows.

$$\langle g_1, g_2 \rangle = \int_0^1 g_1 \overline{g}_2 \mathrm{d}\overline{x}. \tag{3.152}$$

Using the distributive law of the inner product to (3.151) yields

$$\begin{aligned} &\dot{A}_n + \alpha c_{nn} A_n + \gamma_1 d_{nm} \overline{A}_m \mathrm{e}^{\mathrm{i}\sigma\tau_1} = 0, \\ &\dot{A}_m + \alpha c_{mm} A_m + \gamma_1 d_{mn} \overline{A}_n \mathrm{e}^{\mathrm{i}\sigma\tau_1} = 0 \end{aligned} \tag{3.153}$$

where

$$\begin{aligned} c_{kk} &= \frac{\mathrm{i}\omega_k \int_0^1 \varphi_k'''' \overline{\varphi}_k \mathrm{d}\overline{x}}{2\left(\mathrm{i}\omega_k \int_0^1 \varphi_k \overline{\varphi}_k \mathrm{d}\overline{x} + \gamma_0 \int_0^1 \varphi_k' \overline{\varphi}_k \mathrm{d}\overline{x}\right)} \quad (k = n, m), \\ d_{kj} &= -\frac{\left(\omega_j - \omega_k\right) \int_0^1 \overline{\varphi}_j' \overline{\varphi}_k \mathrm{d}\overline{x} + 2\mathrm{i}\gamma_0 \int_0^1 \overline{\varphi}_j'' \overline{\varphi}_k \mathrm{d}\overline{x}}{4\left(\mathrm{i}\omega_k \int_0^1 \varphi_k \overline{\varphi}_k \mathrm{d}\overline{x} + \gamma_0 \int_0^1 \varphi_k' \overline{\varphi}_k \mathrm{d}\overline{x}\right)} \quad (k = n, m;\ j = m, n). \end{aligned} \tag{3.154}$$

These coefficients are derived by the modal parameters calculated from (3.147) and are independent of parametric excitation by cause of the variation of the translating speed. The transformation

$$A_n(\tau_1) = B_n(\tau_1)\mathrm{e}^{\frac{\mathrm{i}\sigma\tau_1}{2}}, \quad A_m(\tau_1) = B_m(\tau_1)\mathrm{e}^{\frac{\mathrm{i}\sigma\tau_1}{2}} \tag{3.155}$$

converts (3.153) into the following autonomous system.

$$\begin{aligned} &\dot{B}_n + \mathrm{i}\frac{\sigma}{2} B_n + \alpha c_{nn} B_n + \gamma_1 d_{nm} \overline{B}_m = 0, \\ &\dot{B}_m + \mathrm{i}\frac{\sigma}{2} B_m + \alpha c_{mm} B_m + \gamma_1 d_{mn} \overline{B}_n = 0. \end{aligned} \tag{3.156}$$

Obviously, (3.156) has a zero solution. We assume that the nonzero solutions of (3.156) take the following form.

$$B_n = b_n \mathrm{e}^{\lambda\gamma_1}, \ B_m = b_m \mathrm{e}^{\overline{\lambda}\gamma_1} \tag{3.157}$$

where b_m and b_n are real coefficients, whereas λ denotes a complex number to be determined. By substituting (3.157) into (3.156) and then taking the complex conjugate of the second resulting equation, we obtain the following results.

$$\left(-\lambda - \frac{\sigma}{2}\mathrm{i} - \alpha c_{nn}\right) b_n + \gamma_1 d_{nm} b_m = 0,$$

$$\gamma_1 \bar{d}_{mn} b_n + \left(-\lambda + \frac{\sigma}{2}\mathrm{i} - \alpha \bar{c}_{mm}\right) b_m = 0. \tag{3.158}$$

The set of homogeneous linear algebraic equations of b_m and b_n in (3.158) has nontrivial solutions if and only if its determinant of coefficient vanishes. Thus, the system is unstable if λ of the following equation

$$\lambda^2 + \alpha(c_{nn} + c_{mm})\lambda + \left(\frac{\sigma}{2}\mathrm{i} + \alpha c_{nn}\right)\left(-\frac{\sigma}{2}\mathrm{i} + \alpha \bar{c}_{mm}\right) - \gamma_1^2 d_{nm}\bar{d}_{mn} = 0 \tag{3.159}$$

has positive real part.

Separate λ, c_{nn}, and c_{mm} into real and imaginary parts as follows.

$$\lambda = \lambda^R + \mathrm{i}\lambda^{\mathrm{I}}, \quad c_{nn} = c_{nn}^{\mathrm{R}} + \mathrm{i}c_{nn}^{\mathrm{I}}, \quad c_{mm} = c_{mm}^{\mathrm{R}} + \mathrm{i}c_{mm}^{\mathrm{I}}. \tag{3.160}$$

Substituting (3.160) into (3.159) and separating the resulting equation into real and imaginary parts lead to

$$\begin{aligned}
&\lambda^{\mathrm{R}2} - \lambda^{\mathrm{I}2} + \alpha\left(c_{nn}^{\mathrm{R}} + c_{mm}^{\mathrm{R}}\right)\lambda^{\mathrm{R}} - \alpha\left(c_{nn}^{\mathrm{I}} + c_{mm}^{\mathrm{I}}\right)\lambda^{\mathrm{I}} + \alpha^2 c_{nn}^{\mathrm{R}} c_{mm}^{\mathrm{R}} \\
&\quad + \left(\frac{\sigma}{2} + \alpha c_{nn}^I\right)\left(\frac{\sigma}{2} + \alpha c_{mm}^I\right) - \gamma_1^2 \mathrm{Re}\left(d_{nm}\bar{d}_{mn}\right) = 0, \\
&2\lambda^{\mathrm{R}}\lambda^{\mathrm{I}} + \alpha\left(c_{nn}^{\mathrm{I}} + c_{mm}^{\mathrm{I}}\right)\lambda^{\mathrm{R}} + \alpha\left(c_{nn}^{\mathrm{R}} + c_{mm}^{\mathrm{R}}\right)\lambda^{\mathrm{I}} \\
&\quad + \alpha\left[c_{mm}^R\left(\frac{\sigma}{2} + \alpha c_{nn}^I\right) - c_{nn}^R\left(\frac{\sigma}{2} + \alpha c_{mm}^I\right)\right] - \gamma_1^2 \mathrm{Im}\left(d_{nm}\bar{d}_{mn}\right) = 0.
\end{aligned} \tag{3.161}$$

For $\alpha = 0$, (3.161) has the solution $\lambda^{\mathrm{R}} = 0$ upon the condition

$$\mathrm{Im}\left(d_{nm}\bar{d}_{mn}\right) = 0, \quad \mathrm{Re}\left(d_{nm}\bar{d}_{mn}\right) > 0, \quad \sigma = \pm\gamma_1\sqrt{\mathrm{Re}\left(d_{nm}\bar{d}_{mn}\right)}. \tag{3.162}$$

For $\alpha \neq 0$, substituting $\lambda^{\mathrm{R}} = 0$ into (3.161) and eliminating λ^{I} in the resulting equation give

$$\begin{aligned}
&\left[\frac{\sigma}{2}\left(c_{nn}^R - c_{mm}^R\right) + \alpha\left(c_{nn}^R c_{mm}^I - c_{mm}^R c_{nn}^I\right) + \frac{\gamma_1^2}{\alpha}\mathrm{Im}\left(d_{nm}\bar{d}_{mn}\right)\right]^2 \\
&\quad + \left(c_{nn}^R + c_{mm}^R\right)\left(c_{nn}^I + c_{mm}^I\right)\left[\frac{\sigma}{2}\left(c_{nn}^R - c_{mm}^R\right)\right. \\
&\quad \left. + \alpha\left(c_{nn}^R c_{mm}^I - c_{mm}^R c_{nn}^I\right) + \frac{\gamma_1^2}{\alpha}\mathrm{Im}\left(d_{nm}\bar{d}_{mn}\right)\right] \\
&\quad + \left(c_{nn}^R + c_{mm}^R\right)^2\left[\frac{\sigma^2}{4} + \frac{\sigma\alpha}{2}\left(c_{nn}^I + c_{mm}^I\right)\right. \\
&\quad \left. + \alpha^2\left(c_{nn}^R c_{mm}^R + c_{nn}^I c_{mm}^I\right) + \gamma_1^2 \mathrm{Re}\left(d_{nm}\bar{d}_{mn}\right)\right] = 0.
\end{aligned} \tag{3.163}$$

Equation (3.163) is the analytical expression of the stability boundary of the summation parametric resonance.

It is noted that the principal parametric resonance may exist when the variation frequency ω is equal to the two times of a natural frequency of the system in (3.145). We denote

$$\omega = 2\omega_n + \varepsilon\sigma. \tag{3.164}$$

Let $m = n$ in (3.163), then the stability boundary in nth principal parametric resonance can be determined via the resulting equation. For $\alpha = 0$, the stability boundary is derived as follows.

$$\sigma = \pm\gamma_1 |d_{nn}|. \tag{3.165}$$

For $\alpha \neq 0$, the stability boundary is expressed by

$$\begin{aligned} &\frac{\gamma_1^4}{\alpha^2}|d_{nn}|^4 + 4\frac{\gamma_1^2}{\alpha}c_{nn}^{\mathrm{R}}c_{nn}^{\mathrm{I}}|d_{nn}|^2 \\ &\quad + 4c_{nn}^{\mathrm{R}\ 2}\left[\frac{\sigma^2}{4} + \sigma\alpha c_{nn}^{\mathrm{I}} + \alpha^2\left(c_{nn}^{\mathrm{R}\ 2} + c_{nn}^{\mathrm{I}\ 2}\right) + \gamma_1^2|d_{nn}|^2\right] = 0 \end{aligned} \tag{3.166}$$

where

$$\begin{aligned} c_{nn} &= \frac{\mathrm{i}\omega_n \int_0^1 \varphi_n''''\overline{\varphi}_n \mathrm{d}\overline{x}}{2\left(\mathrm{i}\omega_n \int_0^1 \varphi_n\overline{\varphi}_n \mathrm{d}\overline{x} + \gamma_0 \int_0^1 \varphi_n'\overline{\varphi}_n \mathrm{d}\overline{x}\right)}, \\ d_{nn} &= -\frac{2\mathrm{i}\gamma_0 \int_0^1 \overline{\varphi}_n''\overline{\varphi}_n \mathrm{d}\overline{x}}{4\left(\mathrm{i}\omega_n \int_0^1 \varphi_n\overline{\varphi}_n \mathrm{d}\overline{x} + \gamma_0 \int_0^1 \varphi_n'\overline{\varphi}_n \mathrm{d}\overline{x}\right)}. \end{aligned} \tag{3.167}$$

Furthermore, the difference parametric resonance may occur if the variation frequency ω approaches the difference of any two natural frequencies of (3.145). The stability in difference parametric resonance is treated similarly. We denote

$$\omega = \omega_n - \omega_m + \varepsilon\sigma. \tag{3.168}$$

The stability boundaries are expressed by (3.165) and (3.166), respectively, for $\alpha = 0$ and $\alpha \neq 0$, while the coefficients in them are given by

$$\begin{aligned} c_{nn} &= \frac{\mathrm{i}\omega_n \int_0^1 \varphi_n''''\overline{\varphi}_n \mathrm{d}\overline{x}}{2\left(\mathrm{i}\omega_n \int_0^1 \varphi_n\overline{\varphi}_n \mathrm{d}\overline{x} + \gamma_0 \int_0^1 \varphi_n'\overline{\varphi}_n \mathrm{d}\overline{x}\right)}, \\ c_{mm} &= \frac{\mathrm{i}\omega_m \int_0^1 \varphi_m'''' \overline{\varphi}_m \mathrm{d}\overline{x}}{2\left(\mathrm{i}\omega_m \int_0^1 \varphi_m\overline{\varphi}_m \mathrm{d}\overline{x} - \gamma_0 \int_0^1 \varphi_m'\overline{\varphi}_m \mathrm{d}\overline{x}\right)}, \end{aligned}$$

$$d_{nm} = \frac{(\omega_m + \omega_n)\int_0^1 \overline{\varphi}'_m \overline{\varphi}_n \mathrm{d}\overline{x} - 2\mathrm{i}\gamma_0 \int_0^1 \overline{\varphi}''_m \overline{\varphi}_n \mathrm{d}\overline{x}}{4\left(\mathrm{i}\omega_n \int_0^1 \varphi_n \overline{\varphi}_n \mathrm{d}\overline{x} + \gamma_0 \int_0^1 \varphi'_n \overline{\varphi}_n \mathrm{d}\overline{x}\right)},$$

$$d_{mn} = \frac{(\omega_n + \omega_m)\int_0^1 \overline{\varphi}'_n \overline{\varphi}_m \mathrm{d}\overline{x} + 2\mathrm{i}\gamma_0 \int_0^1 \overline{\varphi}''_n \overline{\varphi}_m \mathrm{d}\overline{x}}{4\left(\mathrm{i}\omega_m \int_0^1 \varphi_m \overline{\varphi}_m \mathrm{d}\overline{x} - \gamma_0 \int_0^1 \varphi'_m \overline{\varphi}_m \mathrm{d}\overline{x}\right)}. \tag{3.169}$$

The boundary conditions of an axially translating beam with simple supports in dimensionless form are given as follows.

$$\overline{w}(0,\overline{t}) = \overline{w}(1,\overline{t}) = 0, \quad \left.\frac{\partial^2 \overline{w}}{\partial \overline{x}^2}\right|_{\overline{x}=0} = \left.\frac{\partial^2 \overline{w}}{\partial \overline{x}^2}\right|_{\overline{x}=1} = 0. \tag{3.170}$$

Based on the boundary conditions in (3.170), the eigenfunction corresponding to the kth natural frequency ω_k is derived as follows.

$$\begin{aligned}\varphi_k(\overline{x}) = \mathrm{e}^{\mathrm{i}\beta_{1k}\overline{x}} &- \frac{(\beta_{4k}^2 - \beta_{1k}^2)(\mathrm{e}^{\mathrm{i}\beta_{3k}} - \mathrm{e}^{\mathrm{i}\beta_{1k}})}{(\beta_{4k}^2 - \beta_{2k}^2)(\mathrm{e}^{\mathrm{i}\beta_{3k}} - \mathrm{e}^{\mathrm{i}\beta_{2k}})}\mathrm{e}^{\mathrm{i}\beta_{2k}\overline{x}} - \frac{(\beta_{4k}^2 - \beta_{1k}^2)(\mathrm{e}^{\mathrm{i}\beta_{2k}} - \mathrm{e}^{\mathrm{i}\beta_{1k}})}{(\beta_{4n}^2 - \beta_{3n}^2)(\mathrm{e}^{\mathrm{i}\beta_{2k}} - \mathrm{e}^{\mathrm{i}\beta_{3k}})}\mathrm{e}^{\mathrm{i}\beta_{2k}\overline{x}} \\ &- \left[1 - \frac{(\beta_{4k}^2 - \beta_{1k}^2)(\mathrm{e}^{\mathrm{i}\beta_{3n}} - \mathrm{e}^{\mathrm{i}\beta_{1k}})}{(\beta_{4k}^2 - \beta_{2k}^2)(\mathrm{e}^{\mathrm{i}\beta_{3k}} - \mathrm{e}^{\mathrm{i}\beta_{2k}})} - \frac{(\beta_{4k}^2 - \beta_{1k}^2)(\mathrm{e}^{\mathrm{i}\beta_{2k}} - \mathrm{e}^{i\beta_{1k}})}{(\beta_{4k}^2 - \beta_{3k}^2)(\mathrm{e}^{\mathrm{i}\beta_{2k}} - \mathrm{e}^{\mathrm{i}\beta_{3k}})}\right]\mathrm{e}^{\mathrm{i}\beta_{4k}\overline{x}}\end{aligned} \tag{3.171}$$

where β_{jk} ($j = 1,2,3,4$) and ω_k are determined via the following algebraic equations

$$\begin{aligned}&k_\mathrm{f}^4(\beta_{1k} + \beta_{2k} + \beta_{3k} + \beta_{4k}) = \gamma_0^2 - 1, \\ &\beta_{1k}\beta_{2k} + \beta_{1k}\beta_{3k} + \beta_{1k}\beta_{4k} + \beta_{2k}\beta_{3k} + \beta_{2k}\beta_{4k} + \beta_{3k}\beta_{4k} = 0, \\ &k_\mathrm{f}^4(\beta_{1k}\beta_{2k}\beta_{3k} + \beta_{1k}\beta_{2k}\beta_{4k} + \beta_{1k}\beta_{3k}\beta_{4k} + \beta_{2k}\beta_{3k}\beta_{4k}) = 2\gamma_0\omega_k, \\ &k_\mathrm{f}^4\beta_{1k}\beta_{2k}\beta_{3k}\beta_{4k} = -\omega_k^2\end{aligned} \tag{3.172}$$

and the transcendental equation

$$\begin{aligned}&(\beta_{1k}^2 - \beta_{2k}^2)(\beta_{3k}^2 - \beta_{4k}^2)\left[\mathrm{e}^{\mathrm{i}(\beta_{1k}+\beta_{2k})} + \mathrm{e}^{\mathrm{i}(\beta_{3k}+\beta_{4k})}\right] + (\beta_{2k}^2 - \beta_{4k}^2)(\beta_{3k}^2 - \beta_{1k}^2)\left[\mathrm{e}^{\mathrm{i}(\beta_{1k}+\beta_{3k})}\right. \\ &\left.+\mathrm{e}^{\mathrm{i}(\beta_{2k}+\beta_{4k})}\right] + (\beta_{1k}^2 - \beta_{4k}^2)(\beta_{2k}^2 - \beta_{3k}^2)\left[\mathrm{e}^{\mathrm{i}(\beta_{2k}+\beta_{3k})} - \mathrm{e}^{\mathrm{i}(\beta_{1k}+\beta_{4k})}\right] = 0.\end{aligned} \tag{3.173}$$

Consider an axially moving beam with $k_\mathrm{f} = 0.8$ and $\gamma_0 = 2$. The first two natural frequencies and coefficients in corresponding eigenfunctions (3.171), numerically solved from (3.172) and (3.173), are $\omega_1 = 5.3692$, $\beta_{11} = 3.9906$, $\beta_{21} = -1.2424 + 2.4397\mathrm{i}$, $\beta_{31} = -1.2424 - 2.4397\mathrm{i}$, $\beta_{41} = -1.5058$ and $\omega_2 = 30.1200$, $\beta_{12} = 7.4497$, $\beta_{22} = -1.2497 + 6.0726\mathrm{i}$, $\beta_{32} = -1.2497 - 6.0726\mathrm{i}$, $\beta_{42} = -4.9503$.

In the summation parametric resonance, (3.169) gives $c_{11} = 45.8597$, $c_{22} = 709.7023$, $d_{12} = 1.2427 + 0.7843\mathrm{i}$, and $d_{21} = 0.2948 + 0.1860\mathrm{i}$. In the case that c_{kk} is real, (3.166) reduces to

$$\left[\frac{\sigma}{2}\left(c_{nn}^{\mathrm{R}}-c_{mm}^{\mathrm{R}}\right)\right]^2+\left(c_{nn}^{\mathrm{R}}+c_{mm}^{\mathrm{R}}\right)^2\left[\frac{\sigma^2}{4}+\alpha^2\left(c_{nn}^{\mathrm{R}}c_{mm}^{\mathrm{R}}\right)-\gamma_1^2\mathrm{Re}\left(d_{nm}\overline{d}_{mn}\right)\right]=0. \tag{3.174}$$

Therefore, the instability region is given as

$$-2\sqrt{\frac{\gamma_1^2\mathrm{Re}\left(d_{nm}\overline{d}_{mn}\right)-\alpha^2c_{nn}^{\mathrm{R}}c_{mm}^{\mathrm{R}}}{1+k_{\mathrm{c}}^2}}<\sigma<2\sqrt{\frac{\gamma_1^2\mathrm{Re}\left(d_{nm}\overline{d}_{mn}\right)-\alpha^2c_{nn}^{\mathrm{R}}c_{mm}^{\mathrm{R}}}{1+k_{\mathrm{c}}^2}} \tag{3.175}$$

where

$$k_{\mathrm{c}}=\frac{c_{nn}^{\mathrm{R}}-c_{mm}^{\mathrm{R}}}{c_{nn}^{\mathrm{R}}+c_{mm}^{\mathrm{R}}}. \tag{3.176}$$

The instability region exists under the condition that $\mathrm{Re}\left(d_{nm}\overline{d}_{mn}\right)$ and $c_{nn}^{R}c_{mm}^{R}$ have the same sign, and the axial speed variation amplitude is large enough, namely

$$\gamma_1>\alpha\sqrt{\frac{c_{nn}^{\mathrm{R}}c_{mm}^{\mathrm{R}}}{\mathrm{Re}\left(d_{nm}\overline{d}_{mn}\right)}}. \tag{3.177}$$

Now, we investigate the summation resonance of the first two modes in plane σ-γ_1. Figure 3.21 shows the stability boundaries for the summation resonance for $\alpha=0$, 0.0005, 0.001. As shown in this figure, when the viscosity coefficient increases, the stability boundaries move toward the increasing direction of γ_1 in the plane σ-γ_1, and the instability regions become narrow. According to these results, we can conclude that the more significant viscosity coefficient leads to the smaller instability range of σ for a given γ_1 and the larger instability threshold of γ_1 for a given σ.

At the principal parametric resonance, (3.169) gives $d_{11}=-1.0456+1.1879\mathrm{i}$, $d_{22}=-0.4182+0.9776\mathrm{i}$. The instability region is

$$-2\sqrt{\gamma_1^2|d_{nn}|^2-\alpha^2c_{nn}^{\mathrm{R}\,2}}<\sigma<2\sqrt{\gamma_1^2|d_{nn}|^2-\alpha^2c_{nn}^{\mathrm{R}\,2}}. \tag{3.178}$$

It is noted that if the axial speed variation amplitude is beyond a critical value, i.e.,

$$\gamma_1>\frac{\alpha\left|c_{nn}^{\mathrm{R}}\right|}{|d_{nn}|}, \tag{3.179}$$

the instability region exists.

Figure 3.22 shows the stability boundaries for the first principal resonances in the plane σ-γ_1 for $\alpha=0$, 0.02, 0.05, whereas Fig. 3.23 reveals that for the second principal resonances for $\alpha=0$, 0.001, 0.002. These two cases are similar to the summation resonance case: When the viscosity coefficient increases, the stability

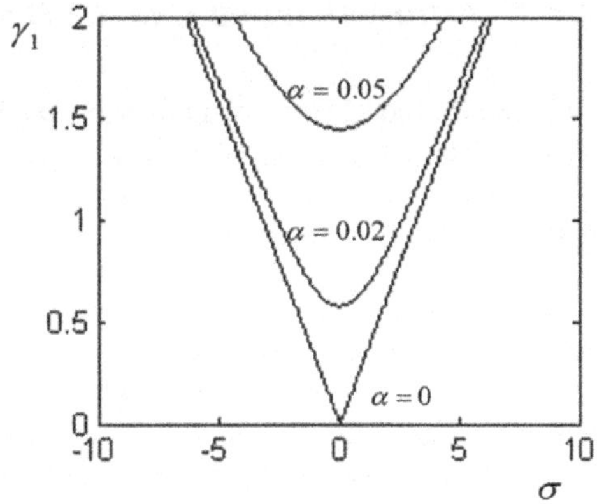

Fig. 3.22 Stability boundaries for the first principal resonance with simple supports

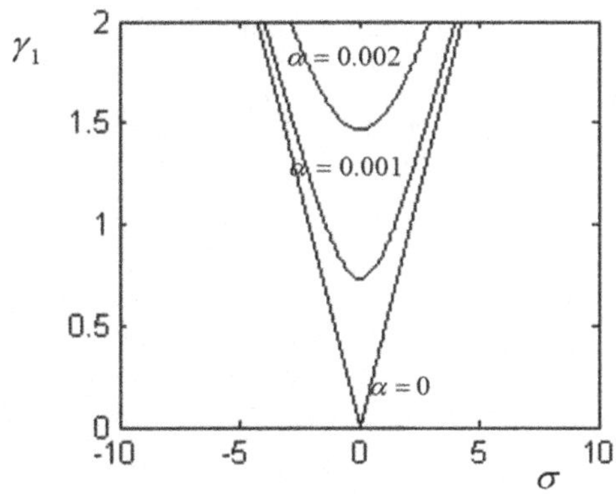

Fig. 3.23 Stability boundaries for the second principal resonance with simple supports

boundaries move toward the increasing direction of γ_1 in the plane σ-γ_1, and the instability regions become narrow.

For the difference parametric resonance, (3.169) gives $c_{11} = 45.8597$, $c_{22} = 741.7379$, $d_{12} = -3.6139 - 2.2809\mathrm{i}$, and $d_{21} = 0.5997 + 0.6081\mathrm{i}$. For real c_{11} and c_{22}, the stability boundary is given by (3.175). In this example, $\mathrm{Re}(d_{nm}\overline{d}_{mn})$ is negative. Therefore, there is no instability region for the different resonance.

In order to demonstrate the stability boundaries on the same scale, the different viscosity coefficients are selected in Figs. 3.21, 3.22 and 3.23. These figures show that the stability boundary in the first principal resonance is most insensitive to the change of the viscosity coefficient. Contrarily, the stability boundary for the summation resonance is most sensitive. The same results can be obtained for the truncated system.

3.4.3 Bifurcation and Chaotic Dynamics

Bifurcation and chaotic dynamics of axially moving Euler–Bernoulli and Timoshenko beams are explored in this subsection. Furthermore, a comparison of the natural supercritical frequencies of Euler–Bernoulli and Timoshenko beams is also performed.

3.4.3.1 Euler–Bernoulli Beam

Bifurcation and chaotic dynamics can be examined by numerical calculations based on the truncated mode. Yang and Chen (2005) and Chen and Yang (2006) explored numerical bifurcations and chaos of axially moving Euler–Bernoulli beams. The governing Eq. (3.128) of the moving Euler–Bernoulli beam can be written in an explicit form without summation notations if the truncation order 4 is chosen.

$$
\begin{aligned}
&\ddot{q}_1 - \pi^2(\overline{v}^2 - 1)q_1 + k_f^2\pi^4 q_1 + \alpha\pi^4\dot{q}_1 - \frac{8}{3}\dot{\overline{v}}q_2 \\
&\quad - \frac{16}{15}\dot{\overline{v}}q_4 - \frac{16}{3}\overline{v}\dot{q}_2 - \frac{32}{15}\overline{v}\dot{q}_4 + k_1^2\pi^4\Big(\frac{8}{3}q_1^3 + 3q_1q_2^2 \\
&\quad + \frac{9}{8}q_1^2q_3 + \frac{9}{2}q_2^2q_3 + \frac{27}{4}q_1q_3^2 + 6q_1q_2q_4 + 18q_2q_3q_4 + 12q_1q_4^2\Big) \\
&\quad + \alpha k_2\pi^4\bigg[\Big(\frac{3}{4}q_1^2 + 2q_2^2 + \frac{3}{2}q_1q_3 + \frac{9}{2}q_3^2 + 4q_2q_4 + 8q_4^2\Big)\dot{q}_1 \\
&\quad + (4q_1q_2 + 6q_2q_3 + 4q_1q_4 + 12q_3q_4)\dot{q}_2 \\
&\quad + \Big(\frac{3}{4}q_1^2 + 3q_2^2 + 9q_1q_3 + 12q_2q_4\Big)\dot{q}_3 \\
&\quad + (4q_1q_2 + 12q_2q_3 + 16q_1q_4)\dot{q}_4\bigg] = 0, \\
&\ddot{q}_2 - \pi^2(\overline{v}^2 - 1)q_2 + 16k_f^2\pi^4 q_2 + 16\alpha\pi^4\dot{q}_2 \\
&\quad + \frac{8}{3}\dot{\overline{v}}q_1 - \frac{24}{5}\dot{\overline{v}}q_3 + \frac{16}{3}\overline{v}\dot{q}_1 + \frac{48}{5}\overline{v}\dot{q}_3 \\
&\quad + k_1^2\pi^4(3q_1q_2^2 + 6q_1^3 + 9q_1q_2q_3 + 3q_1^2q_4 \\
&\quad + 27q_2q_3^2 + 18q_1q_3q_4 + 27q_3^2q_4 + 48q_2q_4^2) \\
&\quad + \alpha k_2\pi^4[(4q_1q_2 + 6q_2q_3 + 4q_1q_4 + 12q_3q_4)\dot{q}_1 \\
&\quad + (2q_1^2 + 12q_2^2 + 6q_1q_3 + 18q_3^2 + 32q_4^2)\dot{q}_2 \\
&\quad + (6q_1q_2 + 36q_2q_3 + 12q_1q_4 + 36q_3q_4)\dot{q}_3 \\
&\quad + (2q_1^2 + 12q_1q_3\dot{q}_4 + 18q_3^2 + 64q_2q_4)\dot{q}_4] = 0, \\
&\ddot{q}_3 - 9\pi^2(\overline{v}^2 - 1)q_3 + 81k_f^2\pi^4 q_3 + 81\alpha\pi^4\dot{q}_3 \\
&\quad + \frac{24}{5}\dot{\overline{v}}q_2 - \frac{48}{7}\dot{\overline{v}}q_4 - \frac{48}{5}\overline{v}\dot{q}_2 - \frac{96}{7}\overline{v}\dot{q}_4
\end{aligned}
$$

$$
\begin{aligned}
&+k_1^2\pi^4\left(\frac{3}{8}q_1^3+\frac{9}{2}q_1q_2^2+\frac{27}{4}q_1^2q_3+27q_3q_2^2\right.\\
&\left.+\frac{243}{8}q_3^3+18q_1q_2q_4+54q_2q_3q_4+108q_3q_4^2\right)\\
&+\alpha k_2\pi^4\left[\left(\frac{3}{4}q_1^2+3q_2^2+9q_1q_3+12q_2q_4\right)\dot{q}_1\right.\\
&+(6q_1q_2+36q_2q_3+12q_1q_4+36q_3q_4)\dot{q}_2\\
&+\left(\frac{9}{2}q_1^2+18q_2^2+\frac{243}{4}q_3^2+36q_2q_4+72q_4^2\right)\dot{q}_3\\
&\left.+(12q_1q_2+36q_2q_3+144q_3q_4)\dot{q}_4\right]=0,\\
&\ddot{q}_4-16\pi^2(\overline{v}^2-1)q_4+256k_f^2\pi^4q_4+256\alpha\pi^4\dot{q}_4\\
&+\frac{16}{15}\dot{\overline{v}}q_1+\frac{48}{7}\dot{\overline{v}}q_3+\frac{32}{15}\overline{v}\dot{q}_1+\frac{96}{7}\overline{v}\dot{q}_3\\
&+k_1^2\pi^4(3q_1^2q_2+18q_1q_2q_3+27q_2^2q_3\\
&+12q_4q_1^2+48q_3^2q_4+108q_3^2q_4+96q_4^3)\\
&+\alpha k_2\pi^4[(4q_1q_2+12q_2q_3+16q_1q_4)\dot{q}_1\\
&+(2q_1^2+12q_1q_3+18q_3^2+64q_2q_4)\dot{q}_2\\
&+(12q_1q_2+36q_2q_3+144q_3q_4)\dot{q}_3\\
&+(8q_1^2+32q_1^2+72q_3^2+192q_4^2)\dot{q}_4]=0.
\end{aligned}
\tag{3.180}
$$

In the sequel, the dimensionless axial speed is assumed a simple harmonic variation about the constant mean speed γ_0, namely

$$
\overline{v}(\overline{t})=\gamma_0+\gamma_1\sin\omega\overline{t},
\tag{3.181}
$$

where γ_1 and ω stand for the amplitude and frequency of the harmonic variation. Under the assumption, (3.180) describes a nonlinear parametrically excited system.

To identify the dynamical behavior, we will use the Poincaré map. The bifurcation diagrams of the Poincaré map are computed to globally view over a range of the values of the parameters and compare chaotic motion and regular motion simultaneously. By calculating the displacement and velocity in the beam's center, we can construct the Poincaré map. The displacement and velocity in the beam's center can be derived via numerical integrations of (3.180), namely

$$
u\left(\frac{1}{2},nT\right)=q_1(nT)-q_3(nT),\ \dot{u}\left(\frac{1}{2},nT\right)=\dot{q}_1(nT)-\dot{q}_3(nT)
\tag{3.182}
$$

where $T=2\pi/\omega$ and $n=1, 2, 3,\ldots$. The bifurcation diagrams are established by changing the viscosity coefficient, the mean axial speed, and the axial speed fluctuation magnitude while keeping other parameters. In order to eliminate the transient vibration, we discard the first 40,000 points of the Poincaré map at each set of parameters. The velocities and displacements for the next 50 points are depicted on the bifurcation diagrams.

The bifurcation diagrams associating with the magnitude of axial speed fluctuation γ_1 are shown in Fig. 3.24, where $k_f = 0.8$, $\alpha = 0.001$, $k_1 = 0.8$, $k_2 = 1.0$, $\gamma_0 = 2.4$, and $\omega = 3.5$. The equilibrium is unstable after γ_1 exceeds a certain value. At the same time, a periodic motion also occurs. When the magnitude of axial speed fluctuation γ_1 keeps increasing, chaotic motions appear after a series of period-doubling bifurcation. The period-doubling bifurcation sequence is demonstrated in Fig. 3.25 (i.e., a local magnification of Fig. 3.24). When the magnitude of axial speed fluctuation γ_1 increases further, chaotic motions suddenly disappear, and a periodic motion appears again. It is observed that the periodic motion finally bifurcates into chaos when the value of γ_1 is sufficiently large. Figure 3.26 shows a local magnification of Fig. 3.24 in the chaotic range. Observably, in the chaotic range, there are different period windows. Thus, the periodic motion and the chaotic motion alternately exchange in both small and large scales.

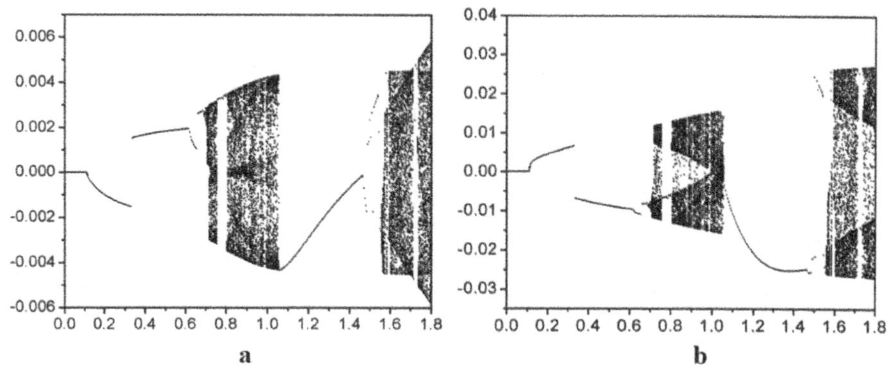

Fig. 3.24 Bifurcation diagrams associating with the magnitude of axial speed fluctuation: **a** lateral displacement, **b** velocity

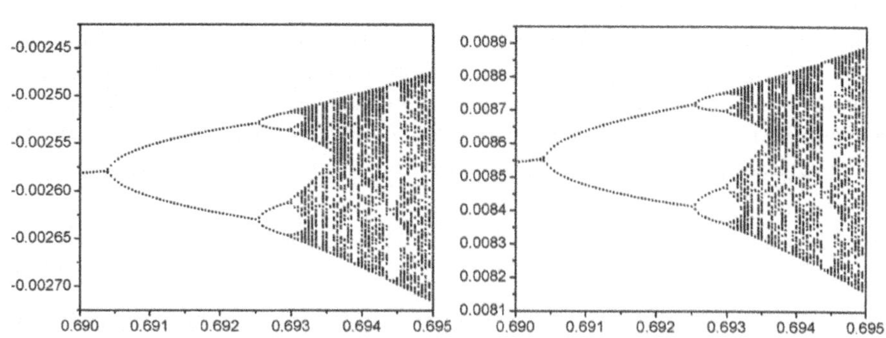

Fig. 3.25 Period-doubling bifurcation sequence: **a** lateral displacement, **b** velocity

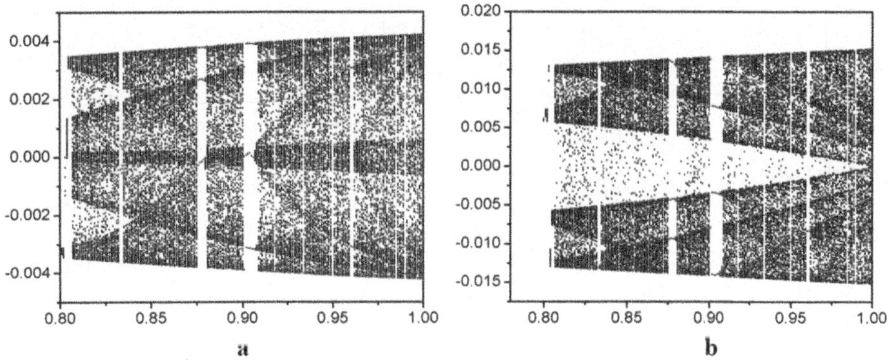

Fig. 3.26 Period windows in the chaotic range: **a** lateral displacement, **b** velocity

The bifurcation diagrams associating with the mean axial speed γ_0 are shown in Fig. 3.27, wherein $k_1 = k_f = 0.8$, $k_2 = 1.0$, $\alpha = 0.001$, $\gamma_1 = 1.0$, and $\omega = 3.5$. The equilibrium is stable for a small mean speed. The equilibrium may lose its stability, and a periodic motion occurs when the mean speed increases. Figure 3.28 reveals that the period-doubling bifurcation appears if the mean speed increases further. Finally, chaotic motions occur when the mean speed is sufficiently large.

The bifurcation diagrams associating with the viscosity coefficient μ are shown in Fig. 3.29, wherein $k_1 = k_f = 0.8$, $k_2 = 1.0$, $\gamma_0 = 1.0$, $\gamma_1 = 1.0$, and $\omega = 3.5$. With a small viscosity coefficient μ, chaotic motions occur. It is observed that there are many period windows in the chaotic range. Figure 3.30 is a local magnification of Fig. 3.29. As shown in this figure, a periodic motion replaces the chaotic motions through the backward period-doubling when μ increases. Finally, the equilibrium may become stable if μ increases to a sufficiently large value; see Fig. 3.29.

A period-doubling bifurcation represented by phase trajectories is shown in Fig. 3.31, wherein $k_1 = k_f = 0.8$, $k_2 = 1.0$, $\alpha = 0.001$, $\gamma_0 = 2.4$, and $\omega = 3.5$.

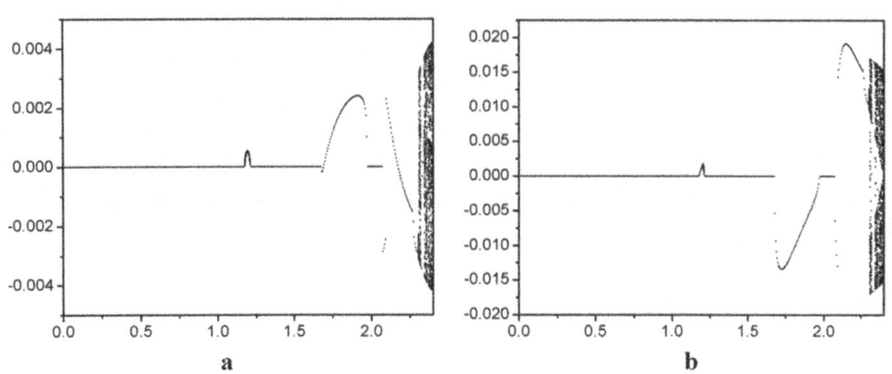

Fig. 3.27 Bifurcation diagrams associating with the mean axial speed: **a** lateral displacement, **b** velocity

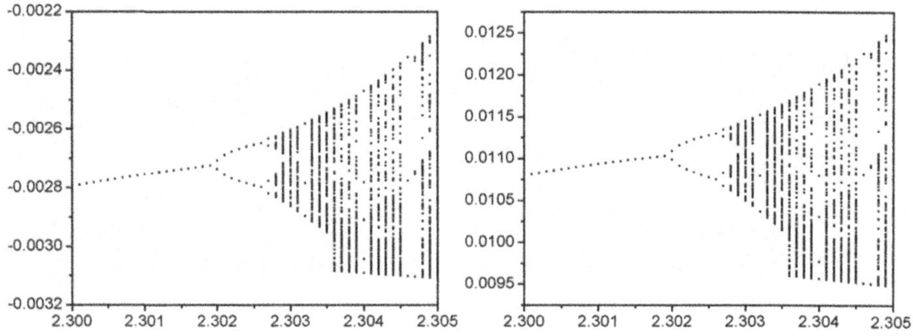

a b

Fig. 3.28 Period-doubling bifurcation sequence: **a** lateral displacement, **b** velocity

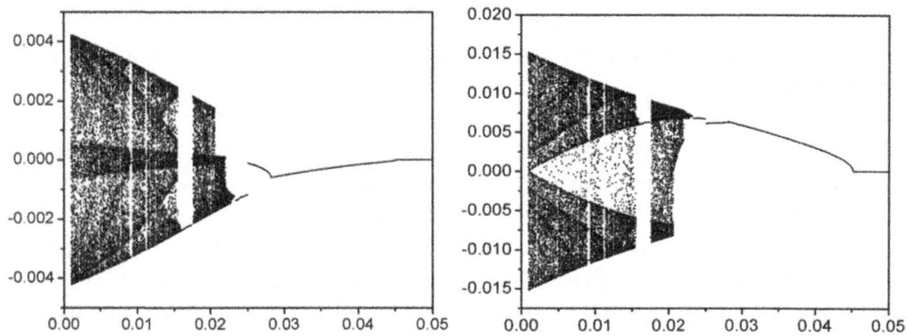

a b

Fig. 3.29 Bifurcation diagrams associating with the viscosity coefficient: **a** lateral displacement, **b** velocity

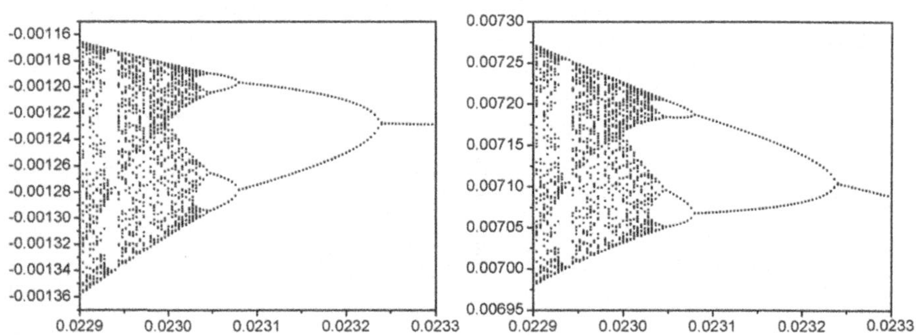

a b

Fig. 3.30 Backward period-doubling bifurcation: **a** lateral displacement, **b** velocity

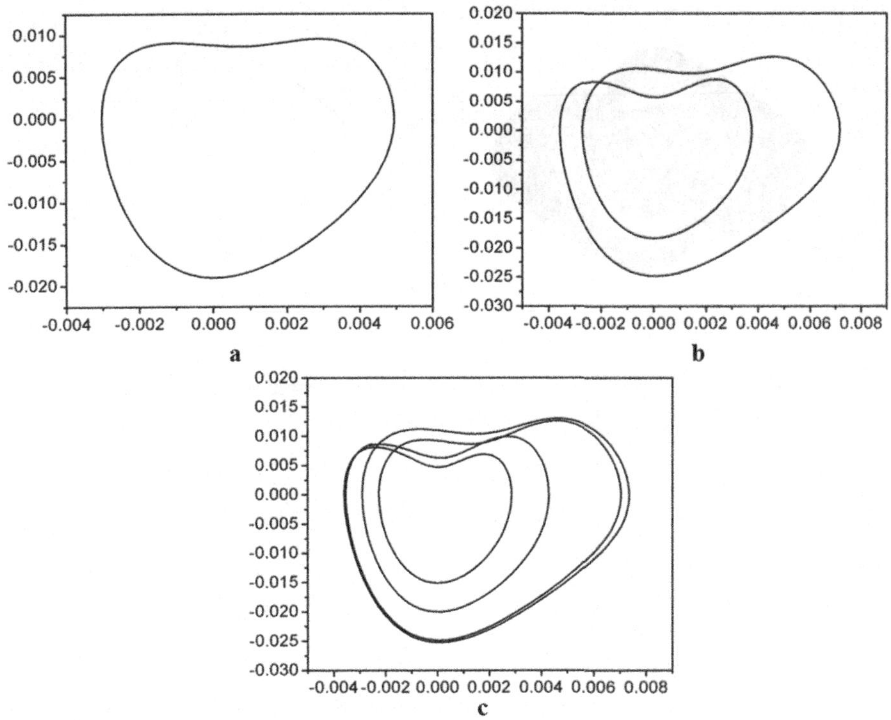

Fig. 3.31 Period-doubling bifurcation shown by phase trajectories: **a** period-1 motion, **b** period-2 motion, **c** period-4 motion

There are period-4 motion, period-2 motion, and period-1 motion for $\gamma_1 = 0.692$, 0.68, 0.5, respectively. Figure 3.32 demonstrates typical chaotic motions identified by the Poincaré map, the phase trajectory, and the largest Lyapunov exponent, wherein $k_f = 0.8$, $\alpha = 0.001$, $k_1 = 0.8$, $k_2 = 1.0$, $\gamma_0 = 2.4$, $\gamma_1 = 0.5$, and $\omega = 3.5$.

The bifurcations and chaos of axially moving beams have shown interesting phenomena. Ding et al. (2014) studied the dynamics in the supercritical regime. Sahoo et al. (2017) considered the responses for the internal resonance case.

3.4.3.2 Timoshenko Beam

The influences of the shear deformation and the rotary inertia of the cross section on the beam's dynamics are ignored in the Euler–Bernoulli model. However, these influences become significant for thick beams. In these situations, the Timoshenko model should be used. The critical speeds, mode shapes, and natural frequencies of an longitudinally translating Timoshenko beam were investigated by Tang et al. (2008). Ghayesh and Balar (2010) explored periodic responses of a translating Timoshenko

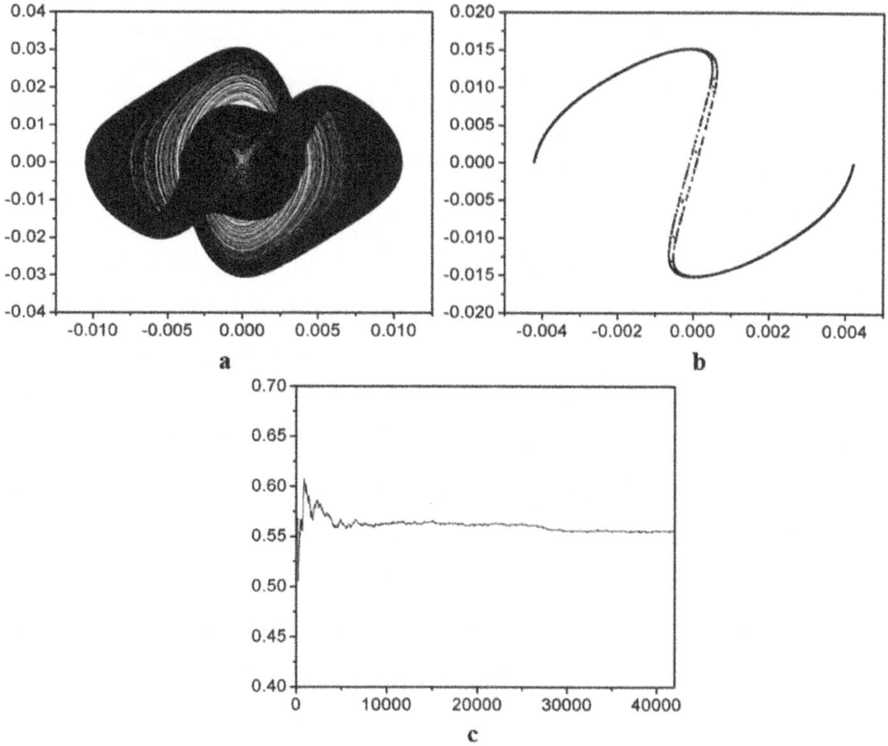

Fig. 3.32 Identification of chaotic motions: **a** phase trajectory, **b** Poincaré map, **c** largest Lyapunov exponent

beam's parametric vibrations. Yan et al. (2014) considered the bifurcations and chaos of longitudinally translating Timoshenko beams.

Consider a uniform Timoshenko beam with length l, area moment of inertia I, cross-sectional area A, mass density ρ, Young's modulus E, shearing modulus G, shear coefficient κ, initial axial tension T_0, and dynamic viscosity α. The beam travels with a time-dependent axial speed $v(t)$ between two simple supports. We assume that the beam's deformation is limited in the vertical plane.

The bending vibration is characterized by two variables dependent on time t and axial coordinate x, namely lateral vibration $w(x, t)$ and the rotation angle $\theta(x, t)$ of the cross section (or slope) by cause of the bending moment. Under the effect of the axial acceleration, the total tension T along the axial direction is given as follows (Chen and Tang 2012).

$$T = T_0 + \zeta \rho A v^2 + (x - l)\rho A \dot{v} \tag{3.183}$$

where $\zeta \in [0, 1]$ denotes the axial support's rigidity parameter (i.e., $\zeta = 0$ corresponds to the infinite rigidity, whereas $\zeta = 1$ corresponds to the zero rigidity). The

Kelvin model for a viscoelastic Timoshenko beam is given as follows.

$$\begin{aligned}\sigma(x,t) &= E[\varepsilon(x,t) + \mu(\varepsilon_t(x,t) + v\varepsilon_x(x,t))],\\ \tau(x,t) &= G\big[\gamma(x,t) + \mu(\gamma_t(x,t) + v\gamma_x(x,t))\big]\end{aligned} \tag{3.184}$$

where $\varepsilon(x,t)$ and $\sigma(x,t)$ are the axial strain and axial stress, respectively, whereas $\tau(x,t)$ and $\gamma(x,t)$ denote the shear stress and shear strain, respectively. According to Newton's second law, the following nonlinear IPDEs are determined for governing the lateral vibration of an axially accelerating viscoelastic Timoshenko beam.

$$\begin{aligned}&\rho A\big(w_{tt} + 2vw_{xt} + v^2 w_{xx}\big) - \big[T_0 + \zeta\rho A v^2 + (x-l)\rho A\dot{v}\big]v_{xx}\\ &\quad - \frac{EA}{2l}w_{xx}\int_0^l w_x^2\mathrm{d}x + \kappa AG[(\theta_x - w_{xx}) + \mu(\theta_{xt} - w_{xxt})\\ &\quad + \mu v(\theta_{xx} - w_{xxx})] = 0,\\ &\rho I\big(\theta_{tt} + 2v\theta_{xt} + \dot{v}\theta_x + v^2\theta_{xx}\big) - EI[\theta_{xx} + \mu(\theta_{xxt} + v\theta_{xxx})]\\ &\quad + \kappa AG[(\theta - w_x) + \mu(\theta_t - w_{xt}) + \mu v(\theta_x - w_{xx})] = 0.\end{aligned} \tag{3.185}$$

The subscripts x or t denote partial differentiation with respect to x or t, and the dot denotes the total derivative with respect to time t. The boundary conditions are given as follows.

$$\begin{aligned}&w(0) = w(l) = 0,\\ &\{EI[\theta_x(0,t) + \mu(\theta_{xt}(0,t) + v\theta_{xx}(0,t))] - \rho I v(\theta_t(0,t) + v\theta_x(0,t))\} = 0,\\ &\{EI[\theta_x(l,t) + \mu(\theta_{xt}(l,t) + v\theta_{xx}(l,t))] - \rho I v(\theta_t(l,t) + v\theta_x(l,t))\} = 0.\end{aligned} \tag{3.186}$$

Introduce the dimensionless parameters and variables as follows.

$$\begin{aligned}&\overline{w} = \frac{w}{l}, \overline{x} = \frac{x}{l}, \overline{t} = \frac{t}{l}\sqrt{\frac{T_0}{\rho A}}, \overline{v} = v\sqrt{\frac{\rho A}{T_0}}, \overline{\alpha} = \frac{\mu}{l}\sqrt{\frac{T_0}{\rho A}},\\ &k_1 = \frac{AG}{\kappa T_0}, k_2 = \frac{I}{Al^2}, k_\mathrm{f} = \sqrt{\frac{EI}{T_0 l^2}}, k_\mathrm{N} = \sqrt{\frac{EA}{T_0}}\end{aligned} \tag{3.187}$$

where k_1 is the dimensionless parameter representing the effect of the shear deformation; k_2 accounts for the effect of the rotary inertia; ${k_\mathrm{f}}^2$ signifies the bending stiffness of the Timoshenko beam; k_N denotes the nonlinear coefficient. By substituting (3.187) into (3.186), the governing equations of the beam can be cast into the dimensionless form

$$\overline{w}_{\overline{t}\overline{t}} + 2\overline{v}\overline{w}_{\overline{x}\overline{t}} + \big[(1-\zeta)\overline{v}^2 - (\overline{x}-1)\dot{\overline{v}} - 1\big]\overline{w}_{\overline{x}\overline{x}} + k_1(\theta_{\overline{x}} - w_{\overline{x}\overline{x}})$$

$$= \frac{1}{2}k_{\mathrm{N}}^2\overline{w}_{\overline{x}\overline{x}}\int_0^1 \overline{w}_{\overline{x}}^2\mathrm{d}\overline{x} - k_1\overline{\alpha}[\theta_{\overline{x}\overline{t}} - \overline{w}_{\overline{x}\overline{x}\overline{t}} + \overline{v}(\theta_{\overline{x}\overline{x}} - \overline{w}_{\overline{x}\overline{x}\overline{x}})]$$

$$k_2\left(\theta_{\overline{t}\overline{t}} + 2\overline{v}\theta_{\overline{x}\overline{t}} + \dot{\overline{v}}\theta_{\overline{x}} + \overline{v}^2\theta_{\overline{x}\overline{x}}\right) - k_{\mathrm{f}}^2\theta_{\overline{x}\overline{x}} + k_1(\theta - \overline{w}_{\overline{x}})$$
$$= \overline{\alpha}\left\{k_{\mathrm{f}}^2(\theta_{\overline{x}\overline{x}\overline{t}} + \overline{v}\theta_{\overline{x}\overline{x}\overline{x}}) - k_1[\theta_{\overline{t}} - \overline{w}_{\overline{x}\overline{t}} + \overline{v}(\theta_{\overline{x}} - \overline{w}_{\overline{x}\overline{x}})]\right\}. \quad (3.188)$$

The dimensionless simply-supported boundary conditions at both ends are given as follows.

$$\overline{w}(0) = \overline{w}(1) = 0,$$
$$\left[\left(k_{\mathrm{f}}^2 - k_2\overline{v}^2\right)\theta_{\overline{x}}\left(0, \overline{t}\right) - k_2\overline{v}\theta_{\overline{t}}\left(0, \overline{t}\right) + k_{\mathrm{f}}^2\overline{\alpha}\left(\theta_{\overline{x}\overline{t}}\left(0, \overline{t}\right) + \overline{v}\theta_{\overline{x}\overline{x}}\left(0, \overline{t}\right)\right)\right] = 0,$$
$$\left[\left(k_{\mathrm{f}}^2 - k_2\overline{v}^2\right)\theta_{\overline{x}}\left(1, \overline{t}\right) - k_2\overline{v}\theta_{\overline{t}}\left(1, \overline{t}\right) + k_{\mathrm{f}}^2\overline{\alpha}\left(\theta_{\overline{x}\overline{t}}\left(1, \overline{t}\right) + \overline{v}\theta_{\overline{x}\overline{x}}\left(1, \overline{t}\right)\right)\right] = 0. \quad (3.189)$$

In the following, we assume that the translating speed is a simple harmonic cosine variation about the constant mean speed γ_0, which is defined as follows.

$$\overline{v} = \gamma_0 + \gamma_1\cos\left(\omega\overline{t}\right) \quad (3.190)$$

where γ_1 and ω are the dimensionless amplitude and dimensionless frequency of the axial speed variation, respectively.

The Galerkin method is used to truncate the nonlinear IPDEs in (3.188), and the resulting truncated system is numerically solved via the fourth-order Runge–Kutta method. According to the boundary conditions in (3.189), the approximate solutions for the lateral vibration of the Timoshenko beam can be written as follows.

$$\overline{w}\left(\overline{x}, \overline{t}\right) = \sum_{n=1}^{N} q_n\left(\overline{t}\right)\sin(n\pi\overline{x}),$$
$$\theta\left(\overline{x}, \overline{t}\right) = \sum_{n=1}^{N} p_n\left(\overline{t}\right)\cos(n\pi\overline{x}) \quad (3.191)$$

in which q_n and p_n ($n = 1, 2, \ldots, N$) are the sets of generalized coordinates for the lateral displacements and the slopes.

Substituting (3.191) into (3.188), one obtains

$$\sum_{n=1}^{N}\ddot{q}_n(\overline{t})\sin(n\pi\overline{x}) + 2\pi\overline{v}\sum_{n=1}^{N}n\dot{q}_n(\overline{t})\cos(n\pi\overline{x})$$
$$-\left[(1-\varsigma)\overline{v}^2 - (\overline{x} - 1)\dot{\overline{v}} - 1\right]\sum_{n=1}^{N}n^2\pi^2 q_n(\overline{t})\sin(n\pi\overline{x})$$

$$
\begin{aligned}
&-\pi k_1 \sum_{n=1}^{N} n p_n(\bar{t}) \sin(n\pi\bar{x}) \\
&+\frac{1}{4}\pi^4 k_{\mathrm{N}}^2 \left[\sum_{n=1}^{N} n^2 q_n(\bar{t}) \sin(n\pi\bar{x})\right]\left[\sum_{n=1}^{N} n^2 q_n^2(\bar{t})\right] \\
&+\pi k_1 \sum_{n=1}^{N} n^2 \pi q_n(\bar{t}) \sin(n\pi\bar{x}) \\
&-n\bar{\alpha}\pi k_1 \left[\sum_{n=1}^{N} \dot{p}_n(\bar{t}) \sin(n\pi\bar{x}) - \sum_{n=1}^{N} n\pi \dot{q}_n(\bar{t}) \sin(n\pi\bar{x})\right] \\
&-\pi^2 n\bar{\alpha} k_1 \bar{v}\left[\sum_{n=1}^{N} n p_n(\bar{t}) \cos(n\pi\bar{x}) - \sum_{n=1}^{N} n^2\pi q_n(\bar{t}) \cos(n\pi\bar{x})\right] = 0,
\end{aligned}
$$

$$
\begin{aligned}
k_2&\left[\sum_{n=1}^{N} \ddot{p}_n(\bar{t}) \cos(n\pi\bar{x}) - 2\pi\bar{v}\sum_{n=1}^{N} n\dot{p}_n(\bar{t}) \sin(n\pi\bar{x})\right. \\
&\left.-\pi\dot{\bar{v}}\sum_{n=1}^{N} n p_n(\bar{t}) \sin(n\pi\bar{x}) - \pi^2\bar{v}^2\sum_{n=1}^{N} n^2 p_n(\bar{t}) \cos(n\pi\bar{x})\right] \\
&+\pi^2 k_{\mathrm{f}}^2 \sum_{n=1}^{N} n^2 p_n(\bar{t}) \cos(n\pi\bar{x}) \\
&+k_1\left[\sum_{n=1}^{N} p_n(\bar{t}) \cos(n\pi\bar{x}) - \sum_{n=1}^{N} n\pi q_n(\bar{t}) \cos(n\pi\bar{x})\right] \\
&+\bar{\alpha}\pi^2 k_{\mathrm{f}}^2\left[\sum_{n=1}^{N} n^2 \dot{p}_n(\bar{t}) \cos(n\pi\bar{x}) - \bar{v}\sum_{n=1}^{N} n^3\pi p_n(\bar{t}) \sin(n\pi\bar{x})\right] \\
&+\bar{\alpha} k_1 \sum_{n=1}^{N} \dot{p}_n(\bar{t}) \cos(n\pi\bar{x}) \\
&+\bar{\alpha} k_1\left\{\pi\bar{v}\left[\sum_{n=1}^{N} n^2\pi q_n(\bar{t}) \sin(n\pi\bar{x}) - \sum_{n=1}^{N} n p_n(\bar{t}) \sin(n\pi\bar{x})\right]\right. \\
&\left.-\sum_{n=1}^{N} n\pi \dot{q}_n(\bar{t}) \cos(n\pi\bar{x})\right\} = 0. \qquad (3.192)
\end{aligned}
$$

Denote the left hand of (3.192), respectively, as R_N and T_N. By choosing the weight function as a trial function itself for the Galerkin truncation, the implementation of the Galerkin method yields

$$\int_0^1 R_N(\overline{x},\overline{t})\sin(m\pi\overline{x})\mathrm{d}\overline{x}=0,$$
$$\int_0^1 T_N(\overline{x},\overline{t})\sin(m\pi\overline{x})\mathrm{d}\overline{x}=0 \tag{3.193}$$

where $m = 1, 2, \ldots, N$.

For given N, substituting (3.192) into (3.193) leads to the second-order ODEs for the N-term Galerkin method of the equations of motion of the translating viscoelastic Timoshenko beam. That is

$$\begin{aligned}
&\ddot{q}_m(\overline{t})-\left[(1-\zeta)\overline{v}^2+\frac{1}{2}\dot{\overline{v}}-1\right]m^2\pi^2q_m(\overline{t})\\
&-m\pi k_1\left[p_m(\overline{t})+\overline{\alpha}\dot{p}_m(\overline{t})\right]+m^2\pi^2k_1\left[q_m(\overline{t})+\overline{\alpha}\dot{q}_m(\overline{t})\right]\\
&+\frac{1}{4}m^2\pi^4k_{\mathrm{N}}^2q_m(\overline{t})\sum_{n=1}^{N}n^2q_n^2(\overline{t})+\sum_{m+n\text{ is odd}}^{N}\left\{\frac{8mn\overline{v}}{m^2-n^2}\dot{q}_n(\overline{t})\right.\\
&\left.-\frac{8mn^3\dot{\overline{v}}}{(m-n)^2(m+n)^2}q_n(\overline{t})-\frac{4mn^2\pi\overline{\alpha}k_1\overline{v}}{m^2-n^2}\left[p_n(\overline{t})-n\pi q_n(\overline{t})\right]\right\}=0\\
&\sum_{m+n\text{ is odd}}^{N}\frac{4k_2}{m^2-n^2}\ddot{p}_n(\overline{t})-\pi^2\left(m^2\pi^2\overline{\alpha}\overline{v}k_{\mathrm{f}}^2+\overline{\alpha}\overline{v}k_1+\dot{\overline{v}}k_2\right)p_m(\overline{t})\\
&-2\pi^2k_2\overline{v}\dot{p}_m(\overline{t})+m\overline{\alpha}\pi^3k_1\overline{v}q_m(\overline{t})\\
&+\sum_{m+n\text{ is odd}}^{N}\frac{4}{m^2-n^2}\left[\left(n^2\pi^2k_{\mathrm{f}}^2+k_1\right)\overline{\alpha}\dot{p}_n(\overline{t})\right.\\
&\left.-\left(n^2\pi^2k_2\overline{v}^2-n^2\pi^2k_{\mathrm{f}}^2-k_1\right)p_n(\overline{t})\right]\\
&-\sum_{m+n\text{ is odd}}^{N}\frac{4n\pi k_1}{m^2-n^2}\left[q_n(\overline{t})+\overline{\alpha}\dot{q}_n(\overline{t})\right]=0.
\end{aligned}\tag{3.194}$$

Now, we analyze the nonlinear dynamics of the axially moving viscoelastic Timoshenko beam via two-term, four-term, and six-term Galerkin truncations. The resulting truncated system is numerically solved based on the fourth-order Runge–Kutta method. Additionally, the following initial conditions are chosen for all cases.

$$\begin{aligned}
&q_1=0.001,\ \dot{q}_1=0,\\
&q_n=0,\ \dot{q}_n=0\quad\text{for } n=2,3,\ldots,N,\\
&p_1=0.001,\ \dot{p}_1=0,\\
&p_n=0,\ \dot{p}_n=0\quad\text{for } n=2,3,\ldots,N.
\end{aligned}\tag{3.195}$$

For $N = 2, N = 4$, and $N = 6$, the lateral displacements and the rotation angle of the cross section in the beam's center are given as follows.

$$\begin{aligned}
\overline{w}_{N=2}(\overline{x},\overline{t}) &= \sum_{n=1}^{2} q_n(\overline{t}) \sin(n\pi\overline{x}) = q_1, \\
\theta_{N=2}(\overline{x},\overline{t}) &= \sum_{n=1}^{2} p_n(\overline{t}) \cos(n\pi\overline{x}) = -p_2, \\
\overline{w}_{N=4}(\overline{x},\overline{t}) &= \sum_{n=1}^{4} q_n(\overline{t}) \sin(n\pi\overline{x}) = q_1 - q_3, \\
\theta_{N=4}(\overline{x},\overline{t}) &= \sum_{n=1}^{4} p_n(\overline{t}) \cos(n\pi\overline{x}) = p_4 - p_2, \\
\overline{w}_{N=6}(\overline{x},\overline{t}) &= \sum_{n=1}^{6} q_n(\overline{t}) \sin(n\pi\overline{x}) = q_1 - q_3 + q_5, \\
\theta_{N=6}(\overline{x},\overline{t}) &= \sum_{n=1}^{6} p_n(\overline{t}) \cos(n\pi\overline{x}) = p_4 - p_2 - p_6.
\end{aligned} \tag{3.196}$$

Consider an axially translating viscoelastic Timoshenko beam with ρ = 7850 kg/m^3, $A = 0.1162 \times 0.07745$ m^2, $l = 0.3$ m, $E = 206 \times 10^9$ Pa, $G = 81 \times 10^9$ Pa, $\kappa = 5/6$, and $T_0 = 1.2 \times 10^7$ N. In the following numerical examples, the dimensionless values are set to $k_1 = 72.8975$, $k_2 = 0.005554$, $k_\mathrm{N} = 12.4296$, $k_\mathrm{f} = 0.9263$, $\mu = 0.001$, and $\zeta = 0.5$ if no other values are assigned. We select parameters, including the magnitude of the translating speed variation, dynamic viscosity, and axial support's rigidity parameter, as the bifurcation parameters. Based on the Galerkin truncation, the bifurcations are investigated.

The frequency of the harmonic variation of the axial speed is set to $\omega = 3.0$. The numerical nonlinear dynamic response in the time interval of $[0, 5000 \times 2\pi/\omega]$ dimensionless time units is obtained. It is noted that the first 96% of the time histories are excluded to remove the transient response. Therefore, the stepping resolution is set so that the last 200 steps are depicted in each bifurcation diagram. Additionally, the lateral vibration and slope of the longitudinally translating beam and their corresponding velocity at the maximum amplitude in the last 4% of the time histories at each step are adopted as the initial conditions of the next step.

We consider the bifurcation diagrams associating with the mean axial speed γ_0 based on the six-term Galerkin truncation. The bifurcation diagrams of the lateral vibration and slope in the beam's center and their corresponding velocity are revealed in Fig. 3.33. This figure shows the significant influence of the mean axial speed γ_0 on the nonlinear vibration of the longitudinally Timoshenko beam. When the axial mean speed increases, the lateral displacement and slope in the beam's center and their corresponding velocity increase. The bifurcation diagrams in Fig. 3.33 reveal

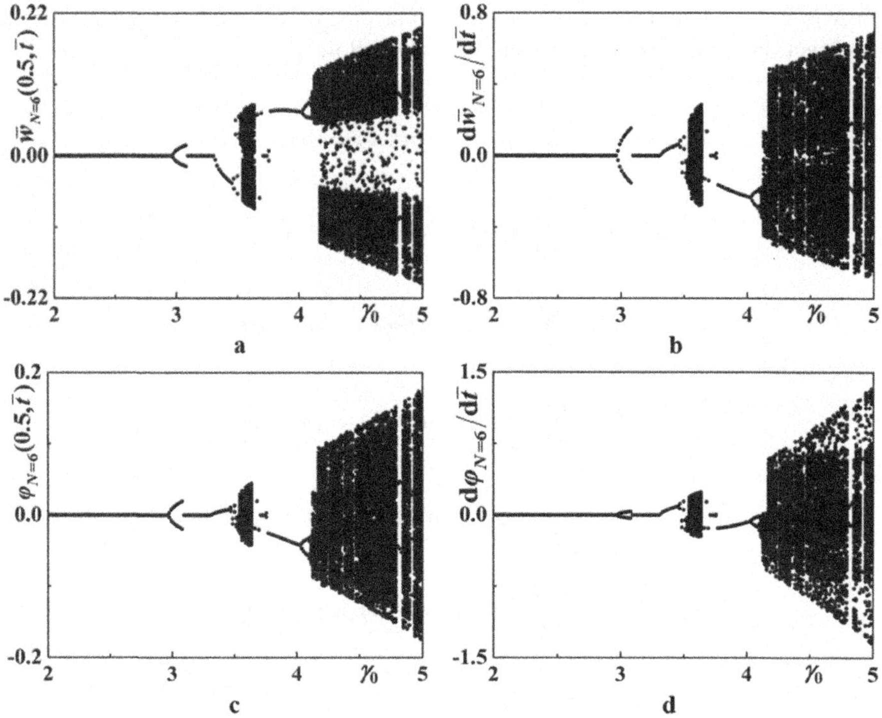

Fig. 3.33 Bifurcation diagrams associating with the mean axial speed (six-term Galerkin truncation): **a** lateral displacement, **b** velocity, **c** slope, **d** angular velocity

that there is no resonant response when the mean axial speed is small. The periodic motion occurs with the increase of the mean axial speed. When the mean axial speed keeps increasing, the complicated motions, such as quasiperiodic or chaotic motions, appear after a series of period-doubling bifurcation. When the mean axial speed increases further, the complicated motions disappear suddenly and are replaced by the periodic motion. It is apparent that the periodic motions and the complicated motions alternately exchange.

Furthermore, Fig. 3.33 shows that the stable steady-state response is dominant in the subcritical regime. Contrarily, quasiperiodic or chaotic motions occur more frequently in the supercritical regime. Observably, the bifurcation diagrams of the lateral displacement and its corresponding velocity predict the same motion, whereas the bifurcation diagrams of the slope and its corresponding velocity are similar. Thus, for simplicity, only the bifurcation diagrams of the lateral displacement and slope are considered in the following numerical examples.

We consider the bifurcation diagrams associating with the mean axial speed γ_0 via two-term Galerkin truncation and four-term Galerkin truncation; see Figs. 3.34 and 3.35. The bifurcation diagrams of the lateral displacement and slope in the center of an longitudinally translating viscoelastic Timoshenko beam are shown in

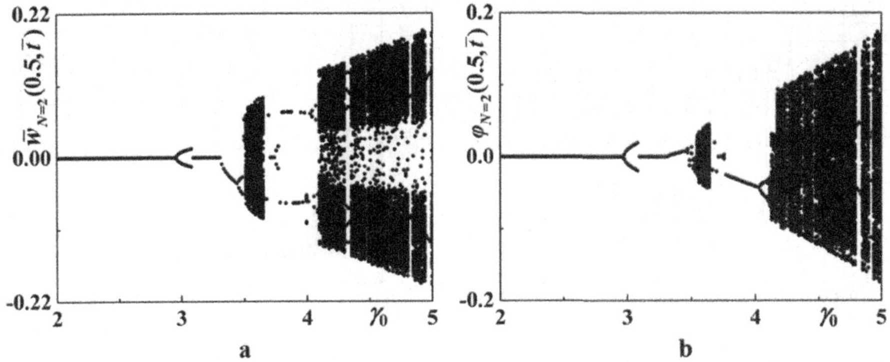

Fig. 3.34 Bifurcation diagrams associating with the mean axial speed (two-term Galerkin truncation): **a** lateral displacement, **b** slope

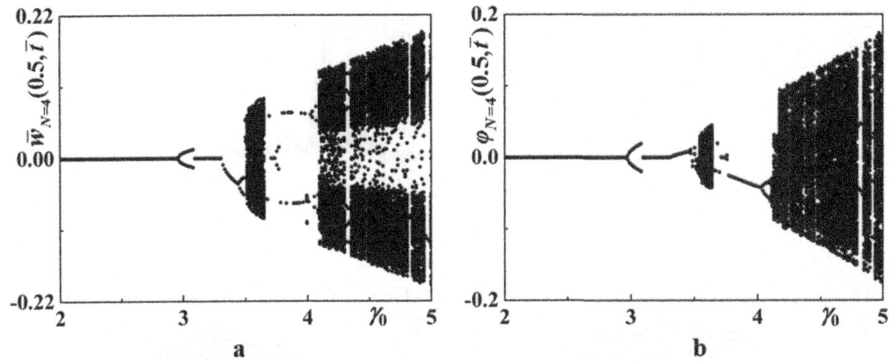

Fig. 3.35 Bifurcation diagrams associating with the mean axial speed (four-term Galerkin truncation): **a** lateral displacement, **b** slope

these figures. Similar to the six-term Galerkin truncation, Figs. 3.34 and 3.35 also show that the lateral displacement and slope in the beam's center increase with the increase of γ_0, and the periodic motions and complicated motions alternately exchange. Compared with the six-term Galerkin truncation in Fig. 3.33, the results in Figs. 3.34 and 3.35 point out that there are apparent differences in predicting bifurcation points among different truncation terms. For the axially moving beam, the more even terms truncation usually leads to better results. Nevertheless, it is difficult to see the tendency in this truncated system in Figs. 3.33, 3.34, and 3.35.

We further investigate the nonlinear dynamical behavior of the complicated motion of a longitudinally translating Timoshenko beam, which is shown in Fig. 3.33. The mean axial speed is fixed by $\gamma_0 = 4.4$. According to the six-term Galerkin truncation, the dynamic characteristics of the lateral vibration are determined and shown in Fig. 3.36. To investigate the long-time nonlinear dynamic, we utilize the following standard indicators: time history, amplitude spectrum, phase portrait, Poincaré map,

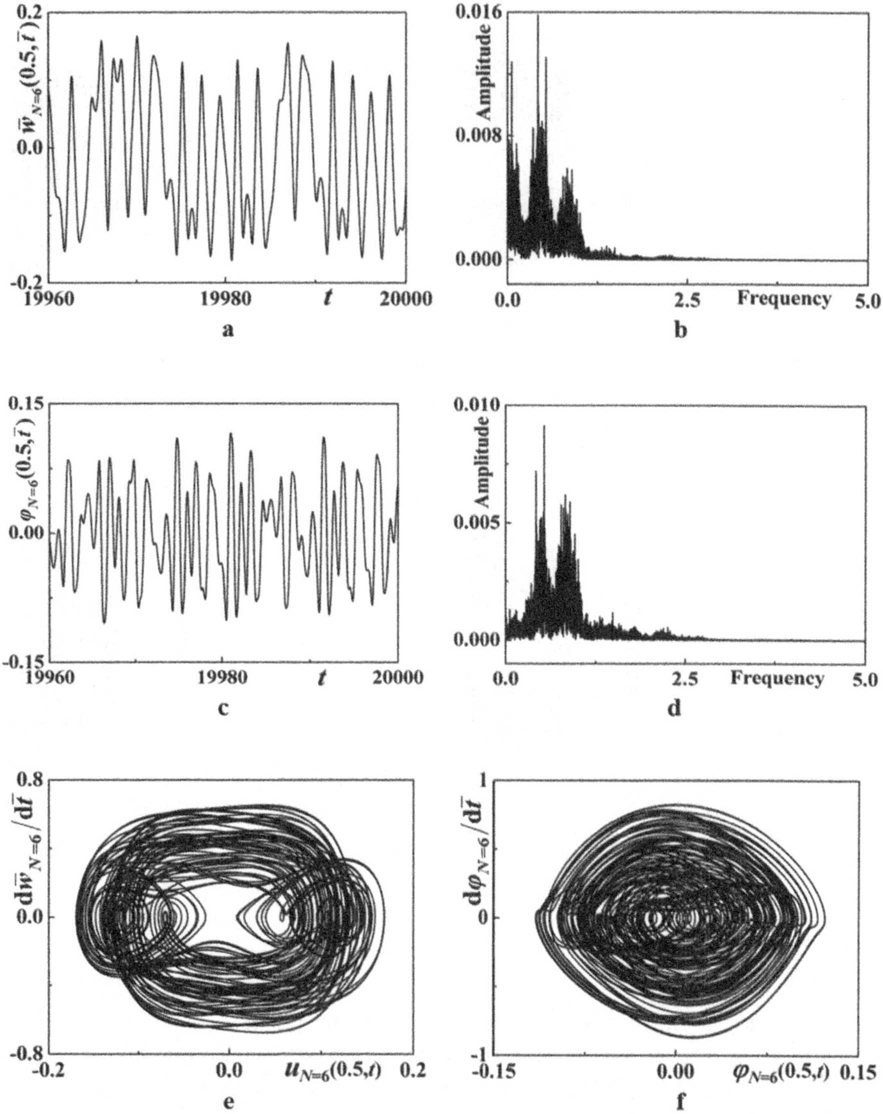

Fig. 3.36 Chaotic motions from the six-term Galerkin truncation: **a** time history of the lateral displacement, **b** amplitude spectrum of the lateral displacement, **c** time history of the slope, **d** amplitude spectrum of the slope, **e** phase portrait of the lateral displacement, **f** phase portrait of the slope, **g** Poincaré map of the lateral displacement, **h** Poincaré map of the slope, **i** sensitive to the initial condition (the initial phase), **j** sensitive to the initial condition (chaotic), **k** sensitive to the initial condition (the initial phase), **l** sensitive to the initial condition (chaotic)

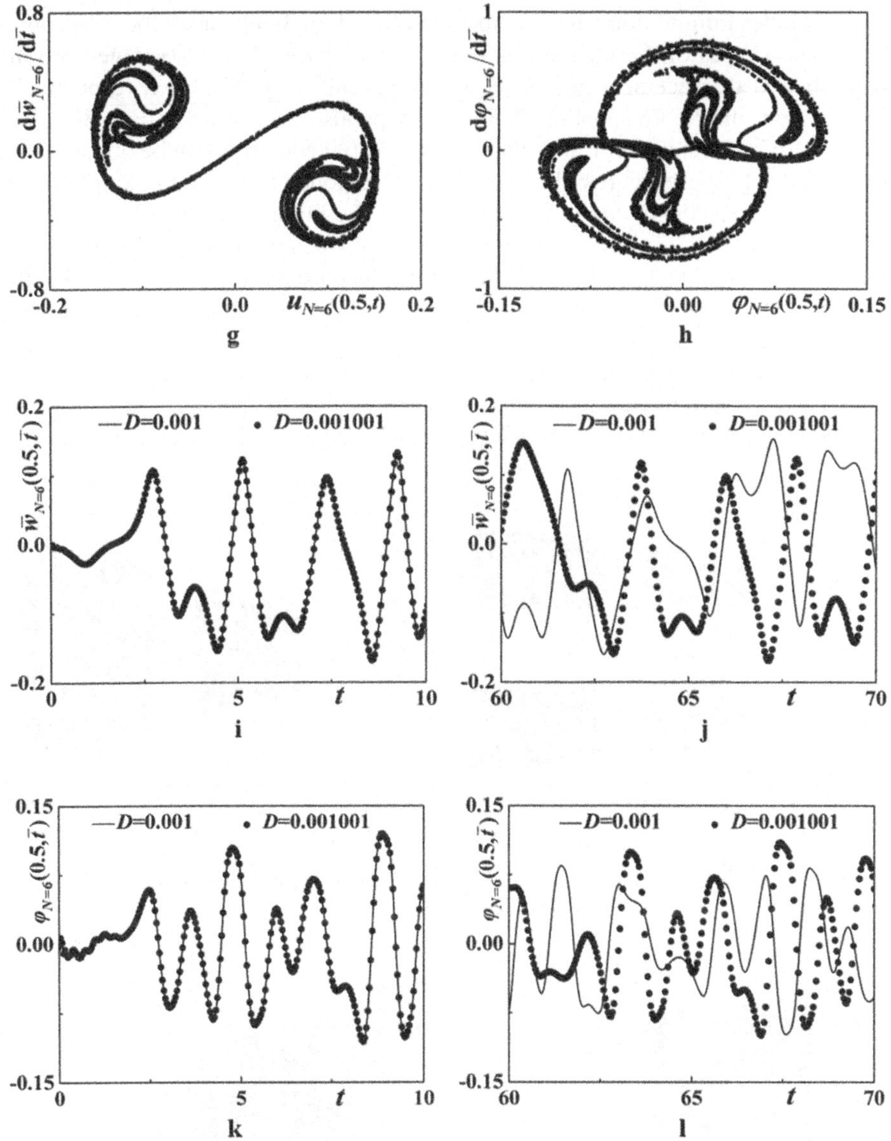

Fig. 3.36 (continued)

and sensitivity to the initial condition. Figure 3.36a presents the time histories of the lateral displacement, whereas Fig. 3.36b shows the corresponding amplitude spectrum based on the fast Fourier transforms. Figure 3.36c, d reveal the time histories of the slope and its amplitude spectrum, respectively. It is observed that the nonlinear vibration does not show the characteristics of the periodic motion from Fig. 3.36a, c. Additionally, the nonlinear vibration of the longitudinally translating Timoshenko beam includes infinite frequency components (i.e., Fig. 3.36b and 3.36d). To appreciating the shapes of the trajectories in the phase space, the phase-plane portraits of the lateral displacement and slope are presented in Fig. 3.36e, f, respectively. Besides that, their corresponding Poincaré maps also are shown in Fig. 3.36g, h. The sensitivity of the nonlinear dynamic to the initial condition can be observed via Fig. 3.36i–l. In these figures, the solid line corresponds to the initial amplitude $D = 0.001$, whereas the dotted line corresponds to the initial amplitude $D = 0.001 + 0.000001$. Figure 3.36i–l reveal that the nonlinear dynamic is very sensitive to the initial conditions. On the whole, all considered standard indicators point out that the complicated motion of the longitudinally translating viscoelastic Timoshenko beam has the characteristics of chaotic motions.

The bifurcation diagrams associating with the magnitude of the axial speed fluctuation γ_1, dynamic viscosity α, and support's rigidity parameter ζ are shown in Figs. 3.37, 3.38, and 3.39, respectively. In these examples, the mean axial speed is

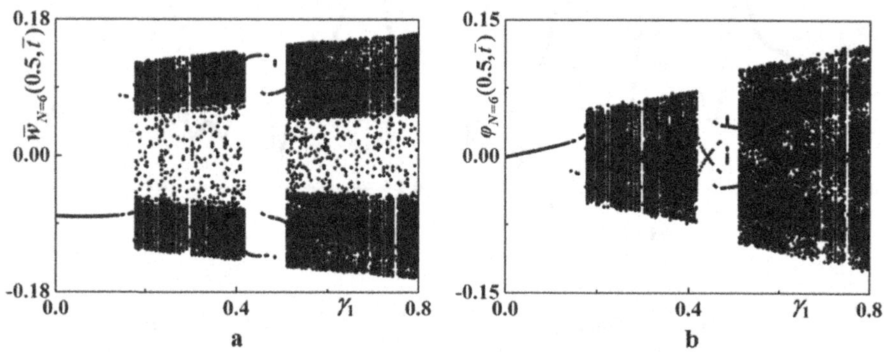

Fig. 3.37 Bifurcation diagrams associating with the magnitude of the axial speed fluctuation (six-term Galerkin truncation): **a** lateral displacement, **b** slope

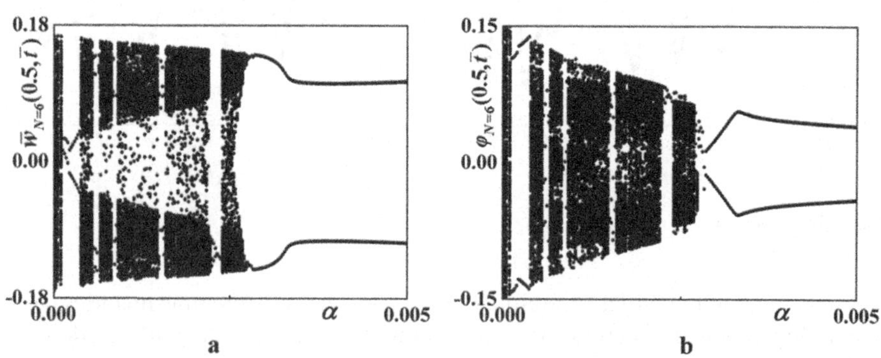

Fig. 3.38 Bifurcation diagrams associating with the dynamic viscosity (six-term Galerkin truncation): **a** lateral displacement, **b** slope

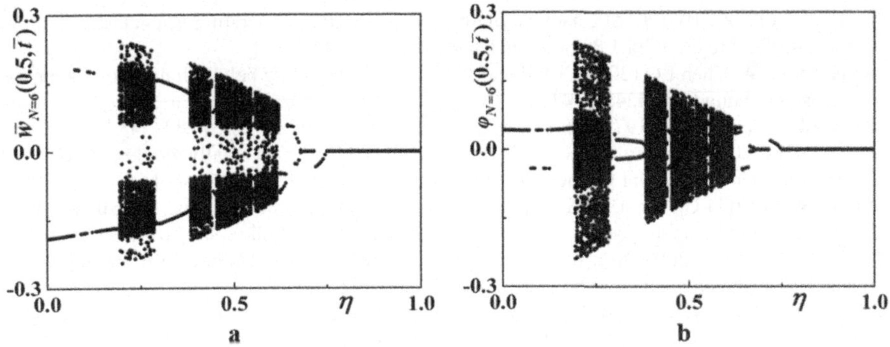

Fig. 3.39 Bifurcation diagrams associating with the axial support's rigidity (six-term Galerkin truncation): **a** lateral displacement, **b** slope

fixed by $\gamma_0 = 4.5$, and the truncation order is $N = 6$ (i.e., six-term Galerkin truncation). Furthermore, in the examples shown in Figs. 3.38 and 3.39, the magnitude of the axial speed fluctuation is fixed by $\gamma_1 = 0.7$. Similar to the bifurcation diagrams associating with the mean axial speed, Figs. 3.37, 3.38, and 3.39 show that the periodic motions and the complicated motions (i.e., quasiperiodic or chaotic motions) exchange alternately. The axially moving beam will enter into the periodic vibration if the dynamic viscosity or axial support's rigidity parameter is large enough. Besides that, when the magnitude of the speed fluctuation increases, the amplitude of the lateral vibration gradually increases. Contrarily, when the dynamic viscosity and axial support's rigidity parameter increases, the amplitude of the lateral vibration gradually decreases.

References

Alshaqaq M, Hawwa MA (2020) Nonlinear behavior of a vibrating axially moving small-size beam under an electrostatic force. ZAMM-Z Angew Math Mech 100(9):e201900104

Chen LQ, Tang YQ (2012) Parametric stability of axially accelerating viscoelastic beams with the recognition of longitudinally varying tensions. J Vib Acoust-Trans ASME 134(1):011008

Chen LQ, Yang XD (2005) Steady-state response of axially moving viscoelastic beams with pulsating speed: comparison of two nonlinear models. Int J Solids Struct 42(1):37–50

Chen LQ, Yang XD (2006) Transverse nonlinear dynamics of axially accelerating viscoelastic beams based on 4-term Galerkin truncation. Chaos Solitons Fractals 27(3):748–757

Ding H (2015) Periodic responses of a pulley-belt system with one-way clutch under inertia excitation. J Sound Vib 353:308–326

Ding H, Li DP (2014) Static and dynamic behaviors of belt-drive dynamical systems with a one-way clutch. Nonlinear Dyn 78(2):1553–1575

Ding H, Zu JW (2013) Effect of one-way clutch on the nonlinear vibration of belt-drive systems with a continuous belt model. J Sound Vib 332:6472–6487

Ding H, Zu JW (2014) Steady-state responses of pulley-belt systems with a one-way clutch and belt bending stiffness. J Vib Acoust-Trans ASME 136(4):041006

Ding H, Yan QY, Zu JW (2014) Chaotic dynamics of an axially accelerating viscoelastic beam in the supercritical regime. Int J Bifurcation Chaos 24(5):1450062

Ding H, Lim CW, Chen LQ (2018) Nonlinear vibration of a traveling belt with non-homogeneous boundaries. J Sound Vib 424:78–93

Eringen AC, Edelen DGB (1972) On nonlocal elasticity. Int J Eng Sci 10(3):233–248

Ghayesh MH (2008) Nonlinear transversal vibration and stability of an axially moving viscoelastic string supported by a partial viscoelastic guide. J Sound Vib 314(3–5):757–774

Ghayesh MH (2011) On the natural frequencies, complex mode functions, and critical speeds of axially traveling laminated beams: parametric study. Acta Mech Solida Sin 24(4):373–382

Ghayesh MH, Balar S (2008) Non-linear parametric vibration and stability of axially moving visco-elastic Rayleigh beams. Int J Solids Struct 45(25–26):6451–6467

Ghayesh MH, Balar S (2010) Non-linear parametric vibration and stability analysis for two dynamic models of axially moving Timoshenko beams. Appl Math Model 34(10):2850–2859

Han SM, Benaroya H, Wei T (1999) Dynamics of transversely vibrating beams using four engineering theories. J Sound Vib 255(5):935–988

Kaw AK (1997) Mechanics of composite materials. CRC Press, New York

Kong LY, Parker RG (2005) Vibration of an axially moving beam wrapping on fixed pulleys. J Sound Vib 280(3–5):1066–1074

Li YH, Wang L, Yang EC (2018) Nonlinear dynamic responses of an axially moving laminated beam subjected to both blast and thermal loads. Int J Nonlinear Mech 101:56–67

Lv HW, Li YH, Li L, Liu QK (2014) Transverse vibration of viscoelastic sandwich beam with time-dependent axial tension and axially varying moving velocity. Appl Math Model 38(9–10):2558–2585

Marynowski K (2012) Dynamic analysis of an axially moving sandwich beam with viscoelastic core. Compos Struct 94(9):2931–2936

Marynowski K, Kapitaniak T (2002) Kelvin-Voigt versus Burgers internal damping in modeling of axially moving viscoelastic web. Int J Non-Linear Mech 37(7):1147–1161

Mase GE, Mase GT (1992) Continuum mechanics for engineers. CRC Press, Boca Raton

Mote CD (1965) A study of band saw vibrations. J Frankl Inst 279(6):430–444

Oz HR, Pakdemirli M (1999) Vibrations of an axially moving beam with time-dependent velocity. J Sound Vib 227(2):239–257

Rezaee M, Lotfan S (2015) Non-linear nonlocal vibration and stability analysis of axially moving nanoscale beams with time-dependent velocity. Int J Mech Sci 96–97:36–46

Sahebkar SM, Ghazavi MR, Khadem SE, Ghayesh MH (2011) Nonlinear vibration analysis of an axially moving drillstring system with time dependent axial load and axial velocity in inclined well. Mech Mach Theory 46(5):743–760

Sahoo B, Panda LN, Pohit G (2017) Stability, bifurcation and chaos of a traveling viscoelastic beam tuned to 3:1 internal resonance and subjected to parametric excitation. Int J Bifurcation Chaos 27(2):1750017

Sarparast H, Ebrahimi-Mamaghan A, Safarpour M, Ouakad HM, Dimitri R, Tornabene F (2020) Nonlocal study of the vibration and stability response of small-scale axially moving supported beams on viscoelastic-Pasternak foundation in a hygro-thermal environment. Math Methods Appl Sci. https://doi.org/10.1002/mma.6859

Shokrieh M (2014) Residual stress in composite material. Woodhead Publishing Limited, Cambridge

Tang YQ, Chen LQ, Yang XD (2008) Natural frequencies, modes and critical speeds of axially moving Timoshenko beams with different boundary conditions. Int J Mech Scim 50(10–11):1448–1458

Timoshenko SP (1921) On the correction for shear of the differential equation for transverse vibrations of bars of uniform cross-section. Philos Mag 41:744–746

Timoshenko SP (1922) On the transverse vibration of bars of uniform cross-section. Philos Mag 43:125

Wang KW, Mote CD (1986) Vibration coupling analysis of band/wheel mechanical systems. J Sound Vib 109(2):237–258

Wang J, Shen H (2019) Nonlinear vibrations of axially moving simply supported viscoelastic nanobeams based on nonlocal strain gradient theory. J Phys Condens Matter 31(48):485403

Wang J, Sun J (2020) Two types of scale effects on the nonlinear forced vibration of axially moving nanobeams. Int J Mod Phys B 34(10):2050095

Wang LH, Hu ZD, Zhong Z, Ju JW (2009) Hamiltonian dynamic analysis of an axially translating beam featuring time-variant velocity. Acta Mech 206(3–4):149–161

Wang L, Yang J, Li YH (2020) Nonlinear vibration of a deploying laminated Rayleigh beam with a spinning motion in hygrothermal environment. Eng Comput. https://doi.org/10.1007/s00366-020-01035-6

Wei MH, Sun L, Hu G (2017) Dynamic properties of an axially moving sandwich beam with magnetor-heological fluid core. Adv Mech Eng 9(2):1687814017693182

Wickert JA (1992) Nonlinear vibration of a traveling tensioned beam. Int J Nonlinear Mech 27(3):503–517

Wickert JA, Mote CD (1989) On the energetics of axially moving continua. J Acoust Soc Am 85(3):1365–1368

Xu T, Lowen GG (1994) A mathematical model of an over-running sprag clutch. Mech Mach Theor 29(1):11–23

Yan QY, Ding H, Chen LQ (2014) Periodic responses and chaotic behaviors of an axially accelerating viscoelastic Timoshenko beam. Nonlinear Dyn 78(2):1577–1591

Yang XD, Chen LQ (2005) Bifurcation and chaos of an axially accelerating viscoelastic beam. Chaos Solitons Fractals 23(1):249–258

Yang TZ, Yang XD, Chen F, Fang B (2013) Nonlinear parametric resonance of a fractional damped axially moving string. J Vib Acoust-Trans ASME 135(6):064507

Zhang NH, Wang JJ, Cheng CJ (2007) Complex-mode Galerkin approach in transverse vibration of an axially accelerating viscoelastic string. Appl Math Mech-Engl 28(1):1–9

Zhao WJ, Chen LQ (2002) A numerical algorithm for non-linear parametric vibration analysis of a viscoelastic moving belt. Int J Nonlinear Sci Numer Simul 3(2):139–144

Zhao WJ, Chen LQ (2008) Iterative algorithm for axially accelerating strings with integral constitutive law. Acta Mech Solida Sin 21(5):449–456

Zhao WJ, Chen LQ, Zu JW (2006) Finite difference method for simulating transverse vibrations of an axially moving viscoelastic string. Appl Math Mech-Engl 27(1):23–28

Zhu WD, Ni J (2000) Energetics and stability of translating media with an arbitrarily varying length. J Vib Acoust-Trans ASME 122(3):295–304

Zhu F, Parker RG (2008) Influence of tensioner dry friction on the vibration of belt drives with belt bending stiffness. J Vib Acoust-Trans ASME 130(1):011002

Chapter 4
Control of Axially Moving Strings and Beams

Abstract Using the models developed in Chaps. 2 and 3, we investigate various control schemes for axially moving strings and beams. First, the control methods using an approximate ODE model, such as the model-based feedback control and variable structure control, are presented. Subsequently, various control schemes using the PDE models, including the transfer function method, wave cancellation method, boundary control based on the Lyapunov method, adaptive control method, and hybrid control methods, are discussed. Numerical simulations are provided to illustrate the performance of the control schemes.

Keywords Vibration control · Sliding mode control · Model-based feedback control · boundary control · Pointwise control · Feedback control

This chapter discusses the design of control schemes for axially moving strings and beams. Dynamic models of axially moving beams are not substantially different from axially moving strings, except for the order in partial differential equations and the boundary conditions. Therefore, we can easily extend most of the control techniques applied for strings to the beams. The control schemes for axially moving systems can be designed based on the approximated ODE models and the original PDE models developed in Chaps. 2 and 3.

4.1 Control Based on an ODE Model

The lateral vibration of an axially moving material is characterized by an infinite number of vibration modes. However, the vibration can be adequately described by considering a sufficiently large number of modes. The system is represented by a truncated discrete model (i.e., ODE model) where only a finite number of vibration modes are considered. Early studies concerning vibration suppression of AMSs were performed based on this model. Since the number of modeled modes may be vast, control of all of these modes requires many actuators and sensors. In practice, it is

© The Author(s), under exclusive license to Springer Nature Singapore Pte Ltd. 2022
K.-S. Hong et al., *Control of Axially Moving Systems*,
https://doi.org/10.1007/978-981-16-2915-0_4

not possible. Balas (1978) proposed a control scheme that considered only a finite number of modes, and the remainder were considered residual modes or uncontrolled modes. This control approach only requires a few actuators and sensors. However, the spillover problem arises, as discussed in Sect. 4.1.1.

4.1.1 Model-Based Feedback Control

Ulsoy (1984) is recognized for his original work in the vibration control of AMSs based on the reduced-order model. The author showed that the lateral vibration could be attenuated via the model-based feedback control proposed by Balas (1978) (Balas did not consider axially moving systems). This control method is discussed in this section.

We design a pointwise controller for suppressing the string's vibration using two sensors and two actuators, as shown in Fig. 4.1. Based on the Galerkin method, the approximate model of the string is given as

$$\mathbf{M}\ddot{\mathbf{q}}(t)+\mathbf{C}\dot{\mathbf{q}}(t)+\mathbf{K}\mathbf{q}(t)=\mathbf{f} \tag{4.1}$$

where $\mathbf{f}$ is the control input vector, and the basis function is chosen as $\varphi_i(x) = \sin(i\pi x/l)$. Equation (4.1) is converted to the state-space form, namely,

$$\dot{\mathbf{x}}=\mathbf{A}\mathbf{x}+\mathbf{B}\mathbf{f} \tag{4.2}$$

where $\mathbf{x} = \begin{bmatrix} \dot{\mathbf{q}}^{\mathrm{T}} & \mathbf{q}^{\mathrm{T}} \end{bmatrix}^{\mathrm{T}}$ is the state-space vector,

$$\mathbf{A}(t)=\begin{bmatrix} -\mathbf{M}^{-1}\mathbf{C} & -\mathbf{M}^{-1}\mathbf{K} \\ \mathbf{I}_{n\times n} & \mathbf{0}_{n\times n} \end{bmatrix}, \mathbf{B}=\begin{bmatrix} \mathbf{M}^{-1} \\ \mathbf{0}_{n\times 1} \end{bmatrix}. \tag{4.3}$$

Figure 4.2 shows the proposed feedback control scheme. The control input is given as

$$\mathbf{f} = -\mathbf{T}_{\mathrm{a}}\mathbf{G}\mathbf{T}_{\mathrm{s}}\mathbf{x} \tag{4.4}$$

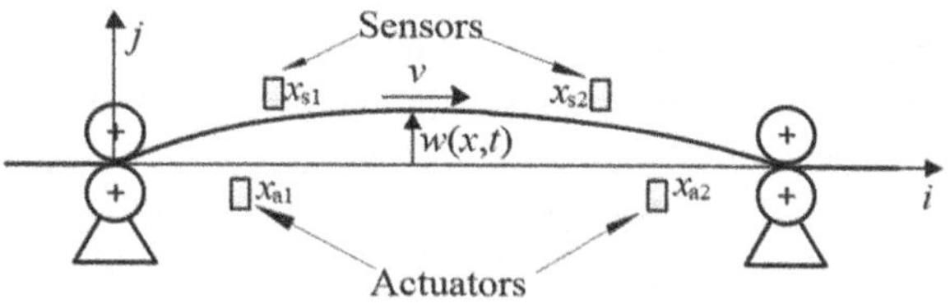

Fig. 4.1 Axially moving string with two sensors and two actuators

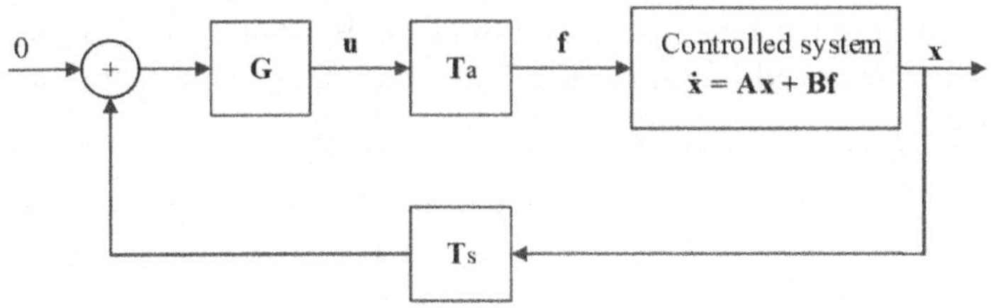

Fig. 4.2 Feedback control scheme for an axially moving string

where $\mathbf{G}$ is the control gain matrix and $\mathbf{T}_a$ and $\mathbf{T}_s$ are the matrices that correspond to the locations of actuators and sensors, respectively. With the control input vector in (4.4), the equation of the closed-loop system can be rewritten as follows:

$$\dot{\mathbf{x}} = (\mathbf{A} - \mathbf{B}\mathbf{T}_a\mathbf{G}\mathbf{T}_s)\mathbf{x} = \mathbf{A}_c\mathbf{x}. \tag{4.5}$$

The desired eigenvalues and eigenvectors of the closed-loop system can be achieved by selecting the matrix $\mathbf{G}$ properly. In addition, we note that matrix $\mathbf{A}$ depends on the axial velocity. Therefore, $\mathbf{G}$ must be a function of the axial velocity to cancel the axial velocity's effects on $\mathbf{A}$.

As previously mentioned, the use of a finite number of actuators and sensors to control an infinite-dimensional system leads to the issue of residual modes. The residual modes cause a problem of vibration suppression of AMSs based on the ODE model: Observation and control spillovers. The spillover phenomenon exists since the design of the controller using the reduced-order model ignores the residual modes; however, the sensors are affected by these modes (i.e., observation spillover), and these modes are excited by the actuators (i.e., control spillover).

Now, we investigate the problems of observation and control spillovers. We assume that the lateral vibration is measured by two sensors located at $x_{s1} = 5l/8$ and $x_{s2} = 3l/4$ and is controlled by two actuators at $x_{a1} = l/4$ and $x_{a2} = 3l/8$. The controller is designed by using the ODE model with $n = 2$. Although the actual axially moving string has an infinite number of vibration modes, the ODE model with $n = 4$ is used to model the system for simulation purposes. This example allows us to observe the spillover phenomenon due to two residual modes (i.e., third and fourth modes). To design the control input vector, we require the matrices for coordinate transformation. Furthermore, since two actuators also excite the residual modes, the transformation matrix $\mathbf{T}_a$ is given as

$$\mathbf{T}_a = \begin{bmatrix} \varphi_1(x_{a1}) & \varphi_1(x_{a2}) \\ \varphi_2(x_{a1}) & \varphi_2(x_{a2}) \\ \varphi_3(x_{a1}) & \varphi_3(x_{a2}) \\ \varphi_4(x_{a1}) & \varphi_4(x_{a2}) \end{bmatrix} \begin{bmatrix} \varphi_1(x_{a1}) & \varphi_1(x_{a2}) \\ \varphi_2(x_{a1}) & \varphi_2(x_{a2}) \end{bmatrix}^{-1} \tag{4.6}$$

where $\varphi_i(x)$ is the basis function of the i-th mode. In addition, two sensors are also affected by the third and fourth modes. Therefore, the transformation matrix $\mathbf{T}_s$ is calculated as

$$\mathbf{T}_s = \begin{bmatrix} \mathbf{T}_{s1} & \mathbf{0} \\ \mathbf{0} & \mathbf{T}_{s1} \end{bmatrix} \tag{4.7}$$

where

$$\mathbf{T}_{s1} = \begin{bmatrix} \varphi_1(x_{s1}) & \varphi_1(x_{s2}) \\ \varphi_2(x_{s1}) & \varphi_2(x_{s2}) \end{bmatrix}^{-1} \begin{bmatrix} \varphi_1(x_{s1}) & \varphi_2(x_{s1}) & \varphi_3(x_{s1}) & \varphi_4(x_{s1}) \\ \varphi_1(x_{s2}) & \varphi_2(x_{s2}) & \varphi_3(x_{s2}) & \varphi_4(x_{s2}) \end{bmatrix}. \tag{4.8}$$

Substituting $x_{s1} = 5l/8$, $x_{s2} = 3l/4$, $x_{a1} = l/4$, and $x_{a2} = 3l/8$ yields the following results.

$$\mathbf{T}_a = \begin{bmatrix} 1 & 0 \\ 0 & 1 \\ \frac{\sin(\frac{\pi}{4})^2 - \sin(\frac{9\pi}{8})}{\sin(\frac{\pi}{4})^2 - \sin(\frac{3\pi}{8})} & \frac{\sin(\frac{\pi}{4})\sin(\frac{9\pi}{8}) - \sin(\frac{\pi}{4})\sin(\frac{3\pi}{8})}{\sin(\frac{\pi}{4})^2 - \sin(\frac{3\pi}{8})} \\ \frac{1}{\sin(\frac{\pi}{4})^2 - \sin(\frac{3\pi}{8})} & \frac{-\sin(\frac{\pi}{4})}{\sin(\frac{\pi}{4})^2 - \sin(\frac{3\pi}{8})} \end{bmatrix}, \tag{4.9}$$

$$\mathbf{T}_s = \begin{bmatrix} \mathbf{T}_a^{\mathrm{T}} & \mathbf{0} \\ \mathbf{0} & \mathbf{T}_a^{\mathrm{T}} \end{bmatrix}. \tag{4.10}$$

The control objective is to suppress the lateral vibration of the string. Therefore, we design the control gain matrix **G** such that the closed-loop system (4.5) has the critically damped response. It is noted that the design of the matrix **G** is performed based on the ODE model with $n = 2$. The desired closed-loop matrix, for example, is chosen as follows:

$$\mathbf{A}_c = \begin{bmatrix} -\frac{2\pi}{l}\sqrt{\frac{T}{\rho A}} & 0 & -\frac{\pi^2 T}{\rho A l^2} & 0 \\ 0 & -\frac{4\pi}{l}\sqrt{\frac{T}{\rho A}} & 0 & -\frac{4\pi^2 T}{\rho A l^2} \\ 1 & 0 & 0 & 0 \\ 0 & 1 & 0 & 0 \end{bmatrix}. \tag{4.11}$$

Based on (4.5), the control gain matrix **G** can be solved, namely,

$$\mathbf{G} = \begin{bmatrix} \frac{\pi^2 v_0^2}{l^2} & 0 & \frac{2\pi}{l}\sqrt{\frac{T}{\rho A}} & -\frac{4v_0}{3l} \\ 0 & \frac{4\pi^2 v_0^2}{l^2} & \frac{4v}{3l} & \frac{4\pi}{l}\sqrt{\frac{T}{\rho A}} \end{bmatrix}. \tag{4.12}$$

We investigate the spillover effects on the system based on the simulation results for four cases:

- *No control or observation spillover*: There is no interaction between the residual modes and the actuators and sensors. The terms in the last two rows of $\mathbf{T}_a$ and the terms in the last two columns of $\mathbf{T}_{s1}$ are set to zero.
- *Control spillover only*: This case assumes that the sensors are not affected by the residual modes (i.e., no observation spillover). We set the terms in the last two columns of $\mathbf{T}_{s1}$ to zero.
- *Observation spillover only*: We assume that the actuators do not excite the residual modes (i.e., no control spillover). Therefore, the terms in the last two rows of $\mathbf{T}_a$ are set to zero.
- *Observation and control spillovers*: The residual modes affect sensors and are excited by actuators. The transformation matrices in (4.9) and (4.10) are utilized to consider both observation spillover and control spillover.

The responses of the controlled modes (i.e., first and second modes) and the residual modes (i.e., the third and four modes) are demonstrated in Figs. 4.3, 4.4, 4.5 and 4.6. The simulation results show that the control spillover's influence on the controlled modes is negligible (i.e., Figs. 4.3 and 4.4). Although the control spillover substantially affects the residual modes, this spillover still does not destabilize the system (i.e., Fig. 4.4). Figures 4.5 and 4.6 show that the observation spillover leads to the instability of both the controlled mode and the residual mode, especially in the case of observation spillover only (i.e., Fig. 4.5). The system's performance without observation spillover can be improved via natural damping, whereas the problem of observation spillover is very complicated. One of the approaches for eliminating the observation spillover is the use of a comb filter (Balas 1978).

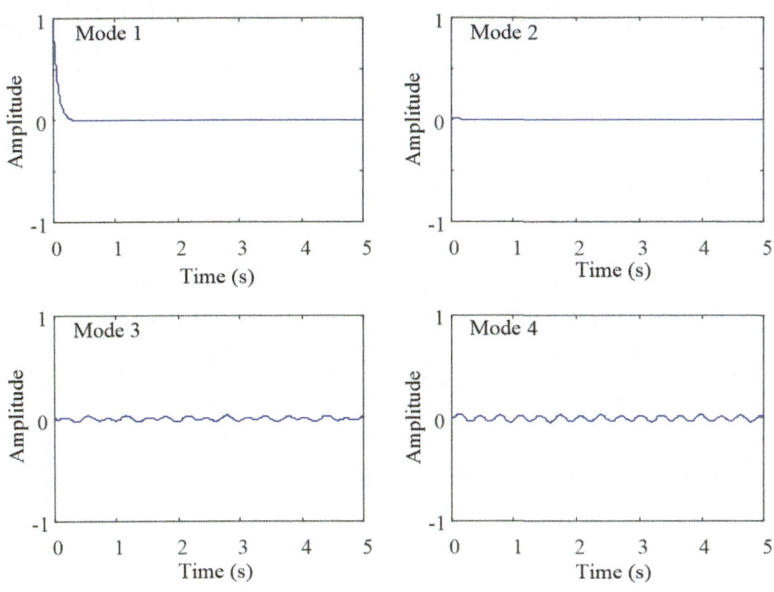

Fig. 4.3 Response of the system without spillover

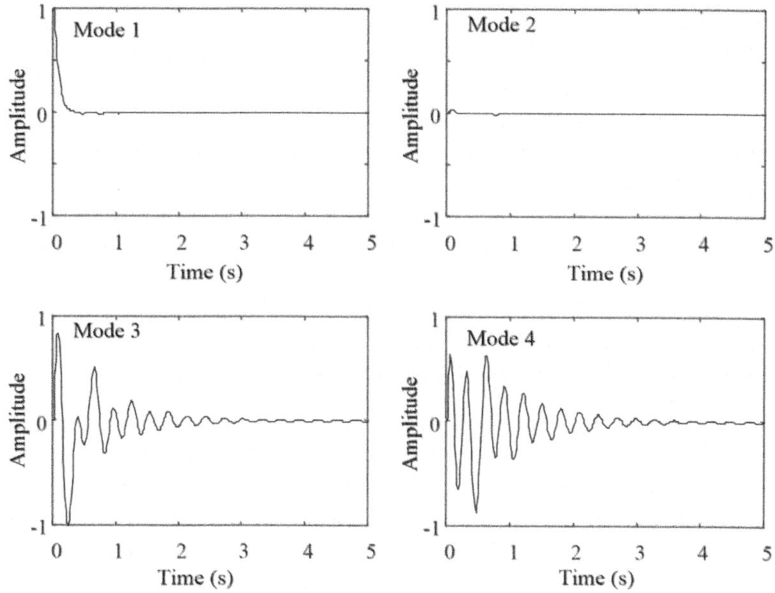

Fig. 4.4 Response of the system with control spillover only

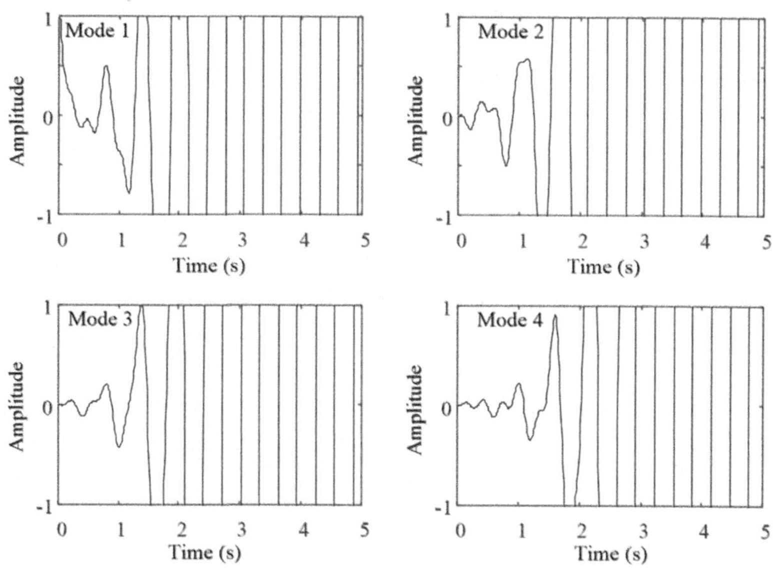

Fig. 4.5 Response of the system with observation spillover only

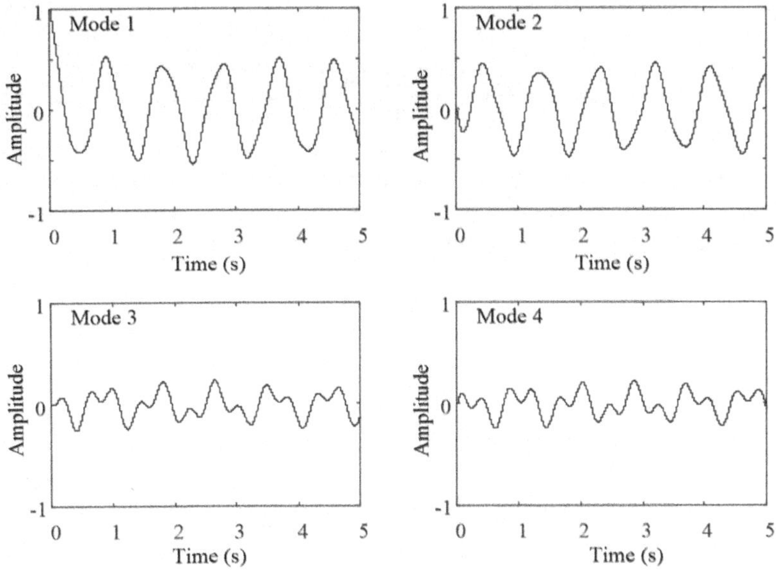

Fig. 4.6 Response of the system with observation and control spillovers

4.1.2 *Sliding Mode Control*

Sliding mode control (SMC)—a particular type of variable structure control (VSC)—is a robust control technique used to handle high-order nonlinear dynamical plants in the presence of model uncertainty. Sliding mode control is characterized by a switching surface and a feedback control law (Edward and Sarah 1998; Hung et al. 1993). First, the switching surface (or the sliding manifold) is designed such that the sliding motion satisfies the required specifications. Subsequently, the control law with varying control gains is defined to move the system's state toward the vicinity of the switching surface. Concerning vibration control of AMSs using the SMC, Fung and Liao (1995) designed a control law based on an ODE model to suppress the lateral vibration of an axially moving string. Later, Fung et al. (1998) improved the sliding mode control scheme via integral and proportional compensation. Furthermore, a sliding mode control for an axially moving string with varying length was presented in Fung et al. (1997).

Now, we discuss the control of axially moving strings by using SMC. The ODE model of an axially moving string is given as follows

$$\mathbf{M}\ddot{\mathbf{q}}(t) + \mathbf{C}\dot{\mathbf{q}}(t) + \mathbf{K}\mathbf{q}(t) = \mathbf{f}(t). \tag{4.13}$$

Let $\mathbf{x} = \begin{bmatrix} \dot{\mathbf{q}}^{\mathrm{T}} & \mathbf{q}^{\mathrm{T}} \end{bmatrix}^{\mathrm{T}}$, the ODE model can be rewritten in the following form:

$$\dot{\mathbf{x}} = \mathbf{A}\mathbf{x} + \mathbf{B}\mathbf{f} \tag{4.14}$$

where

$$\mathbf{A}(t)=\begin{bmatrix} -\mathbf{M}^{-1}\mathbf{C} & -\mathbf{M}^{-1}\mathbf{K} \\ \mathbf{I}_{n\times n} & \mathbf{0}_{n\times n} \end{bmatrix}, \mathbf{B}=\begin{bmatrix} \mathbf{M}^{-1} \\ \mathbf{0} \end{bmatrix}. \tag{4.15}$$

The control objective is to suppress the lateral vibration or to maintain the state in the vicinity of zero. Let the error vectors be $\mathbf{e} = [\ q_1\ q_2\ \ldots\ q_n\ ]^{\mathrm{T}}$. To formulate the SMC scheme, the following switching surface is proposed:

$$\mathbf{s} = \boldsymbol{\lambda}_1\dot{\mathbf{e}} + \boldsymbol{\lambda}_2\mathbf{e} \tag{4.16}$$

where

$$\mathbf{s} = [\ s_1\ s_2\ \ldots\ s_n\ ]^{\mathrm{T}}, \tag{4.17}$$

$$\boldsymbol{\lambda}_1 = \begin{bmatrix} \lambda_{11} & 0 & \cdots & 0 \\ 0 & \lambda_{12} & \cdots & 0 \\ \vdots & \vdots & \ddots & \vdots \\ 0 & 0 & \cdots & \lambda_{1n} \end{bmatrix}, \tag{4.18}$$

$$\boldsymbol{\lambda}_2 = \begin{bmatrix} \lambda_{21} & 0 & \cdots & 0 \\ 0 & \lambda_{22} & \cdots & 0 \\ \vdots & \vdots & \ddots & \vdots \\ 0 & 0 & \cdots & \lambda_{2n} \end{bmatrix}. \tag{4.19}$$

The switching surface in (4.16) includes n scalar switching functions, namely, $s_i = \lambda_{1i}\dot{q}_i + \lambda_{2i}q_i$, $i = 1, 2, \ldots, n$, where $\lambda_{11}, \lambda_{12}, \ldots, \lambda_{1n}$ and $\lambda_{21}, \lambda_{22}, \ldots, \lambda_{2n}$ are designed positive scalars. The requirement in the sliding mode control design is to guarantee that the controlled system satisfies the sliding condition. The sliding condition requires the states to move toward the sliding surface at a finite time and further remain on the sliding surface after this time. We design a differential equation that specifies the dynamics of the switching surface to ensure this condition. This differential equation is referred to as the reaching law. The reaching law is given by

$$\dot{\mathbf{s}} = -\mathbf{R}_{\mathrm{d}}\boldsymbol{\delta} \tag{4.20}$$

where

$$\mathbf{R}_{\mathrm{d}} = \mathrm{diag}\left[\alpha_1|s_1|^{\beta}, \alpha_2|s_2|^{\beta}, \ldots, \alpha_n|s_n|^{\beta}\right], \alpha_i \in \mathbb{R}^+, 0 < \beta < 1, \tag{4.21}$$

$$\boldsymbol{\delta} = \left[\ \mathrm{sgn}(s_1)\ \mathrm{sgn}(s_2)\ \ldots\ \mathrm{sgn}(s_n)\ \right]^{\mathrm{T}}. \tag{4.22}$$

The reaching law in (4.20) guarantees that $\mathbf{s}^{\mathrm{T}}\dot{\mathbf{s}} \leq 0$. It follows that the sliding condition is satisfied, and the existence of the sliding mode on the switching surface is assured. By substituting (4.20) into (4.16), the control law can be derived as follows:

$$\begin{aligned}
&\boldsymbol{\lambda}_1\ddot{\mathbf{e}} + \boldsymbol{\lambda}_2\dot{\mathbf{e}} = -\mathbf{R}_{\mathrm{d}}\boldsymbol{\delta} \\
&\Leftrightarrow \left[\boldsymbol{\lambda}_1\ \boldsymbol{\lambda}_2\right]\begin{bmatrix}\ddot{\mathbf{e}} \\ \dot{\mathbf{e}}\end{bmatrix} = -\mathbf{R}_{\mathrm{d}}\boldsymbol{\delta} \\
&\Leftrightarrow \left[\boldsymbol{\lambda}_1\ \boldsymbol{\lambda}_2\right][\mathbf{A}\mathbf{x} + \mathbf{B}\mathbf{f}] = -\mathbf{R}_{\mathrm{d}}\boldsymbol{\delta} \\
&\Leftrightarrow \mathbf{f} = -\left(\left[\boldsymbol{\lambda}_1\ \boldsymbol{\lambda}_2\right]\mathbf{B}\right)^{-1}\left(\left[\boldsymbol{\lambda}_1\ \boldsymbol{\lambda}_2\right]\mathbf{A}\mathbf{x} + \mathbf{R}_{\mathrm{d}}\boldsymbol{\delta}\right).
\end{aligned} \tag{4.23}$$

Similar to model-based feedback control, the sliding mode control law is designed based on the reduced-order model. Therefore, this approach can lead to the spillover problem due to the residual modes. Spillover is the problem that hinders implementation of the control approaches based on the ODE model in actual axially moving systems.

4.1.3 Control Based on Regulating Axial Velocity

In most existing research on the control of axially moving systems, vibration is suppressed by applying external forces. In several situations, this technique can damage the surface of materials, particularly in roll-to-roll systems that manufacture large-area electronics devices. To overcome this problem, Nguyen and Hong (2011) proposed a vibration control strategy based on axial velocity regulation. Unlike boundary and distributed control techniques, this control technique directly uses the axial velocity as a control input. To quickly decay the vibration energy, the axial velocity is regulated to track an appropriate profile.

Consider an axially moving beam that is described by a PDE as follows:

$$\rho A\left(w_{tt} + \dot{v}w_x + 2vw_{xt} + v^2w_{xx}\right) - T_0w_{xx} + c_{\mathrm{e}}(w_t + vw_x) + EIw_{xxxx} = 0 \tag{4.24}$$

with the following boundary conditions:

$$w(0,t) = w(l,t) = 0,\ w_{xx}(0,t) = w_{xx}(l,t) = 0. \tag{4.25}$$

The equation of motion in (4.24) and boundary conditions in (4.25) can be transformed into a set of ODEs via the Galerkin method, wherein the lateral vibration is approximated by $w(x,t) = \sum_{i=1}^{n} q_i(t)\sin(i\pi x/l)$, that is

$$\mathbf{M}\ddot{\mathbf{q}}(t) + \mathbf{C}(t)\dot{\mathbf{q}}(t) + \mathbf{K}(t)\mathbf{q}(t) = 0 \tag{4.26}$$

where the elements of $\mathbf{M}$, $\mathbf{C}(t)$, and $\mathbf{K}(t)$, which are m_{ij}, c_{ij}, and k_{ij}, respectively, are defined as follows:

$$m_{ij} = \begin{cases} l/2, & \text{if } i = j, \\ 0, & \text{if } i \neq j, \end{cases} \tag{4.27}$$

$$c_{ij} = \begin{cases} c_e l/2\rho A, & \text{if } i = j, \\ 0, & \text{if } i \neq j,\ (i+j) \text{is even}, \\ 4ijv/(j^2 - i^2), & \text{if } i \neq j,\ (i+j) \text{ is odd}, \end{cases} \tag{4.28}$$

$$k_{ij} = \begin{cases} (i\pi/l)^4 EI/\rho A + (i\pi/l)^2 (T/\rho A - v^2), & \text{if } i = j, \\ 0, & \text{if } i \neq j,\ (i+j) \text{ is even}, \\ 2ij\big((c_e/\rho A)^2 v^2 + \dot{v}^2\big)/(j^2 - i^2), & \text{if } i \neq j,\ (i+j) \text{ is odd}. \end{cases} \tag{4.29}$$

Equation (4.26) is converted to the state-space form with the state-space vector $\mathbf{x} = \left[\, \dot{\mathbf{q}}^{\mathrm{T}}\ \mathbf{q}^{\mathrm{T}} \,\right]^{\mathrm{T}}$, namely,

$$\dot{\mathbf{x}} = \mathbf{f}(\mathbf{x}, v, \dot{v}) = \mathbf{A}(t)\mathbf{x}, \tag{4.30}$$

where

$$\mathbf{A}(t) = \begin{bmatrix} -\mathbf{M}^{-1}\mathbf{C} & -\mathbf{M}^{-1}\mathbf{K} \\ \mathbf{I}_{n \times n} & \mathbf{0}_{n \times n} \end{bmatrix}. \tag{4.31}$$

The state-space model given by (4.30) can be rewritten as follows:

$$\dot{\mathbf{x}} = \mathbf{f}(\mathbf{x}, v, \dot{v}) = \mathbf{A}_0\mathbf{x} + v\mathbf{B}_1\mathbf{x} + v^2\mathbf{B}_2\mathbf{x} + (vc_e/\rho A + \dot{v})\mathbf{B}_1\mathbf{x} \tag{4.32}$$

where

$$\mathbf{A}_0 = \begin{bmatrix} -\mathbf{M}^{-1}\mathbf{C}^{\mathrm{s}} & -\mathbf{M}^{-1}\mathbf{K}^{\mathrm{s}} \\ \mathbf{I}_{n \times n} & \mathbf{0}_{n \times n} \end{bmatrix}, \tag{4.33}$$

$$\mathbf{B}_1 = \begin{bmatrix} -\mathbf{M}^{-1}\mathbf{B}_1^{\mathrm{s}} & \mathbf{0}_{n \times n} \\ \mathbf{0}_{n \times n} & \mathbf{0}_{n \times n} \end{bmatrix}, \tag{4.34}$$

$$\mathbf{B}_2 = \begin{bmatrix} \mathbf{0}_{n \times n} & -\mathbf{M}^{-1}\mathbf{B}_2^{\mathrm{s}} \\ \mathbf{0}_{n \times n} & \mathbf{0}_{n \times n} \end{bmatrix}, \tag{4.35}$$

$$\mathbf{B}_3 = \begin{bmatrix} \mathbf{0}_{n \times n} & -\mathbf{M}^{-1}\mathbf{B}_3^{\mathrm{s}} \\ \mathbf{0}_{n \times n} & \mathbf{0}_{n \times n} \end{bmatrix}. \tag{4.36}$$

In (4.33)–(4.36), $\mathbf{K}^{\mathrm{s}}$, $\mathbf{C}^{\mathrm{s}}$, $\mathbf{B}_1^{\mathrm{s}}$, $\mathbf{B}_2^{\mathrm{s}}$, and $\mathbf{B}_2^{\mathrm{s}}$ are $n \times n$ matrices, whose elements are defined as follows:

$$k_{ij}^{\mathrm{s}} = \begin{cases} (i\pi/l)^4 EI/\rho A + (i\pi/l)^2 T/\rho A, & \text{if } i = j, \\ 0, & \text{if } i \neq j, \end{cases} \tag{4.37}$$

$$c_{ij}^{\mathrm{s}} = \begin{cases} c_{\mathrm{e}} l/2\rho A, & \text{if } i = j, \\ 0, & \text{if } i \neq j, \end{cases} \tag{4.38}$$

$$b_{1ij}^{\mathrm{s}} = \begin{cases} 0, & \text{if } i = j, \\ 0, & \text{if } i \neq j,\ (i+j) \text{ is even}, \\ 4ijv/(j^2 - i^2), & \text{if } i \neq j,\ (i+j) \text{ is odd}, \end{cases} \tag{4.39}$$

$$b_{2ij}^{\mathrm{s}} = \begin{cases} -(i\pi/l)^2, & \text{if } i = j, \\ 0, & \text{if } i \neq j, \end{cases} \tag{4.40}$$

$$b_{3ij}^{\mathrm{s}} = \begin{cases} 0, & \text{if } i = j, \\ 0, & \text{if } i \neq j, (i+j) \text{ is even}, \\ 2ij/(j^2 - i^2), & \text{if } i \neq j, (i+j) \text{ is odd}. \end{cases} \tag{4.41}$$

The linear operator $\mathbf{A}(t)$ depends on the axial velocity $v(t)$. Therefore, we can design the eigenvalues of this operator and the convergence speed when $\mathbf{x}(t) \to 0$ by adjusting the velocity $v(t)$. For a specific illustration, we consider the problem of reducing the axial velocity of an axially moving beam from $v(0) = v_0$ to zero at time $t = t_{\mathrm{f}}$. In the case of no control, the axial transport velocity is decreased to zero with a constant deceleration (i.e., Fig. 4.7). With viscous damping, the vibration energy $E(t)$ naturally tends to zero. However, the constant-deceleration velocity profile usually produces considerable residual vibration energy at time t_{f} (i.e., Fig. 4.7). In this situation, vibration suppression depends only on the viscous damping force; therefore, it

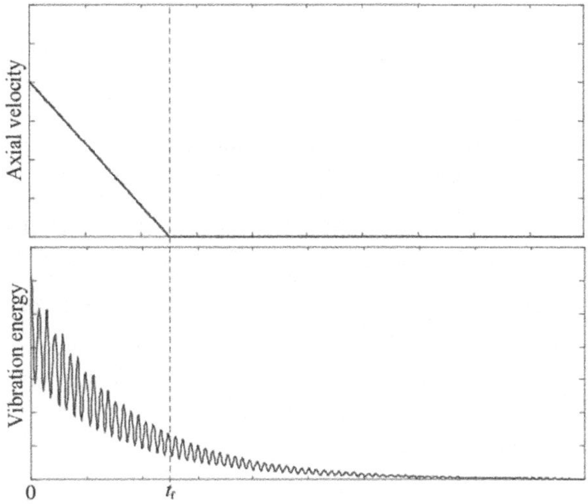

Fig. 4.7 Axial velocity and vibration energy during the deceleration period

requires plenty of time. The control objective is to design an optimal velocity profile to minimize the beam's vibration energy when the axial velocity arrives at zero value.

The vibration energy of the considered beam is computed as follows:

$$E(t) = \frac{1}{2}\int_0^l \{\rho A(w_t + v w_x)^2 + T_0 w_x^2 + EI w_{xx}^2\}\mathrm{d}x. \tag{4.42}$$

By using approximation, $w(x,t) = \sum_{i=1}^n q_i(t)\sin(i\pi x/l)$, the vibration energy can be evaluated as

$$\begin{aligned} E(t) &= \frac{1}{2}\int_0^l \Bigg\{\rho A\left[\sum_{i=1}^n \dot{q}_i(t)\sin\left(\frac{i\pi x}{l}\right) + v\sum_{i=1}^n \frac{i\pi}{l}q_i(t)\cos\left(\frac{i\pi x}{l}\right)\right]^2 \\ &\quad + T_0\left[\sum_{i=1}^n \frac{i\pi}{l}q_i(t)\cos\left(\frac{i\pi x}{l}\right)\right]^2 + EI\left[\sum_{i=1}^n \left(\frac{i\pi}{l}\right)^2 q_i(t)\sin\left(\frac{i\pi x}{l}\right)\right]^2\Bigg\}\mathrm{d}x \\ &\le \frac{1}{2}\int_0^l \Bigg\{\rho A\left[\sum_{i=1}^n \left(\dot{q}_i(t) + v\frac{i\pi}{l}q_i(t)\right)\right]^2 \\ &\quad + T_0\left[\sum_{i=1}^n \frac{i\pi}{l}q_i(t)\right]^2 + EI\left[\sum_{i=1}^n \left(\frac{i\pi}{l}\right)^2 q_i(t)\right]^2\Bigg\}\mathrm{d}x \\ &\le \frac{l}{2}\rho A\left[\sum_{i=1}^{2n}\left(1 + \psi_{\mathrm{H}}(i)v\frac{i\pi}{l}x_i\right)\right]^2 + \frac{l}{2}T_0\left[\sum_{i=1}^{2n}\left(\psi_{\mathrm{H}}(i) + \frac{i\pi}{l}\right)x_i\right]^2 \\ &\quad + \frac{l}{2}EI\left[\sum_{i=1}^{2n}\left(\psi_{\mathrm{H}}(i) + \left(\frac{i\pi}{l}\right)^2\right)x_i\right]^2, \end{aligned} \tag{4.43}$$

where the Heaviside step function $\psi_{\mathrm{H}}(i)$ is given by

$$\psi_{\mathrm{H}}(i) = \frac{1}{2}(\mathrm{sgn}(i-n) + |\mathrm{sgn}(i-n)|). \tag{4.44}$$

Inequality (4.43) reveals that the vibration energy is bounded by $\overline{E}_{\mathrm{b}}(t)$, namely,

$$\overline{E}_{\mathrm{b}}(t) = \frac{1}{2}\mathbf{x}^{\mathrm{T}}\mathbf{Q}(t)\mathbf{x} \tag{4.45}$$

where the elements of $\mathbf{Q}(t)$ are

$$q_{ij} = \frac{l}{2}\rho A\left(1 + \psi_{\mathrm{H}}(i)v\frac{i\pi}{l}\right)\left(1 + \psi_{\mathrm{H}}(j)v\frac{j\pi}{l}\right) + \frac{l}{2}T_0\left(\psi_{\mathrm{H}}(i) + \frac{i\pi}{l}\right)\left(\psi_{\mathrm{H}}(j) + \frac{j\pi}{l}\right)$$

$$+\frac{l}{2}EI\left(\psi_{\mathrm{H}}(i)+\left(\frac{i\pi}{l}\right)^2\right)\left(\psi_{\mathrm{H}}(j)+\left(\frac{j\pi}{l}\right)^2\right). \tag{4.46}$$

It is noted that if the upper bound of the vibration energy $\overline{E}_{\mathrm{b}}(t)$ converges to zero, $E(t)$ also converges to zero. We establish an optimal control problem based on the upper bound $\overline{E}_{\mathrm{b}}(t)$.

Consider the problem of minimizing the functional

$$\overline{E}_{\mathrm{b}}(t_{\mathrm{f}}) = \frac{1}{2}\mathbf{x}^{\mathrm{T}}(t_{\mathrm{f}})\mathbf{Q}(t_{\mathrm{f}})\mathbf{x}(t_{\mathrm{f}}) \tag{4.47}$$

which is subjected to the differential constraint

$$\dot{\mathbf{x}} = \mathbf{f}(\mathbf{x}, v, \dot{v}) = \mathbf{A}(t)\mathbf{x}, \tag{4.48}$$

with the initial and final conditions

$$\mathbf{x}(0) = \mathbf{x}_0, \tag{4.49}$$

$$\mathbf{x}(t_{\mathrm{f}}) = 0, \tag{4.50}$$

where $v(t)$ is the control input subjected to constraints $v(t_{\mathrm{f}}) = 0$ and $v(t) > 0$. To assure the final condition, we combine (4.50) with the cost function given by (4.47); consequently, the following cost function is considered:

$$J = \overline{E}_{\mathrm{b}}(t_{\mathrm{f}}) + \frac{1}{2}\mathbf{x}^{\mathrm{T}}(t_{\mathrm{f}})\mathbf{W}\mathbf{x}(t_{\mathrm{f}}) = \frac{1}{2}\mathbf{x}^{\mathrm{T}}(t_{\mathrm{f}})[\mathbf{Q}(t_{\mathrm{f}}) + \mathbf{W}]\mathbf{x}(t_{\mathrm{f}}), \tag{4.51}$$

where $\mathbf{W}$ is a $2n \times 2n$ weight coefficient matrix. The cost function (4.51) is represented in linear quadratic form, which is a unique function of $v(t)$. To solve this optimal control problem, we can use iterative numerical techniques, such as the steepest descent algorithm (Kirk 2004) and conjugate gradient method (Lasdon et al. 1967).

We discuss the procedure for solving the considered optimal control problem by using the conjugate gradient algorithm. Let H be the Hamiltonian function given by

$$H = \boldsymbol{\lambda}^{\mathrm{T}}\mathbf{f}, \tag{4.52}$$

where $\boldsymbol{\lambda}$ is the vector of adjoint variables that is determined by the following differential equation:

$$\dot{\boldsymbol{\lambda}} = -\frac{\partial H}{\partial \mathbf{x}} = -\boldsymbol{\lambda}^{\mathrm{T}}\mathbf{A}, \tag{4.53}$$

$$\boldsymbol{\lambda}(t_{\mathrm{f}}) = [\mathbf{Q}(t_{\mathrm{f}}) + \mathbf{W}]\mathbf{x}(t_{\mathrm{f}}). \tag{4.54}$$

Using (4.32), the gradient is obtained as follows:

$$g(t) = \frac{\partial H}{\partial v} = \boldsymbol{\lambda}^{\mathrm{T}}\left(\mathbf{B}_1 + 2v\mathbf{B}_2 + \frac{c}{\rho A}\mathbf{B}_1\right)\mathbf{x}. \tag{4.55}$$

Let $v_i(t)$ be the i-th approximation to the optimal velocity profile. The $(i + 1)$-th approximation of the velocity profile is updated by

$$v_{i+1}(t) = v_i(t) + \alpha_i p_i(t), \tag{4.56}$$

where α_i is the step size of the i-th iteration and $p_i(t)$ is the conjugate gradient direction, which is computed by

$$p_i(t) = -g_i(t) + \beta_{i-1} p_{i-1}(t), \tag{4.57}$$

where

$$\beta_{i-1} = \frac{\int_0^{t_f} g_i(t) g_i(t)\mathrm{d}t}{\int_0^{t_f} g_{i-1}(t) g_{i-1}(t)\mathrm{d}t}. \tag{4.58}$$

In (4.57) and (4.58), $g_i(t)$ denotes the gradient of the i-th iteration. The procedure for determining the optimal velocity profile by the conjugate gradient algorithm is summarized as follows:

Step 1: Give t_f, $\mathbf{x}_0$, and v_0 and set $i = 0$.

Step 2: Set $v(t) = v_i(t)$. Solve the state-space Eqs. (4.48)–(4.50) forward, and then solve the adjoint Eq. (4.53)–(4.54) backward.

Step 3: Compute the gradient $g_i(t)$ and the direction $p_i(t)$. If $i = 0$ then $p_i(t) = -g_i(t)$.

Step 4: Choose the step size α_i to minimize the cost function $J(v_i(t) + \alpha_i p_i(t))$. Update the velocity profile $v_{i+1}(t) = v_i(t) + \alpha_i p_i(t)$, wherein $v_{i+1}(t_f) = 0$ and $v_{i+1}(t) = 0$ if $v_{i+1}(t) < 0$.

Step 5: Repeat steps 2, 3, and 4 until $|J(v_{i+1}(t)) - J(v_i(t))| < \delta J(v_i(t))$, where δ is a specified positive number.

The procedure for solving the optimal control problem using the steepest descent algorithm is similar to the conjugate gradient algorithm, except $\beta_i = 0$. The vibration energy substantially decays by applying the optimal velocity profile determined via the steepest descent algorithm or the conjugate gradient algorithm. The lateral vibration is also minimized when the axial velocity reaches zero at time t_f. The inconsiderable residual vibration can be entirely suppressed by viscous damping afterward.

4.2 Control Based on a PDE Model

The use of the ODE model is convenient for applying the conventional control methods, which are well developed for ODEs. However, as mentioned in the previous section, the use of the ODE model has a severe limitation: Spillovers due to the residual modes. To overcome this problem, researchers have investigated the direct use of the PDE model to design a controller for axially moving systems. By utilizing the PDE model, various control methods, such as the transfer function-based method, wave cancellation method, boundary control based on the Lyapunov method, and adaptive control, have been applied for vibration suppression, velocity control, or tension control of axially moving systems. This section presents these PDE-based control methods.

4.2.1 *Transfer Function-Based Method*

The transfer function-based method, which designs a feedback control law for linear strings in the frequency domain, was presented by Yang and Mote (1991). In the book *Structural Theory of Distributed Systems*, which was first published in 1983, Butkovskiy pointed out that the information included in the transfer function can predict the dynamics of a closed-loop system. Accordingly, Yang and Mote analyzed the root loci of the closed-loop transfer function of the linear model of an axially moving string. Stability criteria were established, and then feedback control laws were designed based on these criteria.

To demonstrate this control method, we consider a linear model that describes a moving string, wherein one actuator and one sensor are located at x_a and x_s, respectively, as follows:

$$w_{tt}(x,t) + 2v_0 w_{xt}(x,t) - (1 - v_0^2)w_{xx}(x,t) = f(x,t), \tag{4.59}$$

and the boundary conditions and initial conditions are

$$w(0,t) = 0, \quad w(1,t) = 0, \quad w(x,0) = a(x), \quad w_t(x,0) = b(x) \tag{4.60}$$

where $f(x,t)$ is the control force generated by the actuator (i.e., Fig. 4.8). Note

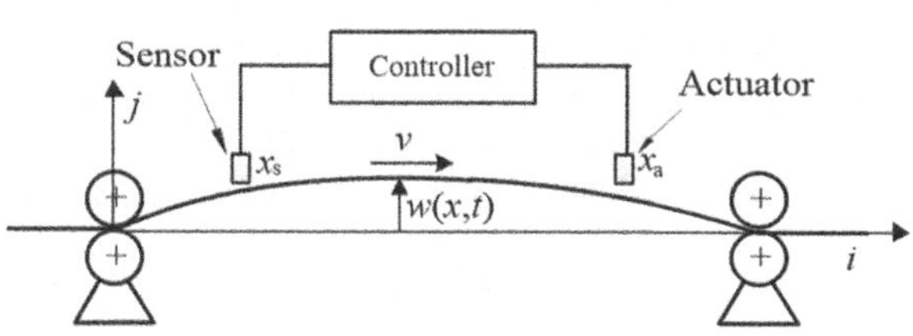

Fig. 4.8 Pointwise control scheme for a moving string using one actuator

that without loss of generality, we assume that the string's length is l and $\rho A = T_0$. Furthermore, we also assume that the string is in the subcritical regime; equivalently, $v_0 < 1$. The control objective is to design the control force f based on the lateral vibration measured by the sensor, such that the lateral vibration $w(x, t)$ is suppressed for any initial conditions. We take the Laplace transform for both sides of (4.59) with respect to t.

$$
\begin{aligned}
&s^2 W(x,s) + 2v_0 s W_x(x,s) - (1 - v_0^2) W_{xx}(x,s) \\
&= F(x,s) + sa(x) + b(x) + 2v_0 a_x(x), \\
&W(0,s) = 0, \quad W(1,s) = 0,
\end{aligned}
\tag{4.61}
$$

where $W(x, s)$ and $F(x, s)$ are the Laplace transforms of $w(x, t)$ and $f(x, t)$, respectively, and s is the Laplace variable. The solution for (4.61) is obtained by using Green's function, namely,

$$
W(x,s) = \int_0^l G_0(x,\xi,s)\left[F(\xi,s) + sa(\xi) + b(\xi) + 2v_0 a_\xi(\xi)\right]\mathrm{d}\xi \tag{4.62}
$$

where $G_0(x, \xi, s)$ is Green's function of the left-hand side of (4.61), which is also the open-loop transfer function of the axially moving string in (4.59) (Butkovskyi 1983). Based on the Green's function theory, $G_0(x, \xi, s)$ takes the following form.

$$
G_0(x,\xi,s) = \frac{N_0(x,\xi,s)}{D_0(s)} \tag{4.63}
$$

where

$$
N_0(x,\xi,s) = \begin{cases} \frac{1}{s}\sinh\left(\frac{sx}{1-v_0^2}\right)\sinh\left(\frac{s(1-\xi)}{1-v_0^2}\right)\mathrm{e}^{\frac{v_0 s(x-\xi)}{1-v_0^2}} & \text{for } 0 \le x \le \xi \le 1, \\ \frac{1}{s}\sinh\left(\frac{s\xi}{1-v_0^2}\right)\sinh\left(\frac{s(1-x)}{1-v_0^2}\right)\mathrm{e}^{\frac{v_0 s(x-\xi)}{1-v_0^2}} & \text{for } 0 \le \xi \le x \le 1, \end{cases} \tag{4.64}
$$

$$
D_0(s) = \sinh\left(\frac{s}{1-v_0^2}\right). \tag{4.65}
$$

The eigenvalues of the moving string are the nonzero roots of $D_0(s) = 0$: $p_k^0 = ik\pi(1 - v_0^2), k = \pm 1, \pm 2, \ldots$ (i.e., $s = 0$ is not an eigenvalue because $\lim_{s\to 0} G_0(x, \xi, s)$ is finite). Note that $\omega_k = k\pi(1 - v_0^2)$ is the natural frequency of the k-th vibration mode of the string and $p_k^0 = i\omega_k$.

Now, we determine the closed-loop transfer function of the system. The control force in the frequency domain is given as

$$
F(x,s) = \delta(x - x_\mathrm{a})G(s)W(x_\mathrm{s}, s) \tag{4.66}
$$

where $G(s)$ is the transfer function of the feedback controller. $G(s)$ also contains the dynamics of the actuator and sensor. We assume that the feedback controller is given by the following form:

$$G(s) = \alpha H(s)e^{-t_d s} = \alpha \frac{N(s)}{D(s)} e^{-t_d s} \tag{4.67}$$

where $N(s)$ and $D(s)$ are the polynomials in s that have no common roots. α is the control gain, and t_d is the time delay in seconds. Substituting (4.66) into (4.62) yields

$$W(x, s) = G_0(x, x_a, s)G(s)W(x_s, s) + \int_0^l G_0(x, \xi, s)\left[sa(\xi) + b(\xi) + 2v_0 a_\xi(\xi)\right]d\xi. \tag{4.68}$$

According to (4.68), the lateral vibration in x_s is shown as follows:

$$W(x_s, s) = [1 - G_0(x_s, x_a, s)G(s)]^{-1} \int_0^l G_0(x_s, \xi, s)\left[sa(\xi) + b(\xi) + 2v_0 a_\xi(\xi)\right]d\xi. \tag{4.69}$$

By substituting (4.63), (4.67), and (4.69) into (4.68), the solution of the closed-loop system is obtained, namely,

$$\begin{aligned} W(x, s) &= \int_0^l \left[\frac{G_0(x, x_a, s)G(s)G_0(x_s, \xi, s)}{1 - G_0(x_s, x_a, s)G(s)} + G_0(x, \xi, s)\right] \left[sa(\xi) + b(\xi) + 2v_0 a_\xi(\xi)\right]d\xi \\ &= \int_0^l G_c(x, \xi, s)\left[sa(\xi) + b(\xi) + 2v_0 a_\xi(\xi)\right]d\xi \end{aligned} \tag{4.70}$$

where $G_c(x, \xi, s)$ is the closed-loop transfer function of the controlled string and is expressed as follows:

$$G_c(x, \xi, s) = \frac{N_c(x, \xi, s)}{D_c(s)}. \tag{4.71}$$

The detailed expression of $G_c(x, \xi, s)$ is shown in Appendix A. The characteristic equation, whose nonzero roots are poles of $G_c(x, \xi, s)$, is

$$D_c(s) = D_0(s)D(s) - \alpha N_0(x_s, x_a, s)N(s)e^{-t_d s} = 0. \tag{4.72}$$

The asymptotic stability of the closed-loop system is guaranteed if all poles of the transfer function $G_c(x, \xi, s)$ have a negative real part. Therefore, the control objective is to design the controller (4.67) to assure that all roots of (4.72) lie in the left-half s-place. The loci of the poles as α varies from zero to infinity are analyzed. The open-loop poles $p_k^0 = i\omega_k$ lie in the imaginary axis. At $\alpha = 0$, the root locus of the k-th vibration mode, which is denoted by $p_k^c(\alpha)$, begins at the open-loop pole p_k^0. As α is increased, the closed-loop poles $p_k^c(\alpha)$ initially move toward the left-half s-place if $R_k = \mathrm{Re}\big(\mathrm{d}p_k^c(\alpha)/\mathrm{d}\alpha\big|_{\alpha=0^+}\big) < 0$ and move toward the right-half s-place if $R_k > 0$. In the former case, the system is asymptotically stable. As the control gain is continuously increased to the critical value α_{cr}, $p_k^c(\alpha)$ crosses the imaginary axis for the first time. Subsequently, the system can be unstable. Note that α_{cr} is either a finite value or an infinite value. Based on the root locus analysis, the following stability criterion is made (i.e., refer to Appendix B for details).

Criterion 4.1 The axially moving string (4.59) under the controller (4.67) is asymptotically stable for $\alpha \in (0, \alpha_{cr})$ if $R_k = \mathrm{Re}\Big(\frac{\mathrm{d}p_k^c(\alpha)}{\mathrm{d}\alpha}\Big|_{\alpha=0^+}\Big) < 0$ for all k.

According to Appendix B, $\frac{\mathrm{d}p_k^c(\alpha)}{\mathrm{d}\alpha}\Big|_{\alpha=0^+}$ and α_{cr} are calculated as follows:

$$\frac{\mathrm{d}p_k^c(\alpha)}{\mathrm{d}\alpha}\bigg|_{\alpha=0^+} = -\frac{\mathrm{i}}{k\pi}\sin(k\pi x_s)\sin(k\pi x_a)H(p_k^0)\mathrm{e}^{-ik\pi\zeta}, \tag{4.73}$$

$$\alpha_{cr} = \inf_n\left\{\frac{\beta_n\sin\big(\beta_n(1-v_0^2)^{-1}\big)}{\sin\big(\xi_{\min}\beta_n(1-v_0^2)^{-1}\big)\sin\big((1-\xi_{\max})\beta_n(1-v_0^2)^{-1}\big)A(\beta_n)}\mathrm{e}^{\mathrm{i}(\zeta\beta_n(1-v_0^2)^{-1}-\Theta(\beta_n))}\right\} \tag{4.74}$$

where

$$\begin{aligned}
&\zeta = v_0(x_a - x_s) + t_d\big(1 - v_0^2\big),\\
&\xi_{\min} = \min(x_a, x_s),\ \xi_{\max} = \max(x_a, x_s),\\
&H(\mathrm{i}\omega) = A(\omega)\mathrm{e}^{\mathrm{i}\Theta(\omega)},\ A(\omega) = \sqrt{\mathrm{Re}(H(\mathrm{i}\omega))^2 + \mathrm{Im}(H(\mathrm{i}\omega))^2},\\
&\Theta(\omega) = \tan^{-1}\left(\frac{\mathrm{Im}(H(\mathrm{i}\omega))}{\mathrm{Re}(H(\mathrm{i}\omega))}\right),
\end{aligned} \tag{4.75}$$

and β_n satisfy $0 < \beta_1 < \beta_2 < \ldots < \beta_n < \ldots$ and β_n are the positive roots of the following equation.

$$\tan\big(\zeta\beta_k(1-v_0^2)^{-1}\big) = \frac{\mathrm{Im}(H(\mathrm{i}\beta_k))}{\mathrm{Re}(H(\mathrm{i}\beta_k))}. \tag{4.76}$$

Based on these stability criteria, the control design is summarized as follows: Design the function $H(s)$ such that $R_k < 0$, and determine the critical value of the control gain. Now, we design the controller for two cases: Collocation and non-collocation of actuators and sensors.

Case 1: Colocation of the actuator and the sensor ($x_a = x_s$). We assume that both the actuator and the sensor are located in x_0, and the feedback controller has no time delay. It follows that $\theta = 0$, $\xi_{\min} = \xi_{\max} = x_0$, and

$$R_k = \mathrm{Re}\left(-\frac{\mathrm{i}}{k\pi}\sin(k\pi x_0)^2 H(p_k^0)\right) = \frac{1}{k\pi}\sin(k\pi x_0)^2 \mathrm{Im}\big(H(p_k^0)\big), \tag{4.77}$$

where $\omega_k = k\pi(1 - v_0^2)$. We further assume that position x_0 is not the node of any vibration mode, i.e., $\sin(k\pi x_0) \neq 0$. Therefore, (4.77) reveals that $R_k < 0$ for all k if and only if $\mathrm{Im}\big(H(p_k^0)\big) < 0$. Next, we derive the range of control gain α by guaranteeing a stable system. Because $\zeta = 0$ and $\mathrm{Im}\big(H(p_k^0)\big) < 0$, (4.76) has no positive root. Therefore, the critical control gain is infinite; equivalently, the controlled system with controller (4.67), where $\mathrm{Im}\big(H(p_k^0)\big) < 0$, is stable for any positive gain α. An example of a stabilizing controller is negative feedback control, $G(s) = -\alpha s$.

Case 2: Non-collocation of the actuator and sensor ($x_a \neq x_s$). In this case, for simplicity, the actuator and sensor are placed in particular positions such that $x_s + x_a = 1$. We design the controller with time delay $t_d = (1 - v_0(x_a - x_s))(1 - v_0^2)^{-1}$. With this time delay, $\zeta = 1$ and the real part of (4.73) becomes

$$\begin{aligned}
R_k &= \mathrm{Re}\left(-\frac{\mathrm{i}}{k\pi}\sin(k\pi x_s)\sin(k\pi x_a)H(p_k^0)e^{-\mathrm{i}k\pi}\right)\\
&= \mathrm{Re}\left(-\frac{\mathrm{i}}{k\pi}\sin(k\pi x_s)^2 A(\omega_k)\mathrm{e}^{\mathrm{i}(\Theta(\omega_k)-k\pi)}\right)\\
&= \frac{(-1)^k}{k\pi}\sin(k\pi x_s)^2 A(\omega_k)\sin(\Theta(\omega_k) - k\pi)\\
&= -\frac{1}{k\pi}\sin(k\pi x_s)^2 A(\omega_k)\sin(\Theta(\omega_k)).
\end{aligned} \tag{4.78}$$

It is observed that $R_k < 0$ for all k if $\Theta(\omega_k) \in (0, \pi)$; equivalently, $\mathrm{Im}\big(H(p_k^0)\big) > 0$. The positive feedback control with time delay, $G(s) = \alpha s e^{s(1-v_0(x_a-x_s))(1-v_0^2)^{-1}}$, satisfies the condition. By substituting $\zeta = 1$ and $H(\mathrm{i}\beta_k) = \mathrm{i}\beta_k$ into (4.76), β_k is obtained as

$$\beta_k = (1 - v_0^2)\left(\frac{\pi}{2} + 2k\pi\right), \tag{4.79}$$

and the critical control gain is derived, namely,

$$\alpha_{\mathrm{cr}} = \inf_k\left\{\sin\left(\xi_{\min}\left(\frac{\pi}{2} + 2k\pi\right)\right)^{-2}\right\} = 1. \tag{4.80}$$

According to the stability criterion, the axially moving string under positive feedback control $G(s) = \alpha s e^{s(1-v_0(x_a-x_s))(1-v_0^2)^{-1}}$ with $\alpha \in (0, 1)$ is asymptotically stable.

Theoretically, the transfer function-based method can overcome the spillover problem since the asymptotical stability of all vibration modes is assured. However, the actuators and sensors must not be set on the nodes of the vibration modes. In practical systems, it seems impossible to avoid the nodes of the high-order modes. Additionally, this method also faces a mathematical challenge when designing the control law of nonlinear strings.

4.2.2 Wave Cancellation Method

Another method used to design the controller for axially moving linear strings in the frequency domain is the wave cancellation method. Its central concept is to suppress the vibration energy of axially moving strings by absorbing traveling waves. This method assures the system's stability by applying control forces to eliminate all reflected waves and prevent the accumulation of vibration energy. From a stability standpoint, these forces cancel all poles of the transfer function. It follows that the controlled system has no pole and is stable. The wave cancellation method was utilized to control vibrations via the pointwise technique (Ying and Tan 1996; Tan and Ying 2000; Zhang and Chen 2006) and the boundary control technique (Chung and Tan 1995).

We design the boundary control of an axially moving string based on the wave cancellation method. Reconsider the linear model of an axially moving string in (4.59), wherein the actuator is located at the right boundary (i.e., $x_a = 1$), whereas the sensor is placed at a position in the vicinity of the actuator (i.e., $x_s < x_a$). The string is further subjected to a bounded external excitation at x_e (i.e., $x_e < x_s$). We assume that the initial conditions are zero. The objective is to derive the control law based on the measured lateral vibration at x_s such that all reflected waves at the right boundary are canceled. According to the previous section, the solution of the lateral vibration in the frequency domain is

$$W(x,s) = \int_0^l G_0(x,\xi,s)[\delta(x-x_e)F_e(s) + \delta(x-1)G(s)W(x_s,s)]d\xi \quad (4.81)$$

where $F_e(s)$ and $G(s)$ are the Laplace transform of the external excitation and the transfer function of the controller, respectively. Because the actuator is located at the boundary, the boundary conditions of the string become

$$W(0,s) = 0, \quad W(1,s) = F(s) = G(s)W(x_s,s). \quad (4.82)$$

With the boundary condition in (4.82), the transfer function of the string is given as follows:

$$G_0(x, \xi, s) = \frac{N_0(x, \xi, s)}{D_0(s)} \tag{4.83}$$

where

$$N_0(x, \xi, s) = \begin{cases} \frac{(-\mathrm{e}^{-s\lambda_1 x} + e^{s\lambda_2 x})(\lambda_1 \mathrm{e}^{s\lambda_1 \xi + s\lambda_2 \xi} + \lambda_2 \mathrm{e}^{s\lambda_1 + s\lambda_2})\mathrm{e}^{-s\lambda_2 \xi}}{2s(\lambda_1 + \lambda_2)} & \text{for } 0 \le x \le \xi \le 1, \\ \frac{(\mathrm{e}^{s\lambda_1 \xi} - \mathrm{e}^{-s\lambda_2 \xi})(\lambda_1 \mathrm{e}^{s\lambda_1 x + s\lambda_2 x} + \lambda_2 \mathrm{e}^{s\lambda_1 + s\lambda_2})\mathrm{e}^{-s\lambda_1 x}}{2s(\lambda_1 + \lambda_2)} & \text{for } 0 \le \xi \le x \le 1, \end{cases}$$

$$\begin{aligned} D_0(s) &= \lambda_1 + \lambda_2 \mathrm{e}^{s(\lambda_1 + \lambda_2)}, \\ \lambda_1 &= (1 + v_0)^{-1}, \quad \lambda_2 = (1 - v_0)^{-1}. \end{aligned} \tag{4.84}$$

By using (4.83), (4.81) is rewritten as follows:

$$\begin{aligned} W(x, s) &= G_0(x, x_\mathrm{e}, s)F_\mathrm{e}(s) + G_0(x, 1, s)G(s)W(x_\mathrm{s}, s) \\ &= G_0(x, x_\mathrm{e}, s)F_\mathrm{e}(s) + \frac{(-\mathrm{e}^{-s\lambda_1 x} + \mathrm{e}^{s\lambda_2 x})\mathrm{e}^{s\lambda_1}}{s\left(\lambda_1 + \lambda_2 \mathrm{e}^{s(\lambda_1 + \lambda_2)}\right)} G(s)W(x_\mathrm{s}, s). \end{aligned} \tag{4.85}$$

By applying the procedure described in the previous section, the closed-loop response of the string is obtained, namely,

$$W(x, s) = G_\mathrm{c}(x, x_\mathrm{e}, s)F_\mathrm{e}(s) = \left[G_\mathrm{cr}(s)\mathrm{e}^{-s\lambda_1 x} + G_\mathrm{cl}(s)\mathrm{e}^{s\lambda_2 x}\right]F_\mathrm{e}(s), \tag{4.86}$$

for $0 \le x \le x_\mathrm{e}$

$$\begin{aligned} G_\mathrm{cr}(s) &= -\frac{(\lambda_1 \mathrm{e}^{s\lambda_1 x_\mathrm{e}} + \lambda_2 \mathrm{e}^{s\lambda_1}\mathrm{e}^{s\lambda_2}\mathrm{e}^{-s\lambda_2 x_\mathrm{e}})}{2s(\lambda_1 + \lambda_2)\left[\lambda_1 + \lambda_2 \mathrm{e}^{s\lambda_1}\mathrm{e}^{s\lambda_2}\right]} - A(s), \\ G_\mathrm{cl}(s) &= -G_\mathrm{cr}(s), \end{aligned} \tag{4.87}$$

for $x_\mathrm{e} \le x \le 1$

$$\begin{aligned} G_\mathrm{cr}(s) &= \frac{\lambda_2 \mathrm{e}^{s(\lambda_1 + \lambda_2)}(\mathrm{e}^{s\lambda_1 x_\mathrm{e}} + \mathrm{e}^{-s\lambda_2 x_\mathrm{e}})}{2s(\lambda_1 + \lambda_2)\left[\lambda_1 + \lambda_2 \mathrm{e}^{s(\lambda_1 + \lambda_2)}\right]} - A(s), \\ G_\mathrm{cl}(s) &= \frac{\lambda_1(\mathrm{e}^{s\lambda_1 x_\mathrm{e}} + \mathrm{e}^{-s\lambda_2 x_\mathrm{e}})}{2s(\lambda_1 + \lambda_2)\left[\lambda_1 + \lambda_2 \mathrm{e}^{s(\lambda_1 + \lambda_2)}\right]} + A(s), \end{aligned} \tag{4.88}$$

where

$$A(s) = \frac{\mathrm{e}^{s\lambda_1} G(s) G_0(x_\mathrm{s}, x_\mathrm{e}, s)}{s\left(\lambda_1 + \lambda_2 \mathrm{e}^{s(\lambda_1 + \lambda_2)}\right) - (-\mathrm{e}^{-s\lambda_1 x_\mathrm{s}} + \mathrm{e}^{s\lambda_2 x_\mathrm{s}})\mathrm{e}^{s\lambda_1} G(s)}. \tag{4.89}$$

The closed-loop transfer function $G_\mathrm{c}(x, x_\mathrm{e}, s)$ in (4.86) is the superposition of the right-propagating wave and left-propagating wave, wherein $G_\mathrm{cr}(s)$ and $G_\mathrm{cl}(s)$ are the transfer functions of the right-propagating wave coefficient and left-propagating wave coefficient, respectively. We assume that the reflected wave at the right boundary is canceled to no accumulated vibration energy. In this case, the left-propagating wave coefficient $G_{\mathrm{cl}}(s)$ in $[x_{\mathrm{e}}, 1]$ must be zero:

$$G_{\mathrm{cl}}(s) = 0, \quad \text{for } x_{\mathrm{e}} \le x \le 1. \tag{4.90}$$

Solving (4.90) yields the transfer function of the controller, which is a negative feedback control law with time delay, namely,

$$G(s) = -\lambda_1 s \mathrm{e}^{s\lambda_1 x_{\mathrm{s}}} \mathrm{e}^{-s\lambda_1}. \tag{4.91}$$

Next, we discuss the stability of the closed-loop system with the controller (4.91). By substituting this controller into (4.87) and (4.88), the closed-loop transfer function of the string is obtained as follows:

$$G_{\mathrm{c}}(x, x_{\mathrm{e}}, s) = \begin{cases} \dfrac{\left(\mathrm{e}^{-s\lambda_1(x-x_{\mathrm{e}})} - \mathrm{e}^{-s(\lambda_1 x + \lambda_2 x_{\mathrm{e}})}\right)(\lambda_1+\lambda_2)^{-1}}{2s}, & x_{\mathrm{e}} \le x \le 1, \\ \dfrac{\left(-\mathrm{e}^{-s(\lambda_1 x + \lambda_2 x_{\mathrm{e}})} + \mathrm{e}^{-s\lambda_2(x_{\mathrm{e}}-x)}\right)(\lambda_1+\lambda_2)^{-1}}{2s}, & 0 \le x \le x_{\mathrm{e}}. \end{cases} \tag{4.92}$$

Equation (4.92) reveals that the closed-loop transfer function of the string has no pole. Therefore, the axially moving string under the designed boundary control force (i.e., $F(s) = -\lambda_1 s \mathrm{e}^{s\lambda_1 x_{\mathrm{s}}} \mathrm{e}^{-s\lambda_1} W(x_{\mathrm{s}}, s)$) is stable and has no resonance. Furthermore, the transfer function of the negative feedback control law is independent, not only the external excitation location but also the type of excitation. The designed control law is also applied to suppress the string's vibration to multiple concentrated and distributed external excitations.

The wave cancellation method can also be applied to the axially moving string via the pointwise control technique. Based on this method, Ying and Tan (1996) designed a controller that consists of a feed-forward controller and feedback controller to suppress the lateral vibration downstream of an axially moving string (i.e., region in the vicinity of the right boundary). The control force is given as follows:

$$\begin{aligned} F(x, s) = &\frac{-2s\left[W(x_{\mathrm{s}1}, s) - W(x_{\mathrm{s}2}, s)\mathrm{e}^{-\lambda_1(x_{\mathrm{s}2}-x_{\mathrm{s}1})}\right]\mathrm{e}^{\lambda_2(x_{\mathrm{a}}-x_{\mathrm{s}1})}}{1 - \mathrm{e}^{(\lambda_2-\lambda_1)(x_{\mathrm{s}2}-x_{\mathrm{s}1})}} \\ &- C(s)\frac{2sW(x_{\mathrm{s}3}, s)}{1 - \mathrm{e}^{(\lambda_2-\lambda_1)(1-x_{\mathrm{s}3})}} \end{aligned} \tag{4.93}$$

where $x_{\mathrm{s}1}$, $x_{\mathrm{s}2}$, and $x_{\mathrm{s}3}$ are the locations of the sensors and x_{a} is the location of the actuator. In this controller, the feed-forward controller [i.e., first term in (4.93)] is used to reduce the string's vibrations, whereas the feedback controller [i.e., second term in (4.93)] acts as a low-pass filter that addresses undesired disturbances in the downstream region. $C(s)$ is designed to minimize the influence of the disturbances. The wave cancellation method is further applied to moving strings with unfixed boundary conditions (Tan and Ying 2000) and moving strings controlled by a tensioner arm (Zhang and Chen 2006).

4.2.3 Boundary Control Based on the Lyapunov Method

Boundary control techniques suppress the vibration of axially moving strings by applying the control force at the system's boundary (i.e., Fig. 4.9). Hence, these techniques do not interfere with the operation of the system. This advantage makes boundary control possible for vibration suppression of practical AMSs. The boundary control scheme based on the Lyapunov stability analysis has received considerable attention in developing controllers for axially moving systems in either linear formulations or nonlinear formulations (Hong et al. 2004; Ngo et al. 2009; Nguyen and Hong 2012; Nguyen et al. 2018). To design a control law, first, we establish an appropriate Lyapunov function candidate that is associated with the system energy. Second, the control law is designed so that the candidate function becomes a Lyapunov function for the system (i.e., the time derivative of the candidate function is negative semidefinite or negative definite). The stability of the system with the designed controller is archived. Furthermore, in many cases, asymptotic and exponential stability can be obtained. Thus far, boundary control based on the Lyapunov method has had an important role in the control of axially moving systems. The control design based on the Lyapunov method can address both linear AMSs and nonlinear AMSs. Furthermore, this technique also handles hybrid systems, such as crane systems (Ngo et al. 2009) and AMSs with mass-dampers (Nguyen et al. 2018).

4.2.3.1 Lyapunov Stability Theorem

We start this subsection by discussing the Lyapunov stability for PDE systems, which is the premise for the control approaches presented in this section. The Lyapunov stability theorem was initially applied to ODE systems. For ODE systems, the basic concepts of stability can be described as follows:

Definition 4.1 (*Khalil* 2002) Consider the ODE system

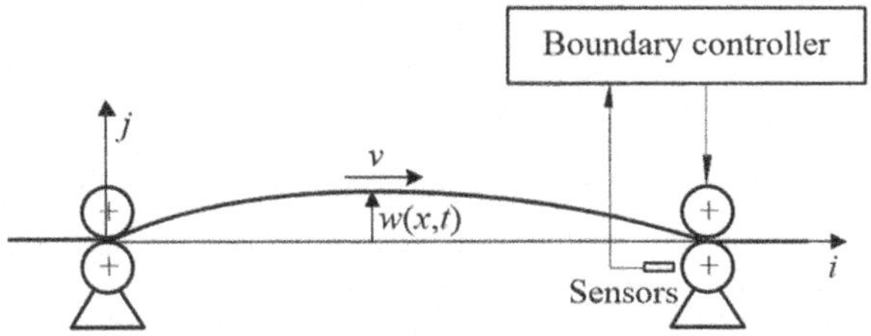

Fig. 4.9 Boundary control scheme for a moving string using one actuator

$$\dot{x} = f(x), \tag{4.94}$$

where $f : D \to \mathbb{R}^n$ is a locally Lipschitz map from the domain $D \subset \mathbb{R}^n$ into $\mathbb{R}^n$. Let $x = 0$ be an equilibrium point of (4.94). The equilibrium point is stable if, for each $\vartheta > 0$, there is $\delta = \delta(\vartheta) > 0$ such that $\|x(t_0)\| < \delta \Rightarrow \|x(t)\| < \vartheta, \forall t > t(0)$. Otherwise, the equilibrium point is unstable. Furthermore, if this point is stable and δ can be chosen such that $\|x(t_0)\| < \delta \Rightarrow \lim_{t\to\infty} x(t) = 0$, the equilibrium point is asymptotically stable.

This concept can be extended to the case where x is a state vector. Notably, the state space is a finite-dimensional Euclidean space; therefore, the state norm is a vector norm. The stability of ODE systems can be concluded based on any norms. Contrary to ODE systems, the state space of PDE systems is infinite-dimensional, and the state norm is a functional norm. Norms of function spaces are not equivalent. Therefore, the conclusion of the stability of PDE systems depends on the norm equipped in the state space. In addition to Definition 4.1, the stability of ODE systems is also examined by analyzing the energy function. According to the energy standpoint, the Lyapunov theory can be extended to PDE systems. The physical observation reveals that the system settles down to an equilibrium point when its total energy is continuously decaying. The relationship between mechanical energy and stability is described as follows:

- The zero-energy is related to the equilibrium solution.
- The system becomes unstable when the mechanical energy grows.
- The system is stable if the mechanical energy is a nonincreasing function.

Therefore, the system stability may be investigated via the variation of a single scalar function associated with the energy.

The procedure of the control design based on the Lyapunov method is to generate a Lyapunov function candidate, which is a scalar energy-like function. Subsequently, we examine the time rate of this function and design a control law to guarantee that this variation is negative semidefinite or negative definite. The key to this approach is to select the Lyapunov function candidate. Unfortunately, there exists no general type of Lyapunov function candidate for AMSs. Although the Lyapunov function candidate is associated with mechanical energy, it varies by case. This section investigates several string models for further discussion about boundary control based on the Lyapunov method.

4.2.3.2 Linear String

The derivation of a boundary control law based on the Lyapunov method for the linear moving string is illustrated in this section. Let us consider the equation of motion of a linear string given as follows:

$$\rho A\left(w_{tt} + 2v_0 w_{xt} + v_0^2 w_{xx}\right) - T_0 w_{xx} = 0. \tag{4.95}$$

The total mechanical energy is computed as

$$E_s(t) = \frac{1}{2}\int_0^l \left[\rho A v_0^2 + \rho A(w_t + v_0 w_x)^2\right]\mathrm{d}x + \frac{1}{2}\int_0^l T_0 w_x^2 \mathrm{d}x. \tag{4.96}$$

We consider vibration control of the linear string in two ways: (i) Right boundary control and (ii) left boundary control.

Right Boundary Control

In this type of control, the control force is applied to the right boundary of the string. The boundary conditions are given as follows:

$$w(0,t) = 0, \quad w_x(l,t) = \frac{1}{T_0} f_r(t), \tag{4.97}$$

where $f_r(t)$ is the control force. The control objective is to design the control force $f_r(t)$ such that the closed-loop system is stable. It should be noted that the aim is to derive a control law to reduce the lateral vibrations. Hence, even though the mechanical energy $E_s(t)$ involves the energy in the longitudinal motion (i.e., $1/2\int_0^l \rho A v_0^2 dx$), we can disregard this energy in the process of the stability analysis. Let us consider the following Lyapunov function candidate, which is associated with the mechanical energy of the string:

$$V(t) = \frac{1}{2}\int_0^l \rho A(w_t + v_0 w_x)^2 \mathrm{d}x + \frac{1}{2}\int_0^l T_0 w_x^2 \mathrm{d}x. \tag{4.98}$$

Notably, the time rate of an energy function of a moving material, which involves a fixed control volume and a control surface with mass flows in and out, can be computed by using Reynold's transport theorem:

$$\begin{aligned}
\frac{\mathrm{d}V(t)}{\mathrm{d}t} &= \frac{1}{2}\rho A \frac{\mathrm{d}}{\mathrm{d}t}\int_0^l (w_t + v_0 w_x)^2 \mathrm{d}x + \frac{1}{2} T_0 v_0 \frac{d}{\mathrm{d}t}\int_0^l w_x^2 \mathrm{d}x \\
&= \frac{1}{2}\rho A \int_0^l \left[\frac{\mathrm{D}}{\mathrm{D}t}(w_t + v_0 w_x)^2\right]\mathrm{d}x + \frac{1}{2} T_0 \int_0^l \frac{\mathrm{D}}{\mathrm{D}t} w_x^2 \mathrm{d}x \\
&= \int_0^l \left\{\rho A\left[(w_t + v_0 w_x)_t + v_0 (w_t + v_0 w_x)_x\right](w_t + v_0 w_x) + T_0 (w_{xt} + v_0 w_{xx}) w_x\right\}\mathrm{d}x \\
&= \int_0^l \left[\rho A\left(w_{tt} + 2v_0 w_x + v_0^2 w_{xx}\right)(w_t + v_0 w_x) + T_0 (w_{xt} + v_0 w_{xx}) w_x\right]\mathrm{d}x.
\end{aligned} \tag{4.99}$$

Substituting (4.95) and the boundary conditions (4.97) into (4.99) yields

$$\begin{aligned}\frac{\mathrm{d}V(t)}{\mathrm{d}t} &= \int_0^l [T_0 w_{xx}(w_t + v_0 w_x) + T_0(w_{xt} + v_0 w_{xx})w_x]\mathrm{d}x \\ &= T_0 \int_0^l (w_{xx} w_t + 2v_0 w_{xx} w_x + w_x w_{xt})\mathrm{d}x \\ &= T_0 \left[w_x w_t + v_0 w_x^2\right]\Big|_{x=0}^{x=l} \\ &= -T_0 v_0 w_x^2(0,t) + f_r(t)\left[w_t(l,t) + \frac{v_0}{T_0} f_r(t)\right]. \end{aligned} \tag{4.100}$$

The control force $f_r(t)$ must be designed such that the rate of change of the Lyapunov function candidate is negative semidefinite. Consider the following control force

$$f_r(t) = -\alpha_0 w_t(l,t),\ \alpha_0 \in \left(0, \frac{T_0}{v_0}\right). \tag{4.101}$$

The control force in (4.101) can assure $\mathrm{d}V(t)/\mathrm{d}t \le 0$. Consequently, the boundedness of the total mechanical energy is guaranteed, and the controlled system is stable.

Left Boundary Control

In contrast to right boundary control, the vibration is suppressed via a control force at the left boundary of the string. The boundary conditions are given as

$$w_x(0,t) = \frac{1}{T_0} f_l(t), \quad w(l,t) = 0. \tag{4.102}$$

According to the foregoing boundary conditions, the rate of change of the Lyapunov function candidate is computed as follows:

$$\frac{\mathrm{d}V(t)}{\mathrm{d}t} = v_0 T_0 w_x^2(l,t) - f_l(t)\left(w_t(0,t) + \frac{v_0}{T_0} f_l(t)\right). \tag{4.103}$$

Although we can design a control force to render the second term in (4.103) negative semidefinite, the first term is always positive. This outcome leads to the local growth of the string energy due to the energy flux caused by waves traveling toward the right boundary. Therefore, right boundary control is recommended for the control of axially moving systems.

Stability of Boundary Control

We further discuss the stability of the right boundary control. The mechanical energy of the linear string under the right boundary control force in (4.101) is bounded.

Therefore, the closed-loop system is stable. The control force in (4.101) can guarantee that the system is asymptotically stable and further exponentially stable. We show the asymptotic and exponential stability of the linear string under the designed control force (4.101) based on the functional analysis and Lyapunov stability theorem.

Asymptotic Stability

For stability analysis, we define the following state-space X.

$$X := \left\{ (w, w_t)^{\mathrm{T}} \middle| w \in H_0^1,\ w_t \in L^2 \right\} \tag{4.104}$$

where the Lebesgue space L^2 and Sobolev space H_0^k are defined as follows:

$$L^2 := \left\{ f : [0, l] \to \mathbb{R} \middle| \int_0^l f(x)^2 \mathrm{d}x < \infty \right\}, \tag{4.105}$$

$$H_0^k := \left\{ f : [0, l] \to \mathbb{R} \middle| f(x), \frac{\mathrm{d}f(x)}{\mathrm{d}x}, \frac{\mathrm{d}^2 f(x)}{\mathrm{d}x^2}, \ldots, \frac{\mathrm{d}^k f(x}{\mathrm{d}x^k}) \in L^2,\ f(0) = 0 \right\}. \tag{4.106}$$

Next, (4.95) can be rewritten in the following distributed state-space form.

$$\dot{\mathbf{z}} = \mathbf{A}\mathbf{z},\ \ \mathbf{z}(0) = \mathbf{z}_0 \in X \tag{4.107}$$

where $\mathbf{z} = [w\ w_t]^{\mathrm{T}} \in X$; the unbounded operator $\mathbf{A} : X \to X$ and its domain $D(\mathbf{A})$ for the controlled system are defined as

$$\mathbf{A} = \begin{bmatrix} 0 & 1 \\ \left(\frac{T_0}{\rho A} - v_0^2\right)\frac{\partial^2}{\partial x^2} & -2v_0\frac{\partial}{\partial x} \end{bmatrix}, \tag{4.108}$$

$$D(\mathbf{A}) := \left\{ [w, w_t]^{\mathrm{T}} \middle| w \in H_0^2,\ w_t \in H_0^1, \alpha w_t(l, t) + w_x(l, t) = 0 \right\}. \tag{4.109}$$

The equilibrium state of (4.107) is $\mathbf{z} = 0$. In space X, we define the inner product of $\mathbf{z}_1 = [w_1\ w_{1t}]^{\mathrm{T}}$ and $\mathbf{z}_2 = [w_2\ w_{2t}]^{\mathrm{T}}$ as

$$\langle \mathbf{z}_1, \mathbf{z}_2 \rangle := \int_0^l \{ \rho A(w_{1t} + v w_{1x})(w_{2t} + v_0 w_{2x}) + T_0 w_{1x} w_{2x} \} \mathrm{d}x. \tag{4.110}$$

The state-space X with the inner product (4.110) forms a Hilbert space. The norm defined by the above inner product is the Lyapunov function candidate in (4.98), namely,

$$V(t) = \|\mathbf{z}\|^2 \triangleq \langle \mathbf{z}, \mathbf{z} \rangle := \int_0^l \left\{ \rho A (w_t + v_0 w_x)^2 + T_0 w_x^2 \right\} \mathrm{d}x \tag{4.111}$$

Since $\dot{V}(t) \leq 0$, operator $\mathbf{A}$ in the space X is dissipative. According to Komura (1967), $\mathbf{A}$ generates a semigroup of contraction $S(t) = \mathrm{e}^{-\mathbf{A}t}$ on X, which implies that (4.107) has a unique solution that can be represented by using the semigroup $S(t)$, namely, $\mathbf{z} = S(t)\mathbf{z}_0$, $t \geq 0$. Now, the asymptotic stability of the system (4.107) will be proven by using the following invariance principle.

Lemma 4.1 (Henry 2006) *Let V be the Lyapunov functional on X such that $\dot{V}(t) \leq 0$, $\forall \mathbf{z} \in X$. Define the set $B = \{\mathbf{z} \in X, \dot{V}(\mathbf{z}) = 0\}$. Let B_{m} be the maximal invariant subset of B and $S(t)$ be the semigroup generated by the closed-loop system. Thus, $\mathbf{z} = S(t)\mathbf{z}_0 \to B_m$ as $t \to \infty$, $\forall \mathbf{z}_0 \in X$.*

Since $V(t)$ in (4.111) is the Lyapunov functional in X, the subset B of X is $B = \{\mathbf{z} \in X, \dot{V}(t) = 0\}$. Based on this lemma, the system is asymptotically stable if the maximal invariant subset B_{m} is the zero solution $\mathbf{z} = 0$. Any nonzero element of B cannot be invariant because waves of any nonzero solution propagate to the right boundary based on the wave equation's characteristics. This finding implies that the origin $\mathbf{z} = 0$ is the unique invariant set; hence, $B_{\mathrm{m}} = \{0\}$. According to Lemma 4.1, $\lim_{t \to \infty} \mathbf{z} = 0$ and the system (4.107) under the right boundary control (4.101) is asymptotically stable.

Exponential Stability To prove that the control force can make the system exponentially stable, we consider the following Lyapunov function candidate:

$$V_{\mathrm{e}} = \alpha_1 V + \alpha_2 \rho A \int_0^l x w_x (w_t + v_0 w_x) \mathrm{d}x, \quad \alpha_1 \in (0, 1), \alpha_2 > 0. \tag{4.112}$$

Lemma 4.2 (Young's inequality) *Let $\varphi_1, \varphi_2 \in \mathbb{R}$. For any $\delta > 0$, we have*

$$\varphi_1 \varphi_2 \leq |\varphi_1 \varphi_2| \leq \frac{1}{2} \frac{\varphi_1^2}{\delta} + \frac{1}{2} \delta \varphi_2^2 \leq \frac{\varphi_1^2}{\delta} + \delta \varphi_2^2. \tag{4.113}$$

Lemma 4.3 (Hardy et al. 1959) *If $\varphi(x, t) \in H_0^1$ and $\varphi(0, t) = 0\ \forall t \in [0, \infty)$, then the following inequality holds*:

$$\varphi(x, t)^2 \leq l \int_0^l (\varphi_x(x, t))^2 \mathrm{d}x. \tag{4.114}$$

Lemma 4.4 *The Lyapunov function candidate in* (4.112) *is bounded as follows*:

$$0 \leq \beta_1 V \leq V_{\mathrm{e}} \leq \beta_2 V \tag{4.115}$$

where β_1 and β_1 are positive constants defined as

$$\beta_1 = (\alpha_1 - \alpha_2 l \max\{1, (\rho A/T_0)\}),$$
$$\beta_2 = (\alpha_1 + \alpha_2 l \max\{1, (\rho A/T_0)\}). \tag{4.116}$$

Proof Applying Lemma 4.2 to the second term of (4.112) yields

$$\begin{aligned}\rho A \int_0^l x w_x (w_t + v_0 w_x) \mathrm{d}x &\le \frac{1}{2} \rho A l \int_0^l \left[w_x^2 + (w_t + v_0 w_x)^2\right] \mathrm{d}x \\ &= l \left[\frac{1}{2} \rho A \int_0^l (w_t + v_0 w_x)^2 \mathrm{d}x + \frac{\rho A}{T_0} \frac{1}{2} T_0 \int_0^l w_x^2 \mathrm{d}x\right] \\ &\le l \max\{1, (\rho A/T_0)\} V.\end{aligned} \tag{4.117}$$

Based on inequality (4.117), the following inequality is established as

$$-l \max\{1, (\rho A/T_0)\} V \le \rho A \int_0^l x w_x (w_t + v_0 w_x) \mathrm{d}x \le l \max\{1, (\rho A/T_0)\} V. \tag{4.118}$$

From (4.112) and (4.118), we can derive the following result:

$$(\alpha_1 - \alpha_2 l \max\{1, (\rho A/T_0)\}) V \le V_e \le (\alpha_1 + \alpha_2 l \max\{1, (\rho A/T_0)\}) V/ \tag{4.119}$$

By letting $\beta_1 = (\alpha_1 - \alpha_2 l \max\{1, (\rho A/T_0)\})$ and $\beta_2 = (\alpha_1 + \alpha_2 l \max\{1, (\rho A/T_0)\})$, Lemma 4.4 is proven.

Lemma 4.5 *If α_1 and α_2 are positive constants satisfying*

$$\alpha_1 > \frac{\rho A l \left(1 - v_0 T_0^{-1} \alpha_0\right)^2 + \alpha_0^2 l T_0^{-1}}{2\alpha_0 \left(1 - v_0 T_0^{-1} \alpha_0\right)} \alpha_2, \tag{4.120}$$

the time derivative of V_e is bounded in the form $\dot{V}_e \le -\alpha_2 V$.

Proof The time derivative of V_e is expressed as

$$\begin{aligned}\frac{\mathrm{d}V_e}{\mathrm{d}t} &= \alpha_1 \frac{\mathrm{d}V}{\mathrm{d}t} + \alpha_2 \rho A \int_0^l x \frac{\mathrm{D}}{\mathrm{D}t} [w_x (w_t + v_0 w_x)] \mathrm{d}x \\ &= \alpha_1 \frac{\mathrm{d}V}{\mathrm{d}t} + \alpha_2 \rho A \int_0^l x [(w_{xt} + v_0 w_{xx})(w_t + v_0 w_x)] \mathrm{d}x\end{aligned}$$

$$
\begin{aligned}
&+\alpha_2\rho A\int_0^l x\big[w_x(w_{tt}+2v_0w_{xt}+v_0^2w_x)\big]\mathrm{d}x\\
&=\alpha_1\frac{\mathrm{d}V}{\mathrm{d}t}+\frac{1}{2}\alpha_2\rho A\int_0^l x\big((w_t+v_0w_x)^2\big)_x\mathrm{d}x+\frac{1}{2}\alpha_2\int_0^l xT_0\big(w_x^2\big)_x\mathrm{d}x\\
&=\alpha_1\frac{\mathrm{d}V}{\mathrm{d}t}+\frac{1}{2}\alpha_2\rho A\left[x(w_t+v_0w_x)^2\Big|_{x=0}^{x=l}-\int_0^l(w_t+v_0w_x)^2\mathrm{d}x\right]\\
&+\frac{1}{2}\alpha_2T_0\left[xw_x^2\Big|_{x=0}^{x=l}-\int_0^l w_x^2\mathrm{d}x\right].
\end{aligned}
\tag{4.121}
$$

Substituting $\mathrm{d}V/\mathrm{d}t$ of (4.100) and the control force of (4.101) into (4.121) yields

$$
\begin{aligned}
\frac{\mathrm{d}V_\mathrm{e}}{\mathrm{d}t}&=-\alpha_1T_0v_0w_x(0,t)^2-\alpha_1\alpha_0w_t(l,t)^2\big(1-v_0T_0^{-1}\alpha_0\big)\\
&+\frac{1}{2}\alpha_2\rho Alw_t(l,t)^2\big(1-v_0T_0^{-1}\alpha_0\big)^2\\
&+\frac{1}{2}\alpha_2\alpha_0^2lT_0^{-1}w_t(l,t)^2-\frac{1}{2}\alpha_2\rho A\int_0^l(w_t+v_0w_x)^2\mathrm{d}x-\frac{1}{2}\alpha_2T_0\int_0^l w_x^2\mathrm{d}x\\
&=-\left[\alpha_1\alpha_0\big(1-v_0T_0^{-1}\alpha_0\big)-\frac{1}{2}\alpha_2\rho Al\big(1-v_0T_0^{-1}\alpha_0\big)^2-\frac{1}{2}\alpha_2\alpha_0^2lT_0^{-1}\right]w_t(l,t)^2\\
&-\alpha_1T_0v_0w_x(0,t)^2-\alpha_2\left[\frac{1}{2}\int_0^l\rho A(w_t+v_0w_x)^2\mathrm{d}x+\frac{1}{2}\int_0^l T_0w_x^2\mathrm{d}x\right].
\end{aligned}
\tag{4.122}
$$

If α_1 and α_2 satisfy (4.120), we can conclude that $\mathrm{d}V_\mathrm{e}/\mathrm{d}t=\dot{V}_\mathrm{e}\le-\alpha_2V$. □

Theorem 4.1 *Consider an axially moving string described by* (4.95) *and the boundary conditions in* (4.97). *Under the right boundary control force* (4.101) *and bounded initial conditions, the mechanical energy of the system decays exponentially.*

Proof According to Lemma 4.4 and Lemma 4.5, we obtain

$$
\dot{V}_\mathrm{e}\le-\alpha_2V\le\frac{-\alpha_2}{\beta_2}V_\mathrm{e}\Leftrightarrow\dot{V}_\mathrm{e}+\frac{\alpha_2}{\beta_2}V_\mathrm{e}\le0
\tag{4.123}
$$

Multiplying (4.123) by $\mathrm{e}^{(\alpha_2/\beta_2)t}$ yields

$$\begin{aligned}
& e^{(\alpha_2/\beta_2)t}\dot{V}_e + \frac{2\alpha_2}{\beta_2}V_e e^{(\alpha_2/\beta_2)t} \le 0 \\
& \Leftrightarrow \frac{d}{dt}\left(V_e e^{(\alpha_2/\beta_2)t}\right) \le 0 \\
& \Leftrightarrow V_e e^{(\alpha_2/\beta_2)t} \le V_e(0) \\
& \Leftrightarrow \beta_1 V e^{(\alpha_2/\beta_2)t} \le V_e e^{\alpha_2 t/\beta_2} \le V_e(0) \le \beta_2 V(0) \\
& \Leftrightarrow V \le \frac{\beta_2}{\beta_1} V(0) e^{-(\alpha_2/\beta_2)t}.
\end{aligned} \tag{4.124}$$

We now use (4.98), (4.124), and Lemma 4.3 to formulate the following inequality.

$$\frac{T_0}{2l}w(x,t)^2 \le \frac{1}{2}\int_0^l T_0 w_x(x,t)^2 dx \le V \le \frac{\beta_2}{\beta_1}V(0)e^{-(2\alpha_2/\beta_2)t}. \tag{4.125}$$

With the bounded initial condition, inequality (4.125) reveals that the lateral vibration of the controlled string decays exponentially.

4.2.3.3 Nonlinear String

In the previous section, we discussed the boundary control of a linear axially moving string based on the Lyapunov method. This approach can be extended to nonlinear strings.

Let us consider an axially moving system with the effect of axial deformation, as shown in Fig. 4.10. The control force is applied at the right boundary via a mass-damper system. The dynamic model of the system is described by a nonlinear PDE and an ODE, namely,

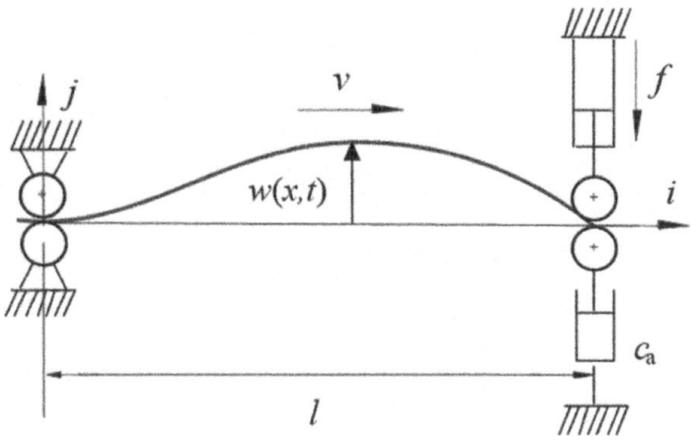

Fig. 4.10 Axially moving system with a hydraulic actuator

$$\rho A\big(w_{tt} + 2v_0 w_{xt} + v_0^2 w_{xx}\big) - T_0 w_{xx} - \frac{3}{2} E A w_x^2 w_{xx} = 0, \tag{4.126}$$

$$m_\mathrm{a} w_{tt}(l,t) + (c_\mathrm{h} - \rho A v_0) w_t(l,t) + (T_0 - \rho A v_0^2) w_x(l,t) = f(t),$$
$$w(0,t) = 0, \tag{4.127}$$

where m_a denotes the mass of the roller and c_h is the damping coefficient of the damper. We consider the Lyapunov function candidate, which is associated with the mechanical energy of the string and mass-damper, as follows.

$$V(t) = \frac{1}{2}\int_0^l \rho A (w_t + v_0 w_x)^2 \mathrm{d}x + \frac{1}{2}\int_0^l T_0 w_x^2 \mathrm{d}x + \frac{1}{8}\int_0^l E A w_x^4 \mathrm{d}x + \frac{1}{2} m_\mathrm{a} w_t(l,t)^2. \tag{4.128}$$

The time rate of change of $V(t)$ is expressed as follows:

$$\frac{\mathrm{d}V(t)}{\mathrm{d}t} = \int_0^l \big[\rho A\big(w_{tt} + 2v_0 w_x + v_0^2 w_{xx}\big)(w_t + v_0 w_x) + T_0 (w_{xt} + v_0 w_{xx}) w_x\big]\mathrm{d}x$$
$$+ \frac{1}{2}\int_0^l \big[E A w_x^3 (w_{xt} + v_0 w_{xx})\big]\mathrm{d}x + m_\mathrm{a} w_t(l,t) w_{tt}(l,t). \tag{4.129}$$

Substituting (4.126) and the boundary condition (4.127) into (4.129) yields

$$\frac{\mathrm{d}V(t)}{\mathrm{d}t} = \big(T_0 w_x w_t + T_0 v_0 w_x^2\big)\Big|_{x=0}^{x=l} + \frac{1}{2}\big(E A w_x^3 w_t + v_0 E A w_x^4\big)\Big|_{x=0}^{x=l}$$
$$+ w_t(l,t)\big[f(t) - (c_\mathrm{h} - \rho A v_0) w_t(l,t) - (T_0 - \rho A v_0^2) w_x(l,t)\big]$$
$$= -\Big[T_0 v_0 w_x(0,t)^2 + \frac{1}{2} v_0 E A w_x(0,t)^4 + (c_\mathrm{h} - \rho A v_0) w_t(l,t)^2\Big]$$
$$+ \rho A v_0^2 w_t(l,t) w_x(l,t) + T_0 v_0 w_x(l,t)^2 + \frac{1}{2} E A w_x(l,t)^3 w_t(l,t)$$
$$+ \frac{1}{2} v_0 E A w_x(l,t)^4 + w_t(l,t) f(t). \tag{4.130}$$

To assure that the system is stable, we propose the following nonlinear feedback control law:

$$f(t) = \begin{cases} -\alpha_1 w_t(l,t) + \alpha_2 \frac{w_x(l,t)^4}{w_t(l,t)}, & \text{for } w_t(l,t) \neq 0, \\ -\alpha_1 w_t(l,t), & \text{for } w_t(l,t) = 0, \end{cases}$$

$$w_x(l,t) = -\beta w_t(l,t),\ \ \beta \in (0,1). \tag{4.131}$$

Under this control law, (4.130) is rewritten as

$$\frac{dV(t)}{dt} = -\Big[T_0 v_0 w_x(0,t)^2 + \frac{1}{2} v_0 E A w_x(0,t)^4$$
$$+(c_h - \rho A v_0) w_t(l,t)^2 + \rho A v_0^2 \beta w_t(l,t)^2\Big]$$
$$-\left[\alpha_1 - T_0 v_0 \beta^2\right] w_t(l,t)^2 - \left[\alpha_2 - \frac{EA}{2}(\beta^{-1} + v_0)\right] w_x(l,t)^4. \quad (4.132)$$

If $\alpha_1 > T_0 v_0 \beta^2$ and $\alpha_2 > EA(\beta^{-1} + v_0)/2$, then $dV(t)/dt \le 0$, which implies that the system under the nonlinear control law in (4.131) is stable. Furthermore, the asymptotic stability of the closed-loop system can be proven by using the approach presented in the previous section.

4.2.3.4 String with Varying Length

We now investigate boundary control of an axially moving string with varying length. For instance, we consider the crane positioning problem (Kim and Hong 2009), wherein the hoisting cable is modeled by a varying-length string (i.e., Fig. 4.11).

In this problem, the control objectives are i) to move the payload to a desired position and ii) to eliminate the lateral vibration of the payload. Let us consider the following mathematical model of the system consisting of the trolley, cable, and payload.

$$\rho A\left(\ddot{y}_{tr} + w_{tt} + \ddot{l} w_x + 2\dot{l} w_{xt} + \dot{l}^2 w_{xx}\right) - (P(x,t) w_x)_x = 0, \quad (4.133)$$

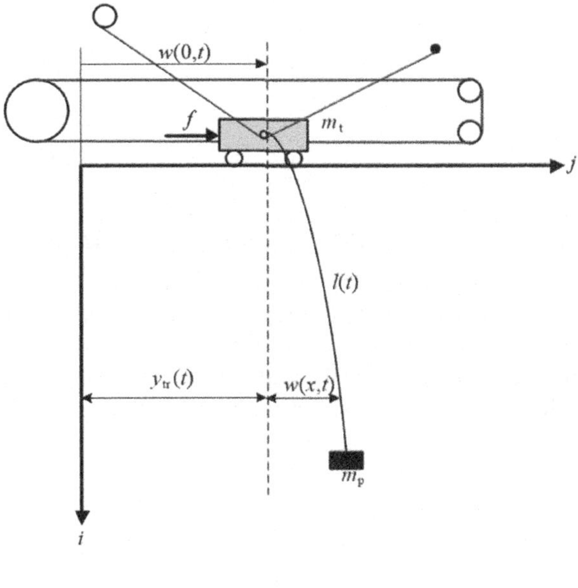

Fig. 4.11 Schematic of an overhead crane

$$m_t\ddot{y}_{tr} + \rho A\dot{l}\dot{y}_{tr} + \rho A\dot{l}\big(w_t(0,t) + \dot{l}w_x(0,t)\big) - P(0,t)w_x(0,t) = f(t), \quad (4.134)$$

$$m_p\big[\ddot{y}_{tr} + w_{tt}(l,t) + \ddot{l}w_x(l,t) + 2\dot{l}w_{xt}(l,t) + \dot{l}^2 w_{xx}(l,t)\big] + P(l,t)w_x(l,t) = 0, \quad (4.135)$$

where y_{tr} is the trolley's position and m_t and m_p are the mass of the trolley and the payload, respectively. The varying axial force is calculated based on (2.23). For convenience, we use P instead of $P(x, t)$. Equations (4.134) and (4.135) are the equations of motion of the trolley and the payload, respectively, which are also the boundary conditions of (4.133). The total mechanical energy, which consists of the energies of the string, trolley, and payload are given as follows:

$$E_s(t) = \frac{1}{2}\int_0^l \left[\rho A\dot{l}^2 + \rho A\big(\dot{y}_{tr} + w_t + \dot{l}w_x\big)^2 + Pw_x^2\right]\mathrm{d}x + \frac{1}{2}m_t\dot{y}_{tr}^2 + \frac{1}{2}m_p\left[\dot{l}^2 + \big(\dot{y}_{tr} + w_t(l,t) + \dot{l}w_x(l,t)\big)^2\right]. \quad (4.136)$$

Based on the mechanical energy, we design a boundary control law to move the trolley along the j-axis and suppress the lateral vibration. For this purpose, the energy corresponding to the motion along the i-axis is disregarded. Let us consider the following Lyapunov function candidate:

$$V(t) = V_1(t) + V_2(t) + V_3(t) \quad (4.137)$$

where

$$V_1(t) = \frac{1}{2}\beta_1\int_0^l [\rho A(\dot{y}_{tr} + w_t + \dot{l}w_x)^2 + Pw_x^2]\mathrm{d}x, \quad (4.138)$$

$$V_2(t) = \frac{1}{2}m_t\dot{y}_{tr}^2 + \frac{1}{2}\beta_2(y_{tr} - y_d)^2, \quad (4.139)$$

$$V_3(t) = \frac{1}{2}\beta_1 m_p\big(\dot{y}_{tr} + w_t(l,t) + \dot{l}w_x(l,t)\big)^2. \quad (4.140)$$

where y_d is the desired position of the payload and β_1 and β_2 are positive control parameters. The time rate of change of $V(t)$ will be calculated. Note that the domain of integration for spatial variable x is a time-varying function. Therefore, the Leibniz rule can be applied. Accordingly, the time rates of change of the three components of $V(t)$ are

$$\frac{dV_1(t)}{dt} = -\beta_1\frac{\dddot{l}}{2}\int_0^l (m_f + \rho A(l-x))w_x^2 dx + \beta_1 P w_x(\dot{y}_{tr} + w_t + \dot{l}w_x)\big|_{x=0}^{x=l}, \tag{4.141}$$

$$\frac{dV_2(t)}{dt} = \dot{y}_{tr}\left[f + Pw_x - \rho A\dot{l}(\dot{y}_{tr} + \dot{l}w_x)\right]\big|_{x=0} + \beta_2\dot{y}_{tr}(y_{tr} - y_d), \tag{4.142}$$

$$\frac{dV_3(t)}{dt} = -\beta_1\left[Pw_x(\dot{y}_{tr} + w_t + \dot{l}w_x)\right]\big|_{x=l}. \tag{4.143}$$

Combining (4.141)-(4.143) yields

$$\begin{aligned}\frac{dV(t)}{dt} &= -\beta_1\frac{\dddot{l}}{2}\int_0^l [m_f + \rho A(l-x)]w_x^2 dx - \beta_1 P\dot{l}w_x^2(0,t)\\ &\quad + \dot{y}_{tr}\left[f - \rho A\dot{l}(\dot{y}_{tr} + \dot{l}w_x) - (\beta_1 - 1)Pw_x + \beta_2(y_{tr} - y_d)\right]\big|_{x=0}.\end{aligned} \tag{4.144}$$

The first term of (4.144) consists of the negative jerk $\dddot{l}$, which can lead to instability. To overcome this problem, we restrict the hoisting cable length $l(t)$ as a second-order polynomial in time, and the remaining destabilizing effects are attenuated by making the second term large enough. With this assumption, the first term vanishes. A feedback control law is proposed as follows:

$$\begin{aligned}f(t) &= (\beta_1 - 1)P(0,t)w_x(0,t) + \rho A\dot{l}(\dot{y}_{tr} + \dot{l}w_x(0,t)) - \beta_2(y_{tr} - y_d) - \beta_3\dot{y}_{tr}\\ &\quad + \frac{2\beta_1}{2\dot{y}_{tr} + \operatorname{sgn}(\dot{l}\dot{y}_{tr})|\dot{y}_{tr}|}\dot{l}P(0,t)w_x(0,t)^2,\end{aligned} \tag{4.145}$$

where β_3 is a positive control parameter. Substituting (4.145) into (4.144) yields

$$\frac{dV(t)}{dt} = -\beta_3\dot{y}_{tr}^2 - \beta_1\left[\frac{|\dot{y}_{tr}\dot{l}|}{2|\dot{y}_{tr}| + \operatorname{sgn}(\dot{l}\dot{y}_{tr})\dot{y}_{tr}}\right]P(0,t)w_x^2(0,t) \le 0. \tag{4.146}$$

According to these results, we conclude that the system described by (4.133)-(4.135) under the control law in (4.145) is uniformly stable.

4.2.4 Adaptive Control

In many practical axially moving strings, the exact determination of system parameters and external disturbances can cause difficulties, similar to the control design problem in the presence of uncertainties. Uncertainties can negatively influence the control performance and cause instability. A widely discussed technique to address this problem is adaptive control. The concept of adaptive control is to estimate the unknown parameters online using the system's signals, followed by the design of a control law based on the estimated parameters. Adaptive control of AMSs has attracted the attention of many investigators for a long time. De Queiroz et al. (1999) pioneered implementing an adaptive technique for AMSs with unknown parameters. They proposed an adaptive control approach for vibration suppression of an axially moving string with a mechanical guide located within the string span. Fung et al. (2002) developed adaptive computed torque control for an axially moving string with a mass-spring-damper system at the right boundary, whereas Yang et al. (2004) designed robust adaptive control for a moving string in the presence of parametric uncertainties and disturbance based on the Lyapunov method. Ngo et al. (2009) further investigated robust adaptive control of a moving string with varying length.

In this section, we discuss the design of adaptive control and stability analysis of axially moving strings with unknown parameters and in the presence of disturbance via an illustration. Consider a moving string system with a mass-damper as shown in Fig. 4.10, wherein the mass density ρ, roller's mass m_{a}, and viscous damping coefficient c_{h} are unknown. Additionally, the string is subjected to an unknown disturbance $d(t)$ at the right boundary. We assume that the disturbance is bounded by the unknown positive constant d_{b}, i.e., $|d(t)| \leq d_{\mathrm{b}}$. We further assume that the lateral vibration is small. The equations of motion of the system are shown as follows:

$$\rho A\left(w_{tt} + 2v_0 w_{xt} + v_0^2 w_{xx}\right) - T_0 w_{xx} = 0, \tag{4.147}$$

$$m_{\mathrm{a}} w_{tt}(l,t) + (c_{\mathrm{h}} - \rho A v_0) w_t(l,t) + (T_0 - \rho A v_0^2) w_x(l,t) + d(t) = f(t), \tag{4.148}$$

and the boundary condition $w(0,t) = 0$. Before designing the adaptive control law for the foregoing system, we discuss a preliminary stability analysis of the closed-loop axially moving system.

Property 4.1 (De Queiroz et al. 2000) *Consider an axially moving string with the kinetic energy and potential energy given as* $K = 1/2 \int_0^l \rho A(w_t + v_0 w_x)^2 \mathrm{d}x$ *and* $U = 1/2 \int_0^l T_0 w_x^2 \mathrm{d}x$, *respectively. If K is bounded, then* w_{xt} *is also bounded. Furthermore, if U is bounded, then* w_x *and* w_{xx} *are bounded.*

Our objective is to design a control law to guarantee the robust stability of the system given by (4.147)–(4.148), where the system parameters, including ρ, m_{a}, and c_{h}, and the bound of the disturbance are unknown. To address this problem, we design robust adaptive boundary control with adaptation laws based on the Lyapunov redesign method. We denote the unknown parameter vector and its estimate vector, respectively, by

$$\boldsymbol{\theta} = \left[\, \rho A \; m_{\mathrm{a}} \; c_{\mathrm{h}} \,\right]^{\mathrm{T}},$$

$$\hat{\boldsymbol{\theta}} = \left[\hat{\rho}(t)A\ \hat{m}_{\mathrm{a}}(t)\ \hat{c}_{\mathrm{h}}(t)\right]^{\mathrm{T}}. \tag{4.149}$$

Let $\tilde{\boldsymbol{\theta}} = \hat{\boldsymbol{\theta}} - \boldsymbol{\theta}$ be the estimated parameter error. The error of the estimated bound value of the disturbance is denoted by $\tilde{d}_{\mathrm{b}} = \hat{d}_{\mathrm{b}} - d_{\mathrm{b}}$, where $\hat{d}_{\mathrm{b}}$ is the estimated value of d_{b}. The following control law is proposed.

$$f = -\mathbf{w}^{\mathrm{T}}\hat{\boldsymbol{\theta}} + T_0 w_x(l,t) - \mathrm{sgn}\left(\frac{\mathrm{D}w(l,t)}{\mathrm{D}t}\right)\hat{d}(t) - \alpha_1\frac{\mathrm{D}w(l,t)}{\mathrm{D}t} + \alpha_2 w_t(l,t) - \alpha_3 w_x(l,t) \tag{4.150}$$

where $\mathbf{w} = \left[v_0 \mathrm{D}w(l,t)/\mathrm{D}t\ \ v_0 w_{xt}(l,t)\ -w_t(l,t)\right]^{\mathrm{T}}$ and α_1 is a positive constant. The estimated values $\hat{\boldsymbol{\theta}}$ and $\hat{d}(t)$ are updated by the following adaptation laws:

$$\begin{aligned} \dot{\hat{\boldsymbol{\theta}}} &= -\xi_1\hat{\boldsymbol{\theta}} + \frac{\mathrm{D}w(l,t)}{\mathrm{D}t}\mathbf{B}^{\mathrm{T}}\mathbf{w}, \\ \dot{\hat{d}}_{\mathrm{b}} &= -\xi_2\hat{d}_{\mathrm{b}} + \frac{1}{\beta_4}\left|\frac{\mathrm{D}w(l,t)}{\mathrm{D}t}\right| \end{aligned} \tag{4.151}$$

where $\mathbf{B} = \mathrm{diag}\{\beta_1\ \beta_2\ \beta_3\}$ and β_1, β_2, β_3, ξ_1, and ξ_2 are positive constants. The terms $-\xi_1\hat{\boldsymbol{\theta}}$ and $-\xi_2\hat{d}_{\mathrm{b}}$ are added to ensure that $\hat{\boldsymbol{\theta}}$ and $\hat{d}_{\mathrm{b}}$ do not become unbounded. Under the proposed control law, the dynamic model of the actuator given by (4.148) becomes

$$\begin{aligned} m_{\mathrm{a}}\frac{\partial}{\partial t}\frac{\mathrm{D}w(l,t)}{\mathrm{D}t} &= -\mathbf{w}^{\mathrm{T}}\tilde{\boldsymbol{\theta}} - \mathrm{sgn}\left(\frac{\mathrm{D}w(l,t)}{\mathrm{D}t}\right)\hat{d}(t) - d(t) \\ &\quad + \alpha_1\frac{\mathrm{D}w(l,t)}{\mathrm{D}t} + \alpha_2 w_t(l,t) - \alpha_3 w_x(l,t). \end{aligned} \tag{4.152}$$

We now analyze the stability of the system given by (4.147)-(4.148) under the proposed adaptive control law. Consider the following energy-based function.

$$V(t) = V_1(t) + V_2(t) + V_3(t) + V_4(t) + V_5(t) \tag{4.153}$$

where

$$V_1(t) = \frac{1}{2}\zeta_1\int_0^l \left[\rho A(w_t + v_0 w_x)^2 + T_0 w_x^2\right]\mathrm{d}x,$$

$$V_2(t) = \frac{1}{2}m_{\mathrm{a}}\frac{Dw(l,t)^2}{Dt},$$

$$V_3(t) = 2\zeta_2 \rho A \int_0^l x w_x (w_t + v_0 w_x) \mathrm{d}x,$$
$$V_4(t) = \frac{1}{2} \widetilde{\boldsymbol{\theta}}^{\mathrm{T}} \mathbf{B} \widetilde{\boldsymbol{\theta}},$$
$$V_5(t) = \frac{1}{2} \beta_4 \tilde{d}_{\mathrm{b}}^2, \tag{4.154}$$

where ζ_1 and ζ_2 are positive control gains. According to Appendix C, the two following lemmas are established.

Lemma 4.6 *The function given by* (4.153) *is bounded as*

$$0 \le \lambda_1 (V_1(t) + V_2(t) + V_4(t) + V_5(t)) \le V(t) \le \lambda_2 (V_1(t) + V_2(t) + V_4(t) + V_5(t)) \tag{4.155}$$

where two positive constants are defined as

$$\lambda_1 = \min\{1, \zeta_1 - \zeta_2 \rho A l / \min\{\rho A, T_0\}\},$$
$$\lambda_2 = \max\{1, \zeta_1 + \zeta_2 \rho A l / \min\{\rho A, T_0\}\}. \tag{4.156}$$

Lemma 4.7 *If the control parameters satisfy*

$$2\zeta_3 \zeta_1 \frac{T_0}{v_0} - \zeta_1 T_0 - \alpha_3 - \zeta_3 \alpha_2 \ge 0,$$
$$2\alpha_3 v_0 - \zeta_2 T_0 l + \alpha_2 v_0^2 - \zeta_1 \zeta_3 T_0 - \zeta_3 \alpha_3 \ge 0,$$
$$\alpha_1 - \frac{1}{2}\alpha_2 - \frac{1}{2}\zeta_2 \rho A l - \zeta_1 \frac{T_0}{v_0} l \ge 0, \tag{4.157}$$

where ζ_1, ζ_2, and ζ_3 are positive constants; then, the time derivative of the function $V(t)$ given by (4.153) *is upper bounded as*

$$\dot{V}(t) \le -\lambda_3 (V_1(t) + V_2(t) + V_4(t) + V_5(t)) + \delta, \tag{4.158}$$

where

$$\lambda_3 = \min\left\{\frac{\zeta_2}{2\zeta_1}, \frac{1}{2}\left(\alpha_1 - \frac{1}{2}\alpha_2 - \frac{1}{2}\zeta_2 \rho A l - \zeta_1 \frac{T_0}{v_0} l\right), \xi_1, \xi_2\right\},$$
$$\delta = \frac{1}{2}\left[\beta_4 \xi_2 d_{\mathrm{b}}^2 + \xi_1 \boldsymbol{\theta}^{\mathrm{T}} \mathbf{B}^{-1} \boldsymbol{\theta}\right]. \tag{4.159}$$

Based on Lemma 4.6 *and* Lemma 4.7, *the following theorem for the stability of an axially moving string under adaptive control is introduced.*

Theorem 4.2 *Consider an axially moving string system described by* (4.147) *and* (4.148) *under the control law given by* (4.150) *and the adaptation law given by*

(4.151). *If the initial conditions are bounded and control gains are selected to satisfy (4.157), then the control law can guarantee the robust stability of the closed-loop system in the sense that the lateral vibration and the estimating error are uniformly bounded.*

Proof According to Lemma 4.6 and Lemma 4.7, we formulate the following inequality:

$$\dot{V}(t) \leq -\lambda V(t) + \delta, \tag{4.160}$$

where $\lambda = \lambda_3/\lambda_2$. Multiplying (4.160) by $e^{\lambda t}$ yields

$$\frac{\partial}{\partial t}\left(V(t)e^{\lambda t}\right) \leq \delta e^{\lambda t} \Leftrightarrow V(t) \leq V(0)e^{-\lambda t} + \frac{\delta}{\lambda}. \tag{4.161}$$

Inequality (4.161) shows that if the boundedness of the initial conditions is ensured, i.e., $V(0)$ is bounded, then $V(t)$ converges to the circle of radius δ/λ as the time approaches infinity. This finding implies the boundedness of $V(t)$ and the boundedness of the kinetic energy and potential energy: $1/2\rho A \int_0^l (w_t + v_0 w_x)^2 dx$ and $1/2 \int_0^l T_0 w_x^2 dx$, respectively. Based on Property 4.1, w_x, w_t, w_{xt}, and w_{xx} are bounded. We use (4.154), (4.161), and Lemma 4.3 to formulate the inequality as follows:

$$\frac{T_0}{2l}w^2 \leq \frac{T_0}{2}\int_0^l w_x^2 dx \leq V(0)e^{-\lambda t} + \frac{\delta}{\lambda}. \tag{4.162}$$

Accordingly, we obtain

$$|w(x,t)| \leq \sqrt{\frac{2l}{T_0}V(0)e^{-\lambda t} + \frac{2l}{T_0}\frac{\delta}{\lambda}}, \forall (x,t) \in [0,l] \times [0,\infty). \tag{4.163}$$

Therefore, the lateral vibration $w(x,t)$ of the string is uniformly bounded. Using these results, we can conclude that the estimated values $\hat{\boldsymbol{\theta}}$ and $\hat{d}_b$ are bounded. Furthermore, the boundedness of the boundary control law given by (4.150) is assured.

Due to the presence of parametric uncertainties and disturbances, the state of the string $w(x,t)$ cannot converge to zero and can only tend to the inside of the domain $D := \{w(x,t) \in \mathbb{R} \| w(x,t)| \leq 2l\delta/T_0\lambda\}$. However, we can minimize the size of domain D by regulating the control gains ξ_1 and ξ_2. If the system parameters are known and the disturbance is disregarded, we can design a control law to guarantee that the lateral vibration is exponentially stable.

4.3 Hybrid Control Methods

The concept of hybrid control methods is the synthesis of different control methods, in which the advantages of each control method are exploited to enhance the control performance of the system. With regard to the vibration control of axially moving strings, some hybrid control approaches have been proposed.

To enhance the control performance of VSC, Fung et al. (1998) presented two hybrid control methods based on proportional and integral compensations: Proportional variable structure control (PVSC) and integral variable structure control (IVSC). According to PVSC, the controller is the synthesis of VSC and a proportional gain, whereas IVSC uses an integral compensator to address time-history errors. By combining these two controllers with the conventional VSC, the authors obtained two new controllers that are referred to as modified PVSC and modified IVSC. The concept associated with using the modified IVSC and PVSC is to switch the controllers at a suitable time to exploit their advantages to enhance the control performance. In modified PVSC, the rising time is reduced by PVSC, whereas the stability of the steady-state responses is assured by VSC. In modified IVSC, VSC and IVSC have a role in minimizing the overshoot and improving the transient response, respectively.

Dai et al. (2014) implemented a hybrid control method based on the combination of sliding mode control and fuzzy logic theory for an axially moving string with varying length. In this work, a control scheme based on the ODE model is proposed to suppress the large amplitude vibrations of the string. Concerning hybrid control based on intelligent control techniques, Chao and Lai (2003) and Huang et al. (2003) combined SMC with fuzzy control and neural networks. Fuzzy control is a powerful tool that uses logical rules and fuzzy set theory to develop a controller when the system information is uncertain, imprecise, or ambiguous. In the neural network method, artificial neural networks inspired by the human neural system have been established. These artificial neural networks are mathematical models with learning capacities that can be employed to determine the input–output relationship of the system. In Huang et al. (2003), a sliding mode control law was developed to control the lateral vibrations of an axially moving string. They also utilized two intelligent control approaches (i.e., fuzzy sliding mode control and fuzzy neural network) to address the control input's discontinuity and non-analyticity when the vibration is small. In these fuzzy logic-based approaches, the switching function and its derivative were used as inputs, while a tension variation was utilized as the output. Chao and Lai (2003) also designed a boundary controller based on these approaches for vibration control of a translating string with a mass-damper-spring system at the boundary.

4.4 Simulation Examples

In this section, various controls of axially moving strings/beams based on both ODE models and PDE models are illustrated via numerical simulations (i.e., Examples 4.1 and 4.2 demonstrate controls of AMSs based on ODE models, whereas Examples 4.3–4.6 illustrate control results based on PDE models. The simulation is performed by using MATLAB. To determine the approximate solutions for the PDEs, we employed finite difference methods. For details, the forward/backward differences are utilized for the elements at the boundaries, whereas the central finite difference is applied to other elements of the string. The small segments that correspond to spatial and temporal variables are set to $\Delta x = 5 \times 10^{-2}$ and $\Delta t = 5 \times 10^{-5}$.

Example 4.1 This example considers the control of an axially moving string based on an ODE model. We use a model-based feedback control law to suppress the vibration of a moving string with two sensors and two actuators, as shown in Fig. 4.1. The string length is 1 m. The locations of actuators are $x_{a1} = 1/4$ and $x_{a2} = 3/8$, whereas the locations of two sensors are $x_{s1} = 5/8$ and $x_{s2} = 3/4$. According to the 4th-order Galerkin truncation, the ODE model of the string is given by (4.1), where

$$\mathbf{M} = \begin{bmatrix} 0.5 & 0 & 0 & 0 \\ 0 & 0.5 & 0 & 0 \\ 0 & 0 & 0.5 & 0 \\ 0 & 0 & 0 & 0.5 \end{bmatrix},$$

$$\mathbf{C} = \begin{bmatrix} 0 & -1.33 & 0 & -0.53 \\ 1.33 & 0 & -2.4 & 0 \\ 0 & 2.4 & 0 & -3.42 \\ 0.53 & 0 & 3.42 & 0 \end{bmatrix},$$

$$\mathbf{K} = \begin{bmatrix} 9.38 & 0 & 0 & 0 \\ 0 & 37.5 & 0 & 0 \\ 0 & 0 & 83.39 & 0 \\ 0 & 0 & 0 & 150.02 \end{bmatrix}, \tag{4.164}$$

and the initial condition is $w(x, 0) = 0.01\sin(\pi x/l)$. The spillover problem was investigated in Sect. 4.1.1. In this section, we consider the effectiveness of the model-based feedback control law $\mathbf{f} = -\mathbf{T}_a\mathbf{G}\mathbf{T}_s\mathbf{x}$, where $\mathbf{x}$, $\mathbf{T}_a$, $\mathbf{T}_s$, and $\mathbf{G}$ are given by (4.2), (4.6), (4.7), and (4.12), respectively, in the case wherein the sensors are not affected by the residual modes (i.e., no observation spillover). The MATLAB simulation scheme is shown in Fig. 4.12, whereas the results are revealed in Fig. 4.13.

Example 4.2 This example illustrates the problem discussed in Sect. 4.1.3: Vibration control of an axially moving beam based on regulating the axial velocity. We consider a beam system with the following parameters: $l = 4$ m, $\rho = 2700$ kg/m^3, $A = 2.5 \times 10^{-4}$ m^2, $T_0 = 300$ N, $I = 5.2 \times 10^{-10}$ m^4, and $c_e = 0.3$ Nm^{2}s. The initial condition is $w(x, 0) = 0.04\sin(\pi x/l)$. At the initial time $t = 0$ s, the beam axially moves with

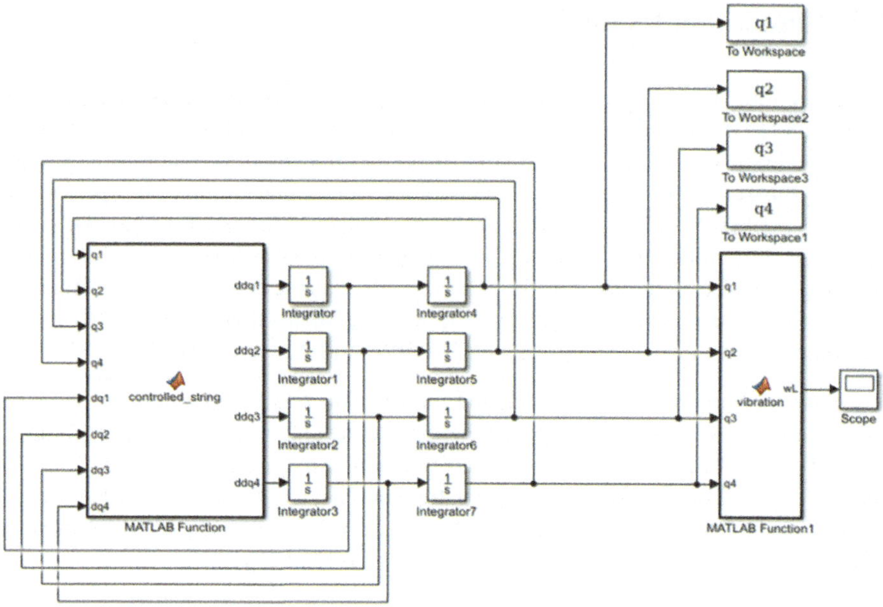

Fig. 4.12 MATLAB simulation scheme of the model-based feedback control of a moving string

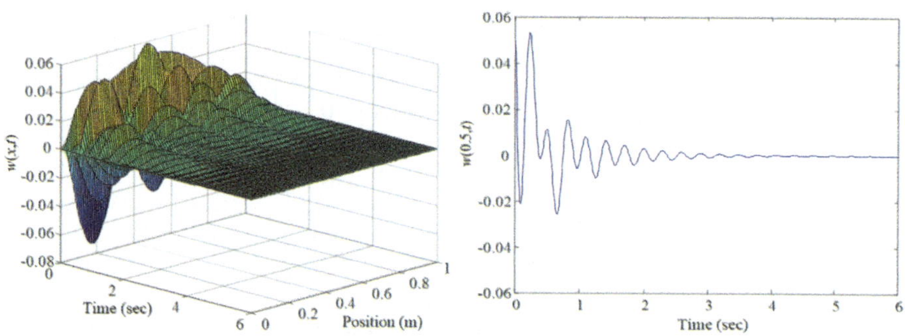

Fig. 4.13 Lateral vibrations of the string **a** $w(x, t)$, **b** $w(0.5, t)$

the velocity $v_0 = 5\text{m/s}$, and then the axial velocity reaches the zero value at time $t_f = 5$ s. The conventional velocity profile is constructed from a slope, wherein the deceleration is 1 (i.e., Fig. 4.14). Figure 4.14 also shows the optimal velocity profile generated by using the steepest descent algorithm. The simulation results in Fig. 4.15 reveal that the vibration energy of the beam quickly converges to zero. Additionally, the lateral vibration using the optimal velocity profile is significantly suppressed at time $t_f = 5$ s, as shown in Fig. 4.16.

Fig. 4.14 Axial velocity profile during the deceleration period

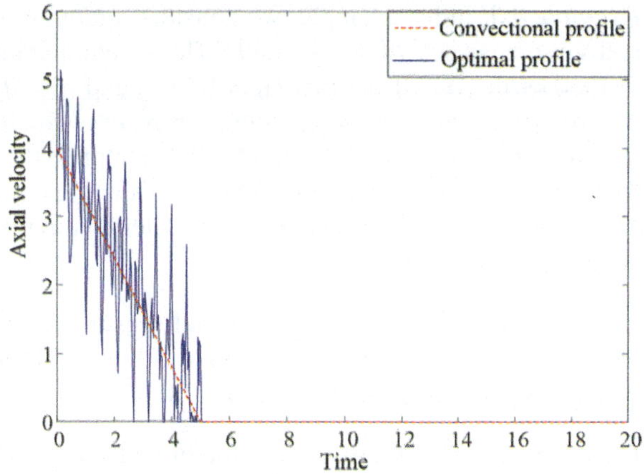

Fig. 4.15 Vibration energy of the beam during the deceleration period

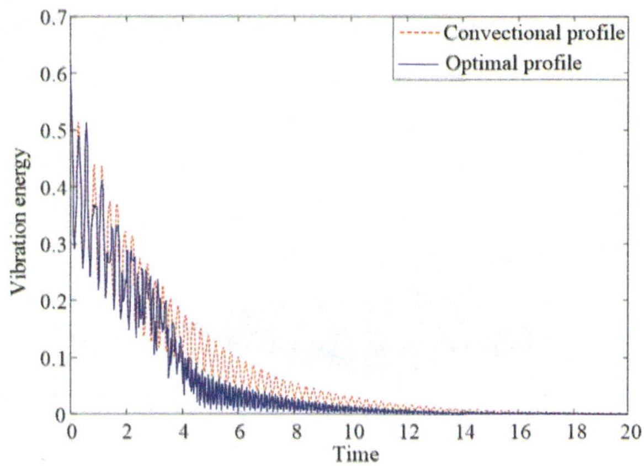

Fig. 4.16 Lateral vibration of the beam in $x = 0.5\ l$

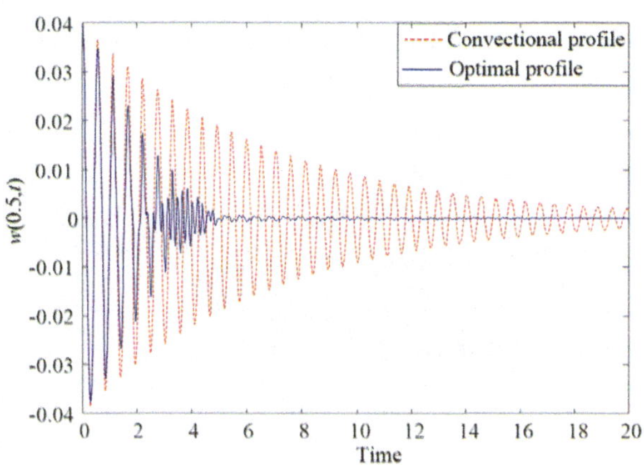

Example 4.3 Inthis example, we consider a moving string with a pair of actuators and sensors, as described by (4.59). The system parameters are given as $l = 1$ m, $\rho = 1530$ kg/m^3, $A = 0.001$ m^2, $T_0 = 1.53$ N, and $v_0 = 0.5$ m/s. The initial condition is $w(x, 0) = 0.05\sin(\pi x/l)$. We use the control laws, which are designed by analyzing the transfer function (Sects. 4.2.1 and 4.2.2) to suppress the string's vibration. The control force in the frequency domain is given by $F(x, s) = -\delta(x - x_a)G(s)W(x_s, s)$, where $W(x_s, s)$ is the Laplace transform of the lateral vibration at the sensor position. Three cases are considered:

Case 1 Colocation of the actuator and sensor (i.e., $x_a = x_s = 0.7$ m): A negative feedback control with $G(s) = -\alpha s$ is applied. The lateral vibration of the string under this control law is shown in Fig. 4.17.

Case 2 Non-collocation of the actuator and sensor (i.e., $x_a = 0.6$ and $x_s = 0.4$ m): We implemented a positive feedback control law with time delay, wherein $G(s) = \alpha s e^{s(1-v_0(x_a-x_s))(1-v_0^2)^{-1}}$. Figure 4.18 reveals the effectiveness of the proposed control law.

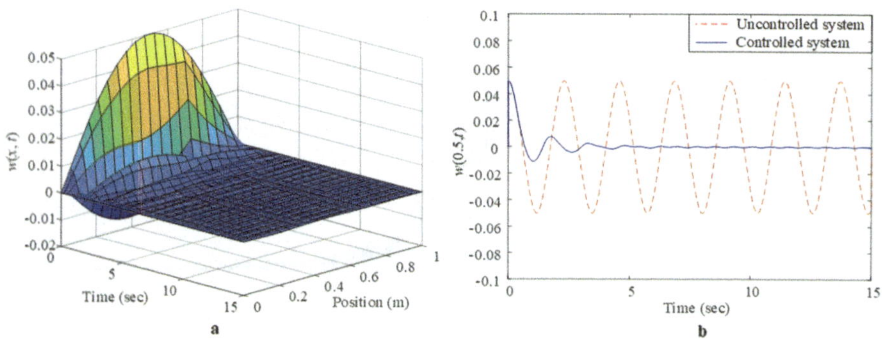

Fig. 4.17 Lateral vibration of the string in case 1: **a** $w(x, t)$, **b** $w(0.5, t)$

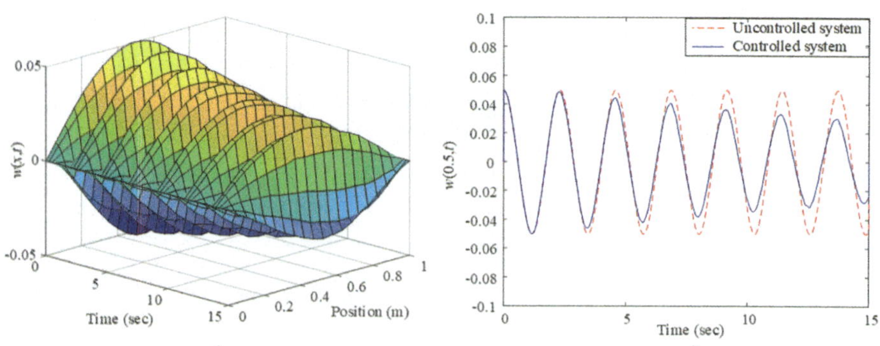

Fig. 4.18 Lateral vibration of the string in case 2: **a** $w(x, t)$, **b** $w(0.5, t)$

Case 3 Non-collocation of the actuator and sensor (i.e., $x_a = 1$ and $x_s = 0.95$ m): We use a control law that is designed based on the wave cancelation method. The boundary control force in the frequency domain is $F(s) = -(\alpha/(1+v_0))se^{sx_s/(1+v_0)}e^{-s/(1+v_0)}W(x_s, s)$. The responses of the controlled string using this boundary control force are shown in Fig. 4.19.

Example 4.4 We verify the effectiveness of the right boundary control law that is designed based on the Lyapunov function. Consider the string described by (4.95) under the control law (4.101). The system parameters are set to $l = 1$ m, $\rho = 1530$ kg/m^3, $A = 0.001$ m^2, $T_0 = 30$ N, and $v_0 = 1$ m/s. The initial condition is $w(x, 0) = 0.05\sin(\pi x/l)$. The exponential stability of the vibration solution of the controlled system is illustrated in Fig. 4.20.

Example 4.5 This example demonstrates the vibration control of an overhead crane hoisting system, wherein the hoisting rope is regarded as an axially moving string with varying length (i.e., Fig. 4.11). The dynamic model of the system is given by

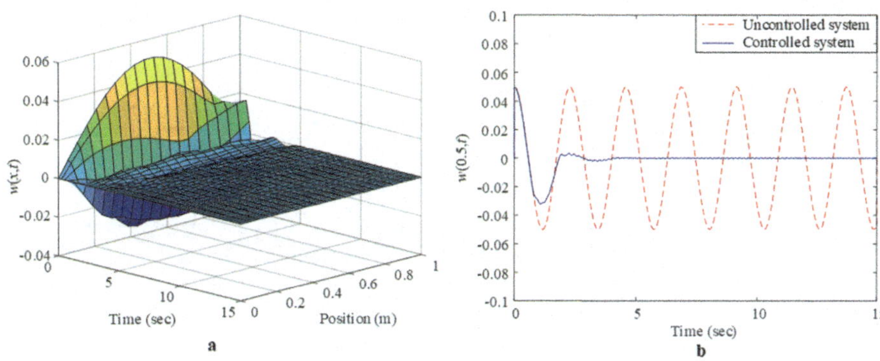

Fig. 4.19 Lateral vibration of the string in case 3: **a** $w(x, t)$, **b** $w(0.5, t)$

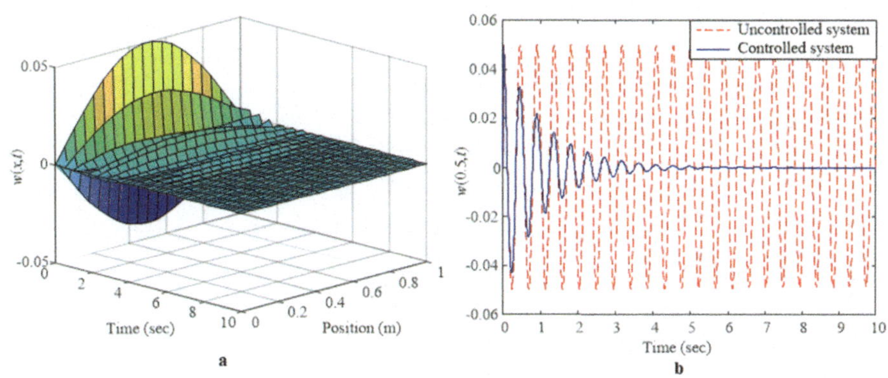

Fig. 4.20 Lateral vibration of the string using the boundary control law based on the Lyapunov method: **a** $w(x, t)$, **b** $w(0.5, t)$

(4.133)–(4.135), where $\rho = 90$ kg/m, $m_t = 10^4$ kg, and $m_p = 10^3$ kg. The trolley carries the payload from 0 mto the desired position at $w_d = 8$ m, whereas the hoisting rope extends from the initial length of 2 mto the final length of 10 m. The MATLAB simulation scheme is shown in Fig. 4.21. Figure 4.22 reveals that the boundary control law given by (4.145) can transport the payload to the desired position and effectively suppress the lateral vibration of the hoisting rope and payload.

Example 4.6 In this example, we consider the adaptive boundary control of an axially moving string shown in Fig. 4.10 in the presence of unknown parameters and disturbance. The mathematical model is given by (4.147) and (4.148), where the system parameters are set to $l = 1$ m, $\rho = 1{,}530$ kg/m^3, $A = 0.001$ m^2, $T_0 = 1.53$ N, and $v_0 = 0.7$ m/s, $c_e = 1.5$ Nm^{2}s, and $m_a = 2$ kg. The initial condition is $w(x, 0) = 0.05\sin(\pi x/l)$. The right boundary of the system is subjected to a bounded disturbance given by $d = 50\sin(45\pi t)$ N. The response of the closed-loop system

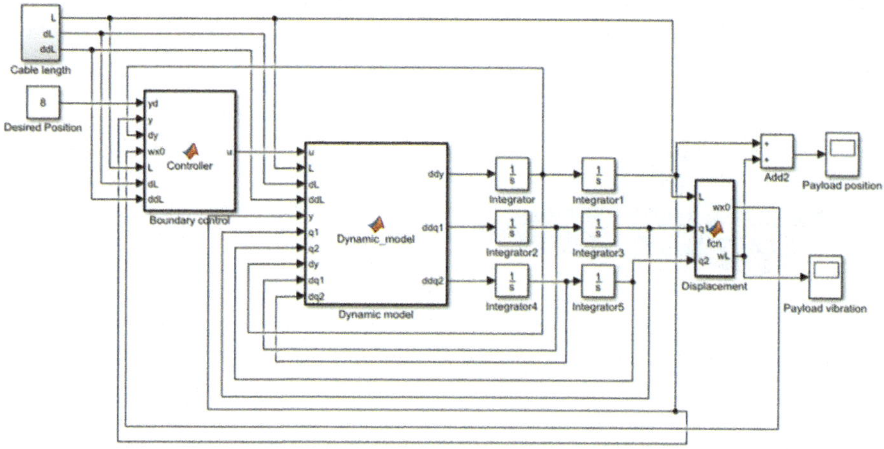

Fig. 4.21 MATLAB simulation scheme of the boundary control of an overhead crane

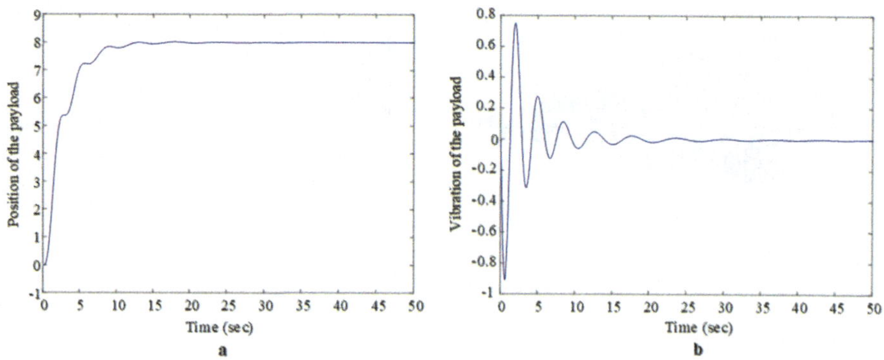

Fig. 4.22 Response of the payload: **a** position of the payload, **b** lateral vibration of the payload

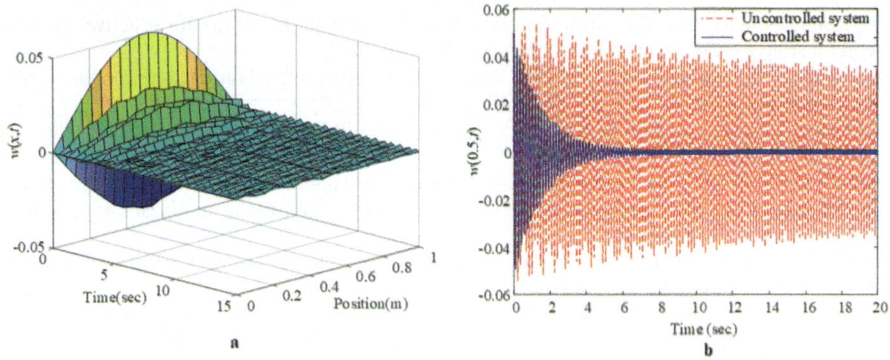

Fig. 4.23 Response of the string under adaptive control: **a** $w(x, t)$, **b** $w(0.5, t)$

Fig. 4.24 Response of the controlled string without disturbances

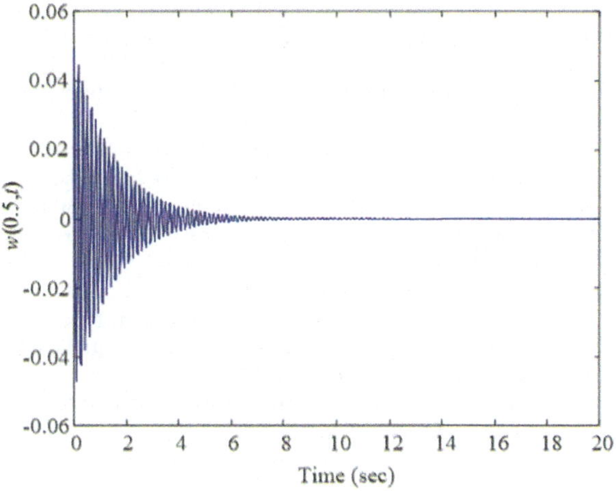

under the adaptive boundary control law in (4.150) and (4.151) is shown in Fig. 4.23. Furthermore, in the case without disturbance, Fig. 4.24 reveals that the vibration solution is asymptotically stable.

References

Balas MJ (1978) Feedfback-control of flexible systems. IEEE T Automat Contr 23(4):673–679

Butkovskiy AG (1983) Structural theory of distributed systems. Prentice Hall, New Jersey

Chao PCP, Lai CL (2003) Boundary control of an axially moving string via fuzzy sliding-mode control and fuzzy neural network methods. J Sound Vib 262(4):795–813

Chung CH, Tan CA (1995) Active vibration control of the axially moving string by wave cancellation. J Vib Acoust 117(1):49–55

Dai L, Sun L, Chen C (2014) Control of an extending nonlinear elastic cable with an active vibration control strategy. Commun Nonlinear Sci 19(10):3901–3912
De Queiroz MS, Dawson DM, Rahn CD, Zhang F (1999) Adaptive vibration control of an axially moving string. J Vib Acoust 121(1):41–49
De Queiroz MS, Dawson DM, Nagarkatti SP, Zhang F (2000) Lyapunov based control mechanical systems. Birkhauser, Boston, USA
Edwards C, Sarah S (1998) Sliding mode control: theory and applications. CRC Press, Boca Raton
Fung RF, Liao CC (1995) Application of variable-structure control in the nonlinear string system. Int J Mech Sci 37(9):985–993
Fung RF, Lin JH, Yao CM (1997) Vibration analysis and suppression control of an elevator string actuated by a PM synchronous servo motor. J Sound Vib 206(3):399–423
Fung RF, Huang JS, Wang YC, Yang RT (1998) Vibration reduction of the nonlinearly traveling string by a modified variable structure control with proportional and integral compensations. Int J Mech Sci 40(6):493–506
Fung RF, Wu JW, Lu PY (2002) Adaptive boundary control of an axially moving string system. J Vib Acoust 124(3):435–440
Hardy GH, Littlewood JE, Polya G (1959) Inequalities. Cambridge University Press, Cambridge
Henry D (2006) Geometric theory of semilinear parabolic equations. Springer, Heidelberg
Hong K-S, Kim CW, Hong KT (2004) Boundary control of an axially moving belt system in a thin-metal production line. Int J Control Autom 2(1):55–67
Hung JY, Gao WB, Hung JC (1993) Variable structure control—a survey. IEEE T Ind Electron 40(1):2–22
Huang JS, Chao PCP, Fung RF, Lai CL (2003) Parametric control of an axially moving string via fuzzy sliding-mode and fuzzy neural network methods. J Sound Vib 264(1):177–201
Khalil HK (2002) Nonlinear systems. Prentice Hall, New Jersey
Kim CS, Hong K-S (2009) Boundary control of container cranes from the perspective of controlling an axially moving string system. Int J Control Autom 7(3):437–445
Kirk DE (2004) Optimal control theory: an introduction. Courier Corporation, New York
Komura Y (1967) Nonlinear semi-groups in Hilbert space. J Math Soc Jpn 19(4):493–507
Lasdon L, Mitter S, Waren A (1967) The conjugate gradient method for optimal control problems. IEEE Trans Autom Control 12(2):132–138
Ngo QH, Hong K-S, Jung IH (2009) Adaptive control of an axially moving system. J Mech Sci Technol 23(11):3071–3078
Nguyen QC, Hong K-S (2011) Stabilization of an axially moving web via regulation of axial velocity. J Sound Vib 330(20):4676–4688
Nguyen QC, Hong K-S (2012) Simultaneous control of longitudinal and transverse vibrations of an axially moving string with velocity tracking. J Sound Vib 331(13):3006–3019
Nguyen QC, Piao M, Hong K-S (2018) Multivariable adaptive control of the rewinding process of a roll-to-roll system governed by hyperbolic partial differential equations. Int J Control Autom 16(5):2177–2186
Tan CA, Ying S (2000) Active wave control of the axially moving string: theory and experiment. J Sound Vib 236(5):861–880
Ulsoy AG (1984) Vibration control in rotating or translating elastic-systems. J Dyn Syst-T Asme 106(1):6–14
Yang K-J, Hong K-S, Matsuno F (2004) Robust adaptive boundary control of an axially moving string under a spatiotemporally varying tension. J Sound Vib 273(4–5):1007–1029
Yang B, Mote CD (1991) Frequency-domain vibration control of distributed gyroscopic systems. J Dyn Syst-T Asme 113(1):18–25
Ying S, Tan CA (1996) Active vibration control of the axially moving string using space feedforward and feedback controllers. J Vib Acoust 118(3):306–312
Zhang W, Chen LQ (2006) Vibration control of an axially moving string system: wave cancellation method. Appl Math Comput 175(1):851–863

Chapter 5
Belt Model

Abstract This chapter focuses on the coupled lateral–longitudinal dynamics of axially moving belts. The mathematical models of axially moving belts under various assumptions are derived, and the dynamic behaviors of the moving belt are analyzed.

Keywords Axially moving belts · Coupled dynamics · Longitudinal vibration · Transverse vibration · Nonhomogenous boundaries · Approximate model · Vibration analysis · Bifurcations

5.1 Introduction

In modeling AMSs, such as beams or strings, the longitudinal displacements are either disregarded or simplified based on the quasi-static stretch assumption (Wickert 1992). In the axially moving belt model, the coupling between the longitudinal vibration and lateral vibration, which becomes significant with an increase in the slenderness ratio of the material (i.e., ratio of length to cross-sectional area), is investigated. In an early study concerning the translating belt, Thurman and Mote (1969) established a comprehensive dynamic model of a translating belt using a closed-form normal strain. Later, simpler models based on approximated strains were presented in the literature to facilitate dynamics analysis of the moving belt (Riedel and Tan 2002; Ghayesh 2012; Ding and Chen 2010). In addition to the Euler–Bernoulli theory, the Timoshenko theory was also utilized to consider the shear deformation and rotary inertia of axially moving belts (Ghayesh and Amabili 2013a, b). Furthermore, Farokhi et al. (2016) and Ghayesh and Farokhi (2016) investigated the dynamical behaviors of the lateral, longitudinal, and transverse vibrations of a moving beam.

Since the belt model considers both longitudinal and lateral vibrations, it is suitable for modeling axially translating components with considerable length, in which the distance between two support points is considerable. Belts of power transmission systems and strips in the zinc galvanization lines are typical examples. Besides that, the axially moving belt model can also be utilized to describe systems in which longitudinal oscillation can affect quality and productivity seriously (e.g., a roll-to-roll printing process producing flexible circuits).

© The Author(s), under exclusive license to Springer Nature Singapore Pte Ltd. 2022
K.-S. Hong et al., *Control of Axially Moving Systems*,
https://doi.org/10.1007/978-981-16-2915-0_5

5.2 Dynamic Models of Axially Moving Belts

5.2.1 Belt Model Using a Closed-Form Strain

A pioneering study on axially moving belts was published in 1969 by Thurman and Mote. Their work established a mathematical model that describes the nonlinear dynamics of a uniform translating strip based on Euler–Bernoulli beam theory. In this chapter, we derive the dynamic models of axially moving belts under various axial strains.

Consider an axially moving belt demonstrated in Fig. 5.1, the displacement and velocity vectors of a unit element of the belt at x are given by

$$\mathbf{r}_\mathrm{d}(x,t) = (x+u)\mathbf{i} + w\mathbf{j}, \tag{5.1}$$

$$\mathbf{r}_\mathrm{v}(x,t) = (v_0 + u_t + v_0 u_x)\mathbf{i} + (w_t + v_0 w_x)\mathbf{j}, \tag{5.2}$$

where w and u denote the lateral and longitudinal displacements, respectively. The kinetic energy consists of the energies associated with the motions of lateral and longitudinal vibrations of the belt which is given as follows:

$$K = \frac{1}{2}\rho A\int_0^l \mathbf{r}_\mathrm{v}\cdot\mathbf{r}_\mathrm{v}\mathrm{d}x = \frac{1}{2}\rho A\int_0^l \left((w_t + v_0 w_x)^2 + (u_t + v_0(1+u_x))^2\right)\mathrm{d}x. \tag{5.3}$$

The potential energy of the axial tension and bending is given by

$$U = \int_0^l \left(T_0 + \frac{1}{2}EA\varepsilon(x,t)\right)\varepsilon(x,t)\mathrm{d}x + \frac{1}{2}\int_0^l EIw_{xx}^2\mathrm{d}x. \tag{5.4}$$

With consideration of the longitudinal displacement, the normal strain [as presented in (2.6)] is expressed as

$$\varepsilon(x,t) = \sqrt{w_x^2 + (1+u_x)^2} - 1. \tag{5.5}$$

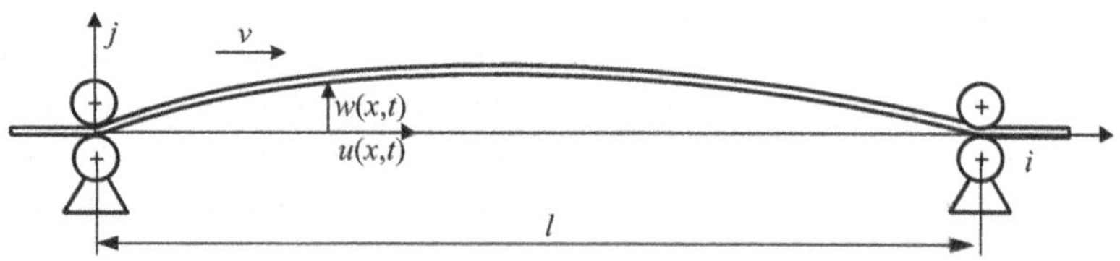

Fig. 5.1 Axially moving belt

The potential energy in (5.4) can be rewritten as

$$U = \int_0^l \left(T_0\left(\sqrt{w_x^2 + (1+u_x)^2} - 1\right) + \frac{1}{2}EA\left(\sqrt{w_x^2 + (1+u_x)^2} - 1\right)^2 \right) dx + \frac{1}{2}\int_0^l EIw_{xx}^2 dx. \tag{5.6}$$

Using the extended Hamilton principle results in

$$\begin{aligned} 0 &= \int_{t_1}^{t_2} \delta\left(\int_0^l \frac{1}{2}\left[\rho A(w_t + v_0 w_x)^2 + \rho A(u_t + v_0(1+u_x))^2 \right.\right. \\ &\quad - T_0\left(\sqrt{w_x^2 + (1+u_x)^2} - 1\right) \\ &\quad \left.\left. -EA\left(\sqrt{w_x^2 + (1+u_x)^2} - 1\right)^2 - EIw_{xx}\right)\right] dx \Bigg) dt \\ &= \int_{t_1}^{t_2}\int_0^l \delta L_{bt} dx dt - \frac{1}{2}EI \int_{t_1}^{t_2}\int_0^l \delta w_{xx}^2 dx dt \end{aligned} \tag{5.7}$$

where

$$\begin{aligned} L_{bt} &= \frac{1}{2}\rho A(w_t + v_0 w_x)^2 + \frac{1}{2}\rho A(u_t + v_0(1+u_x))^2 \\ &\quad - T_0\left(\sqrt{w_x^2 + (1+u_x)^2} - 1\right) \\ &\quad - \frac{1}{2}EA\left(\sqrt{w_x^2 + (1+u_x)^2} - 1\right)^2. \end{aligned} \tag{5.8}$$

By applying the procedure for integration by parts, (5.7) is expressed as follows:

$$\begin{aligned} 0 &= \int_{t_1}^{t_2}\int_0^l \left(\frac{\partial L_{bt}}{\partial w} - \frac{\partial}{\partial x}\frac{\partial L_{bt}}{\partial w_x} - \frac{\partial}{\partial t}\frac{\partial L_{bt}}{\partial w_t}\right)\delta w dx dt + \int_{t_1}^{t_2} \frac{\partial L_{bt}}{\partial w_x}\delta w\bigg|_{x=0}^{x=l} dt \\ &\quad + \int_0^l \frac{\partial L_{bt}}{\partial w_t}\delta w\bigg|_{t=t_1}^{t=t_2} dx \end{aligned}$$

$$
\begin{aligned}
&+ \int_{t_1}^{t_2}\int_0^l \left(\frac{\partial L_{\mathrm{bt}}}{\partial u} - \frac{\partial}{\partial x}\frac{\partial L_{\mathrm{bt}}}{\partial u_x} - \frac{\partial}{\partial t}\frac{\partial L_{\mathrm{bt}}}{\partial u_t}\right)\delta u \mathrm{d}x\mathrm{d}t + \int_{t_1}^{t_2} \frac{\partial L_{\mathrm{bt}}}{\partial u_x}\delta u\Bigg|_{x=0}^{x=l}\mathrm{d}t \\
&+ \int_0^l \frac{\partial L_{\mathrm{bt}}}{\partial u_t}\delta u\Bigg|_{t=t_1}^{t=t_2}\mathrm{d}x \\
&- \int_{t_1}^{t_2}\int_0^l EIw_{xxxx}\delta w \mathrm{d}x\mathrm{d}t - \int_{t_1}^{t_2} EIw_{xx}\delta w_x|_{x=0}^{x=l}\mathrm{d}t \\
&+ \int_{t_1}^{t_2} EIw_{xxx}\delta w|_{x=0}^{x=l}\mathrm{d}t \\
&= \int_{t_1}^{t_2}\int_0^l \left(\frac{\partial L_{\mathrm{bt}}}{\partial w} - \frac{\partial}{\partial x}\frac{\partial L_{\mathrm{bt}}}{\partial w_x} - \frac{\partial}{\partial t}\frac{\partial L_{\mathrm{bt}}}{\partial w_t} - EIw_{xxxx}\right)\delta w \mathrm{d}x\mathrm{d}t \\
&+ \int_{t_1}^{t_2}\int_0^l \left(\frac{\partial L_{\mathrm{bt}}}{\partial u} - \frac{\partial}{\partial x}\frac{\partial L_{\mathrm{bt}}}{\partial u_x} - \frac{\partial}{\partial t}\frac{\partial L_{\mathrm{bt}}}{\partial u_t}\right)\delta u \mathrm{d}x\mathrm{d}t \\
&+ \int_{t_1}^{t_2} \left(\frac{\partial L_{\mathrm{bt}}}{\partial w_x} + 2EIw_{xxx}\right)\delta w\Bigg|_{x=0}^{x=l}\mathrm{d}t + \int_{t_1}^{t_2} \frac{\partial L_{\mathrm{bt}}}{\partial u_x}\delta u\Bigg|_{x=0}^{x=l}\mathrm{d}t \\
&+ \int_0^l \frac{\partial L_{\mathrm{bt}}}{\partial w_t}\delta w\Bigg|_{t=t_1}^{t=t_2}\mathrm{d}x \\
&+ \int_0^l \frac{\partial L_{\mathrm{bt}}}{\partial u_t}\delta u\Bigg|_{t=t_1}^{t=t_2}\mathrm{d}x - \int_{t_1}^{t_2} EIw_{xx}\delta w_x|_{x=0}^{x=l}\mathrm{d}t.
\end{aligned}
\tag{5.9}
$$

The terms in (5.9) can be derived as follows:

$$
\frac{\partial L_{\mathrm{b}}}{\partial w} = \frac{\partial L_{\mathrm{b}}}{\partial u} = 0,
\tag{5.10}
$$

$$
\begin{aligned}
\frac{\partial}{\partial x}\frac{\partial L_{\mathrm{b}}}{\partial w_x} &= \rho A v_0(w_{xt} + v_0 w_{xx}) \\
&\quad + (EA - T_0)\frac{\partial}{\partial x}\left\{w_x\left(w_x^2 + (1+u_x)^2\right)^{-1/2}\right\} - EAw_{xx} \\
&= \rho A v_0(w_{xt} + v_0 w_{xx}) - EAw_{xx} \\
&\quad + (EA - T_0)\left\{\frac{w_{xx}(1+u_x)^2 - w_x u_{xx}(1+u_x)}{\left(w_x^2 + (1+u_x)^2\right)^{3/2}}\right\},
\end{aligned}
\tag{5.11}
$$

$$\frac{\partial}{\partial t}\frac{\partial L_b}{\partial w_t} = \rho A w_{tt} + \rho A v_0 w_{xt}, \tag{5.12}$$

$$\begin{aligned}\frac{\partial}{\partial x}\frac{\partial L_b}{\partial u_x} &= \rho A v_0(u_{xt} + v_0 u_{xx}) \\ &+ (EA - T_0)\frac{\partial}{\partial x}\left\{(1+u_x)\left(w_x^2 + (1+u_x)^2\right)^{-1/2}\right\} - EAu_{xx} \\ &= \rho A v_0(u_{xt} + v_0 u_{xx}) - EAu_{xx} \\ &+ (EA - T_0)\left\{\frac{u_{xx}w_x^2 - (1+u_x)w_x w_{xx}}{\left(w_x^2 + (1+u_x)^2\right)^{3/2}}\right\},\end{aligned} \tag{5.13}$$

$$\frac{\partial}{\partial t}\frac{\partial L_b}{\partial u_t} = \rho A u_{tt} + \rho A v_0 u_{xt}. \tag{5.14}$$

Notably, since the belt is simply supported, $\delta w(0, t) = \delta w(l, t) = 0$ and $\delta w(x, t_1) = \delta w(x, t_2) = 0$. Substituting (5.10)–(5.14) into (5.9), we obtain the PDEs that describe the coupling dynamics of an axially moving belt, namely

$$\begin{aligned}&\rho A\left(u_{tt} + 2v_0 u_{xt} + v_0^2 u_{xx}\right) - EAu_{xx} \\ &+ (EA - T_0)\left\{\frac{u_{xx}w_x^2 - (1+u_x)w_x w_{xx}}{\left(w_x^2 + (1+u_x)^2\right)^{3/2}}\right\} = 0,\end{aligned} \tag{5.15}$$

$$\begin{aligned}&\rho A\left(w_{tt} + 2v_0 w_{xt} + v_0^2 w_{xx}\right) - EAw_{xx} \\ &+ (EA - T_0)\left\{\frac{w_{xx}(1+u_x)^2 - w_x u_{xx}(1+u_x)}{\left(w_x^2 + (1+u_x)^2\right)^{3/2}}\right\} + EIw_{xxxx} = 0,\end{aligned} \tag{5.16}$$

and the boundary conditions are

$$u(0, t) = u(l, t) = 0, \; w(0, t) = w(l, t) = 0, \; w_{xx}(0, t) = w_{xx}(l, t) = 0. \tag{5.17}$$

5.2.2 Belt Model Using Approximate Strains

Equations (5.15)–(5.17) represent a comprehensive dynamic model for axially moving belts, wherein the closed-form strain is used. However, this model is too complicated mathematically. To facilitate dynamic analysis, we develop simpler models using an approximated strain. By using the Taylor series expansion, the strain can be approximated as follows:

$$\varepsilon(x, t) = \sqrt{w_x^2 + (1+u_x)^2} - 1 = \sqrt{\left(w_x^2 + u_x^2 + 2u_x\right) + 1} - 1$$

$$\cong \frac{1}{2}\left(w_x^2 + u_x^2 + 2u_x\right). \tag{5.18}$$

Furthermore, we assume that the longitudinal displacement is sufficiently small such that $u_x^2 \approx 0$. The following approximate strain is derived:

$$\varepsilon(x,t) \approx u_x + \frac{1}{2}w_x^2. \tag{5.19}$$

Under the approximate strain, the term L_{bt} in (5.9) is expressed as

$$\begin{aligned} L_{\mathrm{bt}} = {} & \frac{\rho A}{2}(w_t + v_0 w_x)^2 + \frac{\rho A}{2}(u_t + v_0(1+u_x))^2 \\ & - T_0\left(u_x + \frac{1}{2}w_x^2\right) - \frac{1}{2}EA\left(u_x + \frac{1}{2}w_x^2\right)^2. \end{aligned} \tag{5.20}$$

Substituting (5.20) into (5.9) yields the following dynamic model of an axially moving belt:

$$\rho A\left(u_{tt} + 2v_0 u_{xt} + v_0^2 u_{xx}\right) - EA\left(u_x + \frac{1}{2}w_x^2\right)_x = 0, \tag{5.21}$$

$$\begin{aligned} & \rho A\left(w_{tt} + 2v_0 w_{xt} + v_0^2 w_{xx}\right) - T_0 w_{xx} \\ & - EA\left(w_x\left(u_x + \frac{1}{2}w_x^2\right)\right)_x + EI w_{xxxx} = 0. \end{aligned} \tag{5.22}$$

Interestingly, if we disregard the nonlinear terms in this dynamic model, (5.21) and (5.22) become the governing equations of a traveling rod and a tensioned beam, respectively. The dynamic model of an axially moving material with the coupling of longitudinal and lateral vibrations could be further simplified by using the quasi-static stretch assumption. From a physical point of view, longitudinal waves propagate significantly faster than lateral waves. On the timescales of the lower lateral modes, tension variations propagate almost instantaneously because the influence of longitudinal inertia is small. Hence, we can assume that the time derivative of the longitudinal vibration is disregarded (i.e., $u_{xt} = u_{tt} \approx 0$). Additionally, considering practical axially moving systems, $EA \gg \rho A v_0^2$. Therefore, (5.21) can be approximated as follows:

$$\left(u_x + \frac{1}{2}w_x^2\right)_x \cong 0. \tag{5.23}$$

Equation (5.23) implies that the axial strain is a function of time alone, i.e., because $\varepsilon_x \approx \left(u_x + w_x^2/2\right)_x \approx 0$. Integrating both sides of (5.19) in the spatial interval $[0, x]$ yields

$$\varepsilon(t)x = u(x,t) - u(0,t) + \frac{1}{2}\int_0^x w_\xi^2(\xi,t)\mathrm{d}\xi. \tag{5.24}$$

Considering (5.24) at $x = l$ and using the boundary conditions $u(x,t) = u(0,t) = 0$, we derive the time function of the axial strain

$$\varepsilon(t) = \frac{1}{2l}\int_0^l w_\xi^2(\xi,t)\mathrm{d}\xi. \tag{5.25}$$

By substituting (5.25) into (5.24), the longitudinal vibration can be described by a functional of the lateral vibration as follows:

$$u(x,t) = \frac{x}{2l}\int_0^l w_\xi^2(\xi,t)\mathrm{d}\xi - \frac{1}{2}\int_0^x w_\xi^2(\xi,t)\mathrm{d}\xi. \tag{5.26}$$

Based on (5.22), (5.23), and (5.26), the following dynamic model that describes the lateral vibration only is derived:

$$\begin{aligned} &\rho A\left(w_{tt} + 2v_0 w_{xt} + v_0^2 w_{xx}\right) - T_0 w_{xx} \\ &\quad - \frac{EA}{2l} w_{xx}\int_0^l w_x^2(x,t)\mathrm{d}x + EI w_{xxxx} = 0. \end{aligned} \tag{5.27}$$

Equation (5.27) was presented as the IPDE of an axially moving beam in Chap. 3. In this model, only the "weak" nonlinearity from the dynamics of the longitudinal vibration was incorporated, that is, only the static relationship between the lateral vibration and the longitudinal vibration was considered.

5.2.3 Viscoelastic Belt

If the Kelvin viscoelastic model is considered, Young's modulus E can be replaced by $E + \mu\frac{\partial}{\partial t}$ in (5.21) and (5.22), and the resulting equations are

$$\begin{aligned} &\rho A\left(u_{tt} + 2v_0 u_{xt} + v_0^2 u_{xx}\right) + c_{i1} u_t - EAu_{xx} - EAw_x w_{xx} \\ &\quad - \mu A u_{xxt} - \mu A w_{xt} w_{xx} - \mu A w_x w_{xxt} = 0, \end{aligned} \tag{5.28}$$

$$\begin{aligned} &\rho A\left(w_{tt} + 2v_0 w_{xt} + v_0^2 w_{xx}\right) + c_{i2} w_t - T_0 w_{xx} + EI w_{xxxx} \\ &\quad - EA(u_{xx} w_x + u_x w_{xx}) \end{aligned}$$

$$+\mu I w_{xxxxt} - \mu A(u_{xxt} w_x + u_{xx} w_{xt} + u_{xt} w_{xx} + u_x w_{xxt}) = 0. \tag{5.29}$$

A viscoelastic model with multiple parameters can also be applied to the axially moving belt. However, this leads to the existence of more complicated terms in the governing equations.

5.2.4 Shear Deformation and Rotary Inertia

In this section, we consider the shear force and rotary inertia in addition to the bending moment within the axially moving belt. To take into account these factors, we use the Timoshenko beam theory to develop the belt model. With this theory, the kinetic and potential energies of the belts are given as follows:

$$\begin{aligned} K &= \frac{1}{2}\rho A \int_0^l \left((w_t + v_0 w_x)^2 + (u_t + v_0(1 + u_x))^2\right)\mathrm{d}x \\ &\quad + \frac{1}{2}\rho I \int_0^l (\theta_t + v_0\theta_x)^2 \mathrm{d}x, \end{aligned} \tag{5.30}$$

$$\begin{aligned} U &= \int_0^l \left(T_0\left(u_x + \frac{1}{2}w_x^2\right) + \frac{1}{2}EA\left(u_x + \frac{1}{2}w_x^2\right)^2\right)\mathrm{d}x \\ &\quad + \frac{1}{2}EI\int_0^l \theta_x^2 \mathrm{d}x + \frac{1}{2}\kappa GA \int_0^l (w_x + \theta)^2 \mathrm{d}x, \end{aligned} \tag{5.31}$$

where θ denotes the angle of rotation of the cross section, κ is the shape factor, and G denotes the shear modulus. By using the Hamilton principle, we obtain the following equation:

$$0 = \int_{t_1}^{t_2}\int_0^l \delta L_{\mathrm{bt}} \mathrm{d}x\mathrm{d}t \tag{5.32}$$

where

$$\begin{aligned} L_{\mathrm{bt}} &= \frac{1}{2}\rho A(w_t + v_0 w_x)^2 + \frac{1}{2}\rho A(u_t + v_0(1 + u_x))^2 \\ &\quad + \frac{1}{2}\rho I(\theta_t + v\theta_x)^2 - T_0\left(u_x + w_x^2/2\right) - \frac{1}{2}EA\left(u_x + w_x^2/2\right)^2 \end{aligned}$$

$$-\frac{1}{2}\kappa GA(w_x+\theta)^2-\frac{1}{2}EI\theta_x^2. \tag{5.33}$$

Using the integration by parts formula for (5.32) yields

$$\begin{aligned}
0=&\int_{t_1}^{t_2}\int_0^l\left(\frac{\partial L_{\mathrm{bt}}}{\partial w}-\frac{\partial}{\partial x}\frac{\partial L_{\mathrm{bt}}}{\partial w_x}-\frac{\partial}{\partial t}\frac{\partial L_{\mathrm{bt}}}{\partial w_t}\right)\delta w\mathrm{d}x\mathrm{d}t+\int_{t_1}^{t_2}\frac{\partial L_{\mathrm{bt}}}{\partial w_x}\delta w\bigg|_{x=0}^{x=l}\mathrm{d}t\\
&+\int_0^l\frac{\partial L_{\mathrm{bt}}}{\partial w_t}\delta w\bigg|_{t=t_1}^{t=t_2}\mathrm{d}x\\
&+\int_{t_1}^{t_2}\int_0^l\left(\frac{\partial L_{\mathrm{bt}}}{\partial u}-\frac{\partial}{\partial x}\frac{\partial L_{\mathrm{bt}}}{\partial u_x}-\frac{\partial}{\partial t}\frac{\partial L_{\mathrm{bt}}}{\partial u_t}\right)\delta u\mathrm{d}x\mathrm{d}t+\int_{t_1}^{t_2}\frac{\partial L_{\mathrm{bt}}}{\partial u_x}\delta u\bigg|_{x=0}^{x=l}\mathrm{d}t\\
&+\int_0^l\frac{\partial L_{\mathrm{bt}}}{\partial u_t}\delta u\bigg|_{t=t_1}^{t=t_2}\mathrm{d}x\\
&+\int_{t_1}^{t_2}\int_0^l\left(\frac{\partial L_{\mathrm{bt}}}{\partial \theta}-\frac{\partial}{\partial x}\frac{\partial L_{\mathrm{bt}}}{\partial \theta_x}-\frac{\partial}{\partial t}\frac{\partial L_{\mathrm{bt}}}{\partial \theta_t}\right)\delta \theta\mathrm{d}x\mathrm{d}t+\int_{t_1}^{t_2}\frac{\partial L_{\mathrm{bt}}}{\partial \theta_x}\delta \theta\bigg|_{x=0}^{x=l}\mathrm{d}t\\
&+\int_0^l\frac{\partial L_{\mathrm{bt}}}{\partial \theta_t}\delta \theta\bigg|_{t=t_1}^{t=t_2}\mathrm{d}x.
\end{aligned} \tag{5.34}$$

From (5.33) and (5.34), the following mathematical model for an axially moving belt based on the Timoshenko beam theory is established.

$$\rho A\left(u_{tt}+2v_0u_{xt}+v_0^2u_{xx}\right)-EA(u_{xx}+w_{xx}w_x)=0, \tag{5.35}$$

$$\begin{aligned}
&\rho A\left(w_{tt}+2v_0w_{xt}+v_0^2w_{xx}\right)-T_0w_{xx}-EA\left(w_{xx}u_x+w_xu_{xx}+\frac{3}{2}w_x^2w_{xx}\right)\\
&+\kappa GA(w_{xx}+\theta_x)=0,
\end{aligned} \tag{5.36}$$

$$\rho I(\theta_{tt}+2v_0\theta_{xt}+v_0^2\theta_{xx})-EI\theta_{xx}-\kappa GA(w_x+\theta)=0. \tag{5.37}$$

The boundary conditions for the belt with simply supported boundaries are

$$u(0,t)=u(l,t)=0,\ w(0,t)=w(l,t)=0,\ \theta_x(0,t)=\theta_x(l,t)=0. \tag{5.38}$$

5.2.5 Transverse Vibration

Only longitudinal and lateral vibrations in the plane of the *i*- and *j*-axes have been presented in the previous sections. Now, we further consider the transverse vibration on the *k*-axis. Let $\eta(x, t)$ be the transverse vibration of the belt. The kinetic and potential energies are given as follows:

$$K = \frac{1}{2}\rho A \int_0^l \left((w_t + v_0 w_x)^2 + (\eta_t + v_0 \eta_x)^2 + (u_t + v_0(1 + u_x))^2\right)\mathrm{d}x, \quad (5.39)$$

$$U = \int_0^l \left(T_0\left(u_x + \frac{1}{2}w_x^2 + \frac{1}{2}\eta_x^2\right) + \frac{1}{2}EA\left(u_x + \frac{1}{2}w_x^2 + \frac{1}{2}\eta_x^2\right)^2\right)\mathrm{d}x$$
$$+ \frac{1}{2}EI\int_0^l w_{xx}\mathrm{d}x + \frac{1}{2}EI_z\int_0^l \eta_{xx}\mathrm{d}x. \quad (5.40)$$

According to the Hamilton principle, the following nonlinear PDEs are developed to analyze the vibrations in the *i*-, *j*-, and *k*-axes of an axially accelerating material.

$$\rho A\left(w_{tt} + 2v_0 w_{xt} + v_0^2 w_{xx}\right) - T_0 w_{xx} + EI w_{xxxx}$$
$$- EA\left(u_x w_{xx} + u_{xx} w_x + w_x \eta_x \eta_{xx} + \frac{1}{2}\eta_x^2 w_{xx} + \frac{3}{2}w_x^2 w_{xx}\right) = 0, \quad (5.41)$$

$$\rho A\left(u_{tt} + 2v_0 u_{xt} + v_0^2 u_{xx}\right) - EA(u_{xx} + \eta_x \eta_{xx} + w_x w_{xx}) = 0, \quad (5.42)$$

$$\rho A\left(\eta_{tt} + 2v_0 \eta_{xt} + v_0^2 \eta_{xx}\right) - T_0 \eta_{xx} + EI_z \eta_{xxxx}$$
$$- EA\left(u_x \eta_{xx} + u_{xx}\eta_x + \eta_x w_x w_{xx} + \frac{1}{2}w_x^2 \eta_{xx} + \frac{3}{2}\eta_x^2 \eta_{xx}\right) = 0, \quad (5.43)$$

where I and I_z represent the moments of inertia about the *j*-axis and *k*-axis, respectively. Essentially, the PDEs that describe the transverse and lateral vibrations are similar, except for the moments of inertia.

5.2.6 Belts with Nonhomogenous Boundaries

Similar to the string and beam cases, this subsection focuses on the axially moving belt with nonhomogeneous boundaries. The dynamic model of a band-wheel mechanical system, a typical example presented by Wang and Mote (1986), is investigated in this subsection. Ding et al. (2018) and Ding and Zu (2014) have revealed the dynamics of axially moving belts in detail. A two-pulley–belt-drive system flexibly coupled with a one-way clutch is developed considering the bending stiffness. A mechanism consisting of a wrap spring and a one-way rigid clutch is used to connect the accessory and the driven pulley. In this system, the belt span can be treated as a slender Euler–Bernoulli beam with a high slenderness ratio (i.e., with $I/(Al^2) < 0.001$). Therefore, two spans of the belt are modeled by longitudinally translating viscoelastic belts undergoing an in-plane vibration. The viscoelastic property of the belt is considered via the Kelvin model.

The belt is an one-dimensional continuum with coupled longitudinal and lateral vibrations, and the Eulerian equation gives (Kong and Parker, 2003, 2004)

$$\begin{aligned}&\rho A\left(w_{itt}+2v_0w_{ix_it}+v_0^2w_{ix_ix_i}\right)-\left[(T_0+T_{di})w_{ix_i}\right]_{x_i}+M_{ixx}(x_i,t)=0,\\&\rho A\left(u_{itt}+2v_0u_{ix_it}+v_0^2u_{ix_ix_i}\right)-(T_0+T_{di})_{x_i}=0\end{aligned}\tag{5.44}$$

where T_0 denotes the axial static tension of the axially moving belt, T_{di} $(i = 1, 2)$ is the dynamic tension of the i-th span, and $M(x_i, t)$ accounts for the bending moment. Additionally, the viscoelastic property of the belt is considered based on the Kelvin model. Thus, the linear moment–curvature relationship of the belt is sufficiently accurate as

$$M_i(x,t)=\left(E+\mu\frac{\partial}{\partial t}\right)Iw_{ix_ix_i}.\tag{5.45}$$

In (5.45), the dynamic tension T_{di} of the i-th belt span is given as follows:

$$T_{\mathrm{d}i}=EA\left(u_{ix_i}+\frac{1}{2}w_{ix_i}^2\right)\tag{5.46}$$

where $w_{ix_i}^2 >> u_{ix_i}^2$ is assumed. According to the quasi-static stretch assumption (Wickert 1992), we assume that the dynamic tension in each span is uniform throughout the span. Accordingly, the dynamic tensions can be defined as follows:

$$T_{\mathrm{d}i}=\frac{EA}{l}[u_i(l,t)-u_i(0,t)]+\frac{EA}{2l}\int_0^l w_{ix_i}^2\mathrm{d}x_i,\quad (i=1,2).\tag{5.47}$$

Substitution of (5.45) and (5.46) into (5.44) yields

$$\begin{aligned}&\rho A\left(w_{itt}+2v_0w_{ix_it}+v_0^2w_{ix_ix_i}\right)-T_0w_{ix_ix_i}+EIw_{ix_ix_ix_ix_i}+\mu Iw_{ix_ix_ix_ix_it}\\&=\frac{EA}{l}\left\{[u_i(l,t)-u_i(0,t)]+\frac{1}{2}\int_0^l w_{ix_i}^2\mathrm{d}x_i\right\}w_{ix_ix_i},\\&\rho A\left(u_{itt}+2v_0u_{ix_it}+v_0^2u_{ix_ix_i}\right)=0,\end{aligned}\tag{5.48}$$

with the boundary conditions given by

$$w_i(0,t) = 0,\ w_i(l,t) = 0,\ w_{ix_ix_i}(0,t) = 1/r,\ w_{ix_ix_i}(l,t) = 1/r, \quad (i = 1,2) \tag{5.49}$$

where r is the pulley's radius and EI denotes the bending stiffness. In (5.48), the lateral vibration and longitudinal vibration of axially moving belts are decoupled.

The rotational motions for the pulleys and the accessory are derived as follows:

$$J\ddot{\theta}_1 + c_b\dot{\theta}_1 = (T_{d1} - T_{d2})r - f(\theta_1 - \theta_3)\left[K_d(\theta_1 - \theta_3) + c_d\left(\dot{\theta}_1 - \dot{\theta}_3\right)\right] + M_1,$$

$$J_a\ddot{\theta}_3 + c_a\dot{\theta}_3 = f(\theta_1 - \theta_3)\left[K_d(\theta_1 - \theta_3) + c_d\left(\dot{\theta}_1 - \dot{\theta}_3\right)\right] - M_1,$$

$$J\ddot{\theta}_2 + c_b\dot{\theta}_2 = (T_{d2} - T_{d1})r + M_2, \tag{5.50}$$

where c_a, c_b, and c_d are the damping coefficients of the rotating accessory, the friction between belt and pulley, and the wrap spring, respectively. $f(\theta_1 - \theta_3)$ is expressed as the following piecewise function

$$f(\theta_1 - \theta_3) = \begin{cases} 1,\ \theta_1 - \theta_3 > 0, \\ 0,\ \theta_1 - \theta_3 \le 0. \end{cases} \tag{5.51}$$

The one-way clutch disengages or engages depending on the relative angular displacement, $\theta_1 - \theta_3$. The clutch is engaged if the angular displacement θ_1 exceeds θ_3 (i.e., $\theta_1 - \theta_3 > 0$). The pulley–belt system transmits the clutch torque $f(\theta_1 - \theta_3)K_d(\theta_1 - \theta_3)$ to the accessory. In the zero clutch torque case, the accessory's motion is decoupled from the pulley–belt system. It is noted that the accessory's motion is a damped free oscillation of one degree-of-freedom (DOF) if $\theta_1 - \theta_3 \le 0$.

The nondimensional parameters and variables are given as follows:

$$\overline{x}_i = \frac{x_i}{l},\ \overline{w}_i = \frac{w_i}{l},\ \overline{u}_i = \frac{u_i}{l},\ \overline{t} = t\sqrt{\frac{T_0}{\rho A l^2}},\ \overline{v}_0 = v_0\sqrt{\frac{\rho A}{T_0}},\ \overline{K}_d = \frac{K_d}{T_0 r_a},\ \overline{M}_i = \frac{M_i}{T_0 r},$$

$$\overline{J}_a = \frac{J_a}{\rho A r_a l^2},\ \overline{J} = \frac{J}{\rho A r l^2},\ \overline{c}_b = \frac{c_b}{l r}\sqrt{\frac{1}{\rho A T_0}},\ \overline{c}_a = \frac{c_a}{l r_a}\sqrt{\frac{1}{\rho A T_0}},\ \overline{c}_d = \frac{c_d}{l r_a}\sqrt{\frac{1}{\rho A T_0}},$$

$$k_1 = \frac{EA}{T_0},\ \alpha = \frac{I\mu}{l^3\sqrt{\rho A T_0}},\ k_f = \sqrt{\frac{EI}{T_0 l^2}}, \tag{5.52}$$

where k_1 is the nonlinear coefficient representing the effect of nonlinearity of the axially moving belt, α accounts for the dynamic viscosity, and k_f^2 corresponds to the beam's flexural rigidity. The nondimensional equations of motion governing the lateral vibration of the longitudinally moving spans and the rotational motions of the pulleys and accessory are derived as follows:

$$\bar{w}_{1\bar{t}\bar{t}} + 2\bar{v}_0\bar{w}_{1\bar{x}_1\bar{t}} + \left(\bar{v}_0^2 - 1\right)\bar{w}_{1\bar{x}_1\bar{x}_1} + k_f^2\bar{w}_{1\bar{x}_1\bar{x}_1} + \alpha\bar{w}_{1\bar{x}_1\bar{x}_1\bar{t}}$$
$$= k_1\left[\frac{r}{l}(\theta_2 - \theta_1) + \frac{1}{2}\int_0^1 \bar{w}_{1\bar{x}_1}^2 \mathrm{d}\bar{x}_1\right]\bar{w}_{1\bar{x}_1\bar{x}_1},$$
$$\bar{w}_{2\bar{t}\bar{t}} + 2\bar{v}_0\bar{w}_{2\bar{x}_2\bar{t}} + \left(\bar{v}_0^2 - 1\right)\bar{w}_{2\bar{x}_2\bar{x}_2} + k_f^2\bar{w}_{2\bar{x}_2\bar{x}_2} + \alpha\bar{w}_{2\bar{x}_1\bar{x}_1\bar{t}}$$
$$= k_1\left[\frac{r}{l}(\theta_1 - \theta_2) + \frac{1}{2}\int_0^1 \bar{w}_{2\bar{x}_2}^2 \mathrm{d}\bar{x}_2\right]\bar{w}_{2\bar{x}_2\bar{x}_2}, \tag{5.53}$$

$$\bar{J}\ddot{\theta}_1 = -\bar{c}_b\dot{\theta}_1 + \frac{k_1}{2}\left[\frac{4r}{l}(\theta_2 - \theta_1) + \int_0^1 \overline{w}_{1\overline{x}_1}^2 \mathrm{d}\overline{x}_1 - \int_0^1 \overline{w}_{2\overline{x}_2}^2 \mathrm{d}\overline{x}_2\right]$$
$$+ M_1 - f(\theta_1 - \theta_3)\left[\overline{K}_d(\theta_1 - \theta_3) + \overline{c}_d\left(\dot{\theta}_1 - \dot{\theta}_3\right)\right]\frac{r_a}{r},$$

$$\overline{J}_a\ddot{\theta}_3 = -\overline{c}_a\dot{\theta}_3 - \frac{r\overline{M}_1}{r_a} + f(\theta_1 - \theta_3)\left[\overline{K}_d(\theta_1 - \theta_3) + \overline{c}_d\left(\dot{\theta}_1 - \dot{\theta}_3\right)\right], \tag{5.54}$$

$$J\ddot{\theta}_2 = -c_b\dot{\theta}_2 + \frac{k_1}{2}\left[\frac{4r}{l}(\theta_1 - \theta_2) + \int_0^1 \overline{w}_{2\overline{x}_2}^2 \mathrm{d}\overline{x}_2 - \int_0^1 \overline{w}_{1\overline{x}_1}^2 \mathrm{d}\overline{x}_1\right] + M_2 \tag{5.55}$$

with the dimensionless boundary conditions given by

$$\overline{w}_i(0, \overline{t}) = \overline{w}_i(1, \overline{t}) = 0, \overline{w}_{i\overline{x}_i\overline{x}_i}(0, \overline{t}) = \overline{w}_{i\overline{x}_i\overline{x}_i}(1, \overline{t}) = \frac{l}{r}, \quad (i = 1, 2). \tag{5.56}$$

It is assumed that the displacement fluctuation of the driving pulley's crankshaft is a factor exciting the pulley–belt system. The displacement fluctuation is given by $\theta_2 = A_m\cos\omega\overline{t}$, in which the nondimensional parameter A_m denotes the magnitude of the driving pulley's motion. That is, A_m is the excitation amplitude (Zhu and Parker 2008). Additionally, the translating velocity is determined via the driving pulley motion: $v_0 = r\omega/n_e l$, where n_e is the number of cylinder pairs of the engine (Mockensturm and Balaji 2005). In this section, we assume that the considered system is operated by an engine with six cylinders; therefore, $n_e = 3$.

The determination of the exact solution of the nonlinear hybrid discrete–continuous equations in (5.53)–(5.56) is impossible. Therefore, the numerical solutions are pursued in the sequel. The numerical methods, such as the differential quadrature and integral quadrature methods (DQM and IQM), can be utilized to investigate the nonlinear vibration of AMSs. Both two methods exactly satisfy many boundary conditions and have very high accuracy. Additionally, they do not require much formulating and programming efforts. In order to accelerate the convergence rate of DQM and IQM, we consider unequally spaced Chebyshev–Gauss–Lobatto sampling points with δ points immediate adjacent at two ends for the i-th belt span (Bert and Malik 1996). That is,

$$\overline{x}_{i1} = 0, \overline{x}_{i2} = \delta, x_{i(N-1)} = 1 - \delta, \overline{x}_{iN} = 0, \quad i = 1, 2,$$

$$\overline{x}_{ij} = \frac{1}{2}\left[1 - \cos\frac{(j-2)\pi}{N-3}\right], \quad i = 1, 2;\ j = 3, 4, \ldots, N-2. \tag{5.57}$$

Let us consider an arbitrary function $f(\overline{x}_i, \overline{t})$ on a set of preselected sampling points $\overline{x}_{ij}$. Based on the integral quadrature rule, the value of an integral over the whole domain is given by a linear weighted sum of $f(\overline{x}_{ij}, \overline{t})$. Based on the differential quadrature rule, the values of the function derivatives at sampling points are given by a linear weighted sum of the values $f_{ij} = f(\overline{x}_{ij}, \overline{t})$. Accordingly, an n-th order derivative at $\overline{x}_{ij}$ is written as follows:

$$\left.\frac{\partial^n f(\overline{x}_i, \overline{t})}{\partial \overline{x}_i^n}\right|_{\overline{x}_i = \overline{x}_{ij}} = \sum_{k=1}^{N} A_{jk}^{(n)} f(\overline{x}_{ik}, \overline{t}) = \sum_{k=1}^{N} A_{jk}^{(n)} f_k, \tag{5.58}$$

and the integral of the function over the whole domain is given as follows:

$$\int_0^1 f(\overline{x}_i, \overline{t}) \mathrm{d}\overline{x}_i = \sum_{k=1}^{N} I_k f(\overline{x}_{ik}, \overline{t}) = \sum_{k=1}^{N} I_k f_{ik} \tag{5.59}$$

where the integral weighting coefficients I_k (k = 1,2,..., N) for integrals are solved from

$$\begin{pmatrix} I_1 \\ I_2 \\ \vdots \\ I_{N-1} \\ I_N \end{pmatrix} = \begin{pmatrix} 1 \\ 1/2 \\ \vdots \\ 1/(N-1) \\ 1/N \end{pmatrix} \begin{pmatrix} 1 & 1 & \cdots & 1 & 1 \\ \overline{x}_{i1} & \overline{x}_{i2} & \cdots & \overline{x}_{iN-1} & \overline{x}_{iN} \\ \vdots & \vdots & & \vdots & \vdots \\ \overline{x}_{i1}^{N-2} & \overline{x}_{i2}^{N-2} & \cdots & \overline{x}_{i(N-1)}^{N-2} & \overline{x}_{iN}^{N-2} \\ \overline{x}_{i1}^{N-1} & \overline{x}_{i2}^{N-1} & \cdots & \overline{x}_{i(N-1)}^{N-1} & \overline{x}_{iN}^{N-1} \end{pmatrix}^{-1}. \tag{5.60}$$

The differential quadrature weighting coefficients in (5.58) are determined for the first-order derivative in the neutral axis coordinate x as follows:

$$A_{jk}^{(1)} = \begin{cases} \dfrac{\prod_{m=1, m\neq i}^{N} (\overline{x}_{ij} - \overline{x}_{im})}{(\overline{x}_{ij} - \overline{x}_{ik}) \prod_{m=1, m\neq k}^{N} (\overline{x}_{ik} - \overline{x}_{im})} & \text{for } k \neq j;\ j, k = 1, 2, \ldots, N, \\ \sum\limits_{k=1; k\neq j}^{N} \dfrac{1}{\overline{x}_{ij} - \overline{x}_{ik}} & \text{for } k = j. \end{cases} \tag{5.61}$$

The second and high-order derivative weighting coefficients are determined through the following recurrence relationship.

$$A_{jk}^{(2)} = \sum_{m=1}^{N} A_{jm}^{(1)} A_{mk}^{(1)} \text{ for } j, k = 1, 2, \ldots, N \tag{5.62}$$

and

$$A_{jk}^{(r)} = \sum_{m=1}^{N} A_{jm}^{(1)} A_{mk}^{(r-1)} = \sum_{m=1}^{N} A_{jm}^{(r-1)} A_{mk}^{(1)} \quad \text{for } r = 3, 4, \ldots, \text{ and } j, k = 1, 2, \ldots, N. \tag{5.63}$$

Using (5.58) yields

$$\begin{aligned} \overline{w}_{i\overline{x}_i}(\overline{x}_{ij}, \overline{t}) &= \sum_{k=1}^{N} A_{jk}^{(1)} \overline{w}_i(\overline{x}_{ik}, \overline{t}), \\ \overline{w}_{i\overline{x}_i\overline{x}_i}(\overline{x}_{ij}, \overline{t}) &= \sum_{k=1}^{N} A_{jk}^{(2)} \overline{w}_i(\overline{x}_{ik}, \overline{t}), \\ \overline{w}_{i\overline{x}_i\overline{x}_i\overline{x}_i\overline{x}_i}(\overline{x}_{ij}, \overline{t}) &= \sum_{k=1}^{N} A_{jk}^{(4)} \overline{w}_i(\overline{x}_{ik}, \overline{t}). \end{aligned} \tag{5.64}$$

Furthermore, the substitution of (5.64) into (5.53) leads to the following series of ordinary differential equations.

$$\begin{aligned} &\overline{w}_i(\overline{x}_{i1}, \overline{t}) = 0, \\ &\overline{w}_i(\overline{x}_{i2}, \overline{t}) = \frac{l/r - A_{2(N-1)}^{(2)} \overline{w}_{i(N-1)} - \sum_{k=3}^{N-2} A_{2k}^{(2)} \overline{w}_i(\overline{x}_{ik}, \overline{t})}{A_{22}^{(2)}}, \\ &\overline{w}_{i\overline{t}\overline{t}}(\overline{x}_{ij}, \overline{t}) - \sum_{k=1}^{N} \left[(\overline{v}_0^2 - 1) A_{jk}^{(2)} + k_{\mathrm{f}}^2 \tilde{A}_{jk}^{(4)} \right] \overline{w}_i(\overline{x}_{ik}, \overline{t}) \\ &\quad + \sum_{k=1}^{N} \left[2\overline{v}_0 A_{jk}^{(1)} + \overline{\alpha} A_{jk}^{(4)} \right] \overline{w}_{i\overline{t}}(\overline{x}_{ik}, \overline{t}) \\ &= \frac{k_1^2}{2} \left[\sum_{k=1}^{N} A_{jk}^{(2)} \overline{w}_i(\overline{x}_{ik}, \overline{t}) \right] \left\{ \lambda_i + \sum_{g=1}^{N} I_g \left[\sum_{k=1}^{N} A_{gk}^{(1)} \overline{w}_i(\overline{x}_{ik}, \overline{t}) \right]^2 \right\}, \\ &j = 3, 4, \ldots, N-2, \\ &\overline{w}_i(\overline{x}_{i(N-1)}, \overline{t}) \\ &= \frac{l/r - \sum_{k=3}^{N-2} A_{(N-1)k}^{(2)} \overline{w}_i(\overline{x}_{ik}, \overline{t}) - A_{(N-1)2}^{(2)} \frac{l/r - \sum_{k=3}^{N-2} A_{2k}^{(2)} \overline{w}_i(\overline{x}_{ik}, \overline{t})}{A_{22}^{(2)}}}{A_{(N-1)(N-1)}^{(2)} - A_{(N-1)2}^{(2)} \frac{A_{2(N-1)}^{(2)}}{A_{22}^{(2)}}}, \\ &\overline{w}_i(\overline{x}_{iN}, \overline{t}) = 0 \end{aligned} \tag{5.65}$$

where $i = 1, 2$ and

$$\lambda_1 = \frac{2r}{l}(\theta_2 - \theta_1), \lambda_2 = \frac{2r}{l}(\theta_1 - \theta_2). \tag{5.66}$$

Applying (5.59) to (5.54) yields

$$\begin{aligned}
\overline{J}\ddot{\theta}_1 = &-\overline{c}_b\dot{\theta}_1 + \frac{k_1}{2}\left\{\frac{4r}{l}(\theta_2 - \theta_1) + \sum_{g=1}^{N} I_g\left[\sum_{k=1}^{N} A_{gk}^{(1)}\overline{w}_1(\overline{x}_{1k}, \overline{t})\right]^2\right. \\
&\left. - \sum_{g=1}^{N} I_g\left[\sum_{k=1}^{N} A_{gk}^{(1)}\overline{w}_2(\overline{x}_{2k}, \overline{t})\right]^2\right\} \\
&+ \overline{M}_1 - f(\theta_1 - \theta_3)\left[\overline{K}_d(\theta_1 - \theta_3) + \overline{c}_d(\dot{\theta}_1 - \dot{\theta}_3)\right]\frac{r_a}{r}, \\
\overline{J}_a\ddot{\theta}_3 = &-\overline{c}_a\dot{\theta}_3 - \frac{r\overline{M}_1}{r_a} + f(\theta_1 - \theta_3)\left[\overline{K}_d(\theta_1 - \theta_3) + \overline{c}_d(\dot{\theta}_1 - \dot{\theta}_3)\right].
\end{aligned} \tag{5.67}$$

With given dimensionless parameters, we can apply the four-stage Runge–Kutta method to (5.65) and (5.67) to derive the numerical solutions of $\overline{w}_i(\overline{x}_{ij}, \overline{t})$ ($i = 1, 2$ and $j = 2, 3, \ldots, N - 1$), $\theta_1(t)$, and $\theta_3(t)$. In all numerical examples, the following initial conditions for the first calculation are used.

$$\overline{w}_i(\overline{x}_i, 0) = 0, \overline{w}_{i\overline{t}}(\overline{x}_i, 0) = 0.001,$$

$$\theta_1 = \theta_3 = 0, \dot{\theta}_1 = 0.001, \dot{\theta}_3 = 0. \tag{5.68}$$

If the number of sampling points N is odd, the lateral displacement of the center of the i-th span will be $\overline{w}_i(\overline{x}_{i(N+1)/2}, \overline{t})$. In the numerical examples, we set $N = 15$.

Table 5.1 shows the geometrical and physical properties of the pulley–belt system. We set $\theta_2 = 0$. The time histories of the belt's lateral vibration and rotation of the driven pulley are shown in Fig. 5.2. An engaged state refers to the case that a wrap spring without one-way clutch is used to connect the driven pulley and accessory directly. The frequency spectrum of a time-dependent function can be determined via the fast Fourier transform. Accordingly, the frequencies of the nonlinear vibration can be computed based on the time series of the belt's vibration and the pully's rotation. Figure 5.2a demonstrates the time histories of the rotation of the driven pulley, whereas Fig. 5.2b shows the frequencies of the rotational angle. Additionally, Fig. 5.2c, d exhibits the time histories of the lateral vibration of the belt's span one and its frequency spectrum, respectively. Based on the numerical solutions and the fast Fourier transform, we can determine frequencies of the motion of the pulley–belt system. As shown in Fig. 5.2, ω_B is the natural frequency of the belt's lateral vibration, whereas ω_1, ω_2, and ω_3 are the first, second, and third rotational natural frequencies of the driven pulley.

The influence of the belt's axial speed on the system's natural frequencies is illustrated in Fig. 5.3. As shown in this figure, if the belt's axial speed increases,

Table 5.1 System parameters of the pulley–belt system with one-way clutch

Notation	Parameter	Value
Pulleys		
r	Pulley's radius	0.0452 m
J	Pulley's rotation inertia	0.001607 kg·m^2/rad
r_a	Accessory's radius	0.0889 m
J_a	Accessory's rotation inertia	1.62 J
M_0	Preload on the driven pulley	5 N m
A_m	Excitation amplitude	0.001264
c_a	Friction coefficient, accessory	0.02 N·m·s/rad
c_b	Friction coefficient, pulley	0.02 N·m·s/rad
Belts		
b	Width	0.02 m
h	Height	0.005 m
l	Span length	0.5518 m
E	Young's modulus	2×10^9 N/m^2
I	Moment of inertia	2.08×10^{-10} m^4
ρ	Mass density	1150 kg/m^3
T_0	Static tension	350 N
μ	Viscous damping	2×10^6 N s/m^2
(k_1)	Nonlinear coefficient	571.43
$(k_f{}^2)$	Bending stiffness	0.0039
One-way clutch		
c_d	Damping coefficient	2 N m s/rad
K_d	Torsional stiffness	4000 N m/rad

the natural frequencies will decrease. This conclusion was also pointed out by Wang and Mote (1986). Figure 5.3 further shows that if the excitation frequency, which is computed by $\omega = v_0 n_e l / r$, approaches twice the belt's natural frequency, the parametric resonance of the axially moving belt will exist. Additionally, when the excitation frequency closes to the pulley's natural frequency, the primary resonances of the pulleys will appear. As shown in Fig. 5.3, the prototypical system occurs in the resonance phenomenon when the translating speed is close to v_n ($n = 1, 2, 3, 4$). It is observed that $\omega = 226$ rad/s, $\omega = 377$ rad/s, $\omega = 626$ rad/s, and $\omega = 1110$ rad/s are the approximate values of the excitation frequencies corresponding to v_n.

It is noted that $\omega = 1110$ rad/s, $\omega = 626$ rad/s, $\omega = 377$ rad/s, and $\omega = 226$ rad/s correspond to the engine speeds 3533.24 rpm, 1992.62 rpm, 1200.03 rpm, and 719.38 rpm, which are in the range of practical importance. We investigate the dynamic responses of the clutch torque (or transmission torque) at these resonance frequencies via numerical simulations shown in Fig. 5.4. In the principal parametric resonance of the belt's vibration ($\omega = 1110$ rad/s) and the primary resonances of the

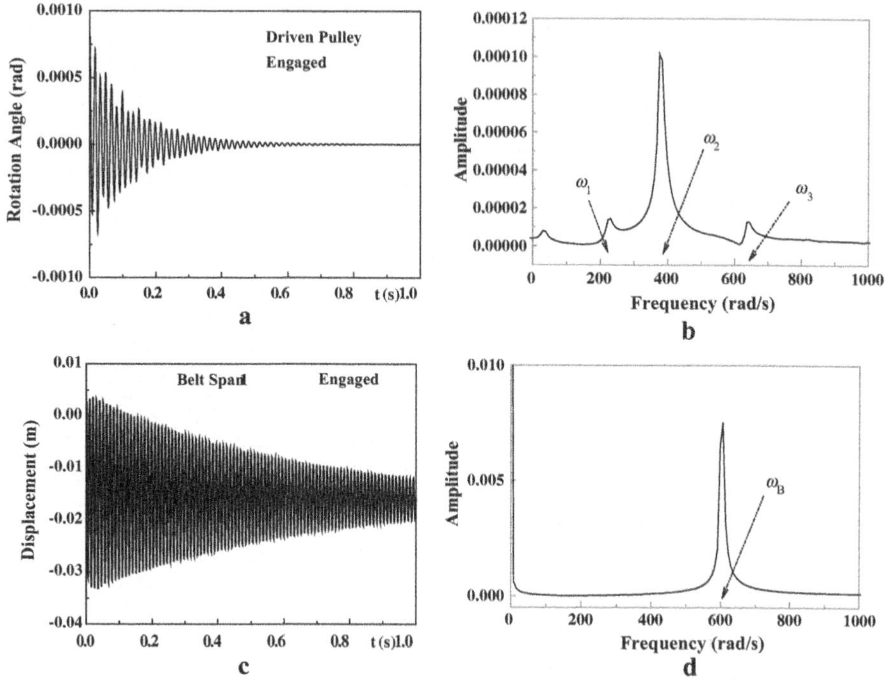

Fig. 5.2 Free vibration of the pulley–belt system: **a** time history of the driven pulley's rotation, **b** amplitude spectrum the driven pulley's rotation, **c** time history of the lateral vibration of the belt, and **d** amplitude spectrum of the lateral vibration of the belt

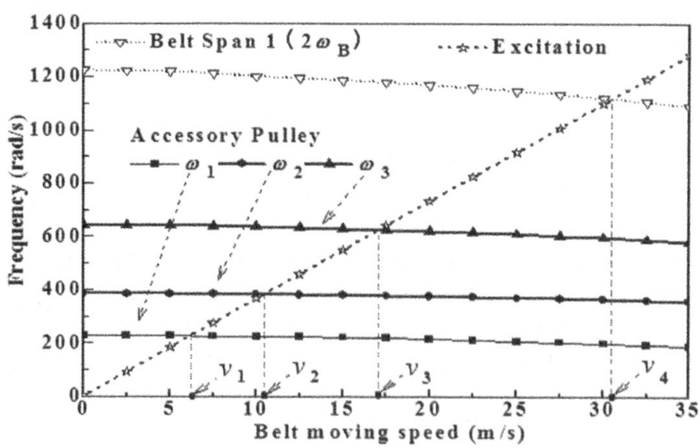

Fig. 5.3 Frequencies of the free vibration of the pulley–belt system versus the belt's translating speed

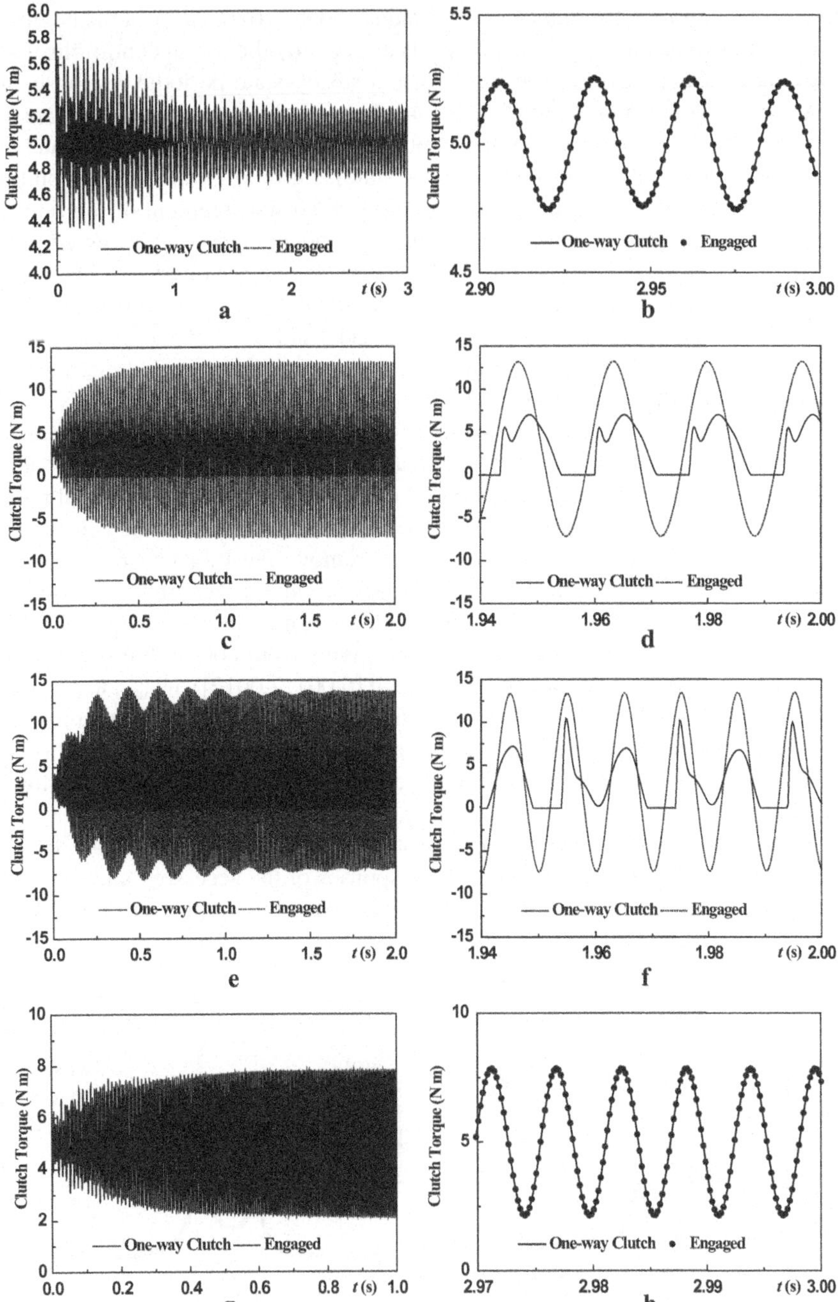

Fig. 5.4 Dynamic response of the clutch torque of the pulley–belt system: **a** time history ($\omega = 226$), **b** steady-state phase ($\omega = 226$), **c** time history ($\omega = 377$), **d** steady-state phase ($\omega = 377$), **e** time history ($\omega = 626$), **f** steady-state phase ($\omega = 626$), **g** time history ($\omega = 1110$), and **h** steady-state phase ($\omega = 1110$)

pulley's rotation ($\omega = 226$ rad/s, $\omega = 377$ rad/s, and $\omega = 626$ rad/s), at the beginning phase, the transmission torque's response is affected by the initial conditions. Later, the response of the transmission torque has a steady-state periodic response form after undergoing the transition. Figure 5.4d, f shows the response of the transmission torque for the accessory and driven pulley engaged with a one-way clutch and without any one-way clutch (i.e., engaged state). Furthermore, in the second and third primary resonances of the pulley rotation, the disengagement of the one-way clutch is from the zero transmission torque. This phenomenon was pointed out by Zhu and Parker (2005, 2006). Figure 5.4b, h reveals that the one-way clutch maintains the engaged state in the case of $\omega = 226$ and $\omega = 1110$. Nevertheless, in the second and third primary resonances ($\omega = 377$ rad/s and $\omega = 626$ rad/s), the one-way clutch removes the torque transmission in the opposite direction and decreases the resonance's magnitude (i.e., Fig. 5.4d, f).

The system's responses in the time interval of [0, 700] seconds are considered to investigate the magnitude of the stable steady-state periodic response. The nondimensional time interval is calculated based on the geometrical and physical parameters given in Table 5.1. That is [0, 7.0012]. It is noted that the data in the time interval of [0, 600) seconds are excluded to remove the transient response. The steady-state response's magnitudes of the accessory and driven pulley are obtained via recording local maximums of θ_1 and θ_3 in the time interval of [600, 700]. On the other hand, the local maximums and minimums from 600 to 700 s (i.e., corresponding to the nondimensional time interval of [6.001, 7.0012]), $w_{1\max}$, $w_{2\max}$, $w_{1\min}$, and $w_{2\min}$, are recorded. Accordingly, the magnitudes of the steady-state response of the lateral vibration in the belt's center are derived by $(w_{i\max} - w_{i\min})/2$ ($i = 1, 2$). These magnitudes are utilized to represent the belt's response for the following investigation.

The sweep frequency responses of the prototypical pulley–belt system are depicted in Figs. 5.5 and 5.6. A comparison of the responses of the accessory and that of the driven pulley is shown in Fig. 5.5a, whereas Fig. 5.5b compares the periodic response

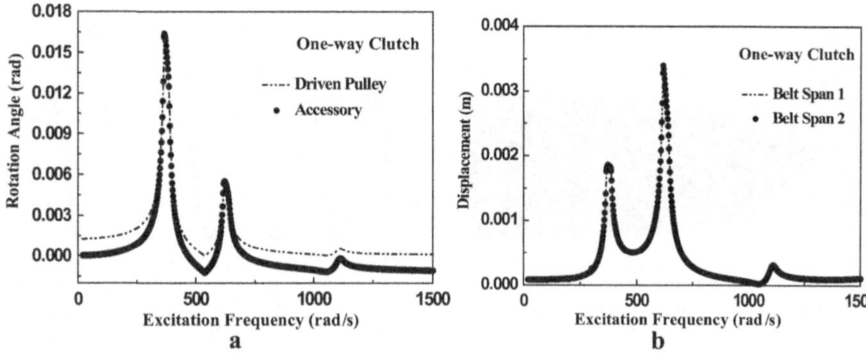

Fig. 5.5 Sweep frequency–response of the prototypical pulley–belt system with a one-way clutch: **a** comparison of the driven pulley with the accessory, **b** comparison of belt spans

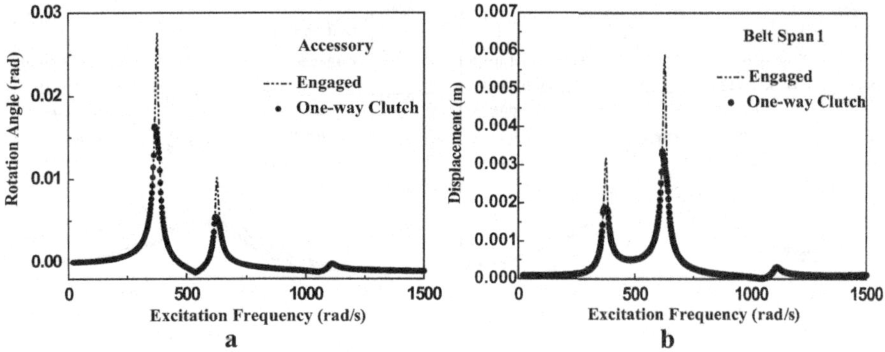

Fig. 5.6 Sweep frequency of the prototypical pulley–belt system to compare the steady-state response with and without one-way clutch: **a** comparison of pulley rotations, **b** comparison of lateral vibrations

of belt's spans. It is observed that when the excitation frequency approximates the natural frequency of the driven pulley (377 rad/s, 626 rad/s, or 1110 rad/s), the resonance phenomenon exists in all components of the pulley–belt system. It shows that the belt's lateral vibration is coupled with the pulley's rotations and the accessory's rotations.

Figure 5.6 also shows that the softening nonlinearities occur in the resonance of the system with the one-way clutch. This figure compares the sweep frequency of the system in two cases: with a one-way clutch and without one-way clutch. The sweep frequency corresponding to the accessory's rotation is shown in Fig. 5.6a, whereas that of the belt's vibration is revealed in Fig. 5.6b. In all cases, the resonance phenomenon exists in all components of the system. It is noted that the one-way clutch decreases the resonant magnitudes in the case of the second and third primary resonances. Nevertheless, the one-way clutch does not work in the case of subharmonic resonance. Observably, it does not significantly differ in responses near the second and the third primary resonances. Thus, the subsequent study will concentrate on the first primary resonance.

The influence of the belt's aspect ratio (b/h) on the system's resonance is shown in Fig. 5.7. In this investigation, the cross-section area is fixed. Therefore, the aspect ratio only affects the moment of inertia; see Table 5.2. In other words, this ratio only affects the belt's flexural rigidity EI. For a small moment of inertia, the typical hardening nonlinearity more clearly displays (Fig. 5.7). Figure 5.8 shows the considerable influence of the aspect ratio on the steady-state of the pulley–belt system. In two cases, with and without one-way clutch, the system's resonance is more significant when the aspect ratio is large. It is noted that the resonance of the spans is more affected by the aspect ratio. It is noted that when the aspect ratio is large, the flexural rigidity will be slight. In this case, the beam-belt model is closer to the string-belt model. Therefore, the dynamic model considering the belt's flexural rigidity for the axially moving belt (i.e., the beam-belt model) is more stable than the moving

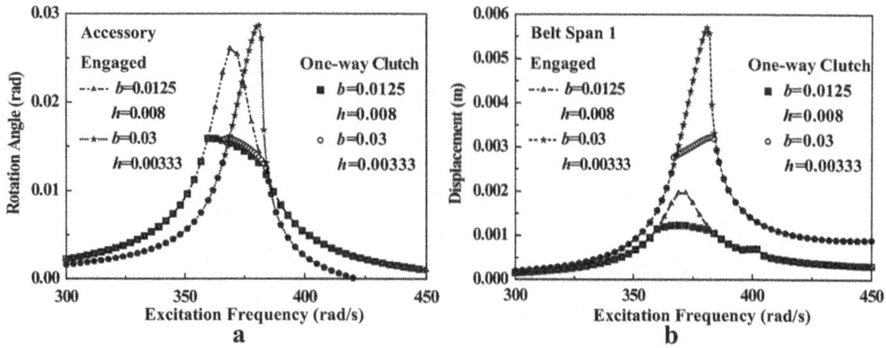

Fig. 5.7 Steady-state response for the different aspect ratio of the belt's cross section: **a** accessory, **b** belt span 1

Table 5.2 Influence of the aspect ratio on the moment of inertia

b (m)	h (m)	b/h	I
0.0125	0.008	1.5626	5.333×10^{-10} m^4
0.03	0.00333	9.009	0.926×10^{-10} m^4

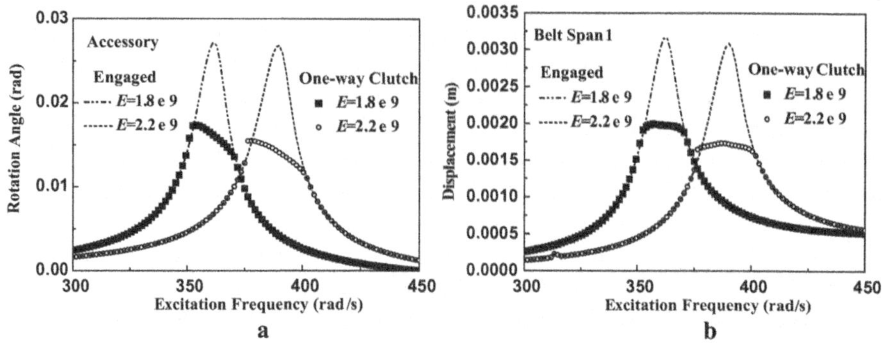

Fig. 5.8 Steady-state response for different Young's moduli of the belt: **a** the accessory, **b** the belt span 1

belt without flexural rigidity (the string-belt model). Additionally, the amplitude–frequency curve moves to the left with the increase of the aspect ratio. From this aspect, Ding and Zu (2014) and Ding et al. (2018) clarified the modeling problem of axially moving belts.

The steady-state responses for different Young's moduli are shown in Fig. 5.8, whereas that for different static tensions are revealed in Fig. 5.9. When Young's modulus and static tension increase, the resonance regions move to a higher frequency. It results in the increase of the natural frequency of the pulley's rotation. This reveals that the belt's lateral vibration and the rotation motion of the pulley

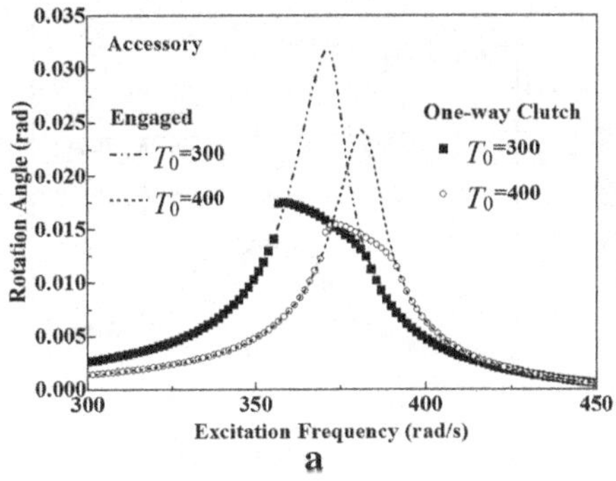

Fig. 5.9 Steady-state response for the different initial static tension of the belt: **a** the accessory, **b** the belt span 1

are coupled. Figure 5.8 shows that the resonance region is remarkably affected by Young's modulus, whereas the resonant magnitude is slightly affected. Contrarily, Fig. 5.9 reveals the considerable effect of the static tension on both magnitude and region of the resonance: The resonant magnitude will decrease with the increase of the static tension. Figure 5.10 demonstrates the steady-state responses for different viscous damping coefficients. It is observed that the viscous damping coefficient significantly affects the magnitude of the system's resonance. The resonance region becomes wider for larger Young's modulus and smaller static tension. Meanwhile, a bigger static tension, a bigger viscous damping coefficient, and a smaller Young's modulus generate the periodic response with less softening nonlinearity.

The influence of the stiffness and damping coefficient of the wrap spring on the steady-state response is shown in Figs. 5.11 and 5.12. The jumping phenomenon only exists in the response of the pulley–belt system with a one-way clutch when the stiffness is significant and the damping coefficient is small. Thus, the softening

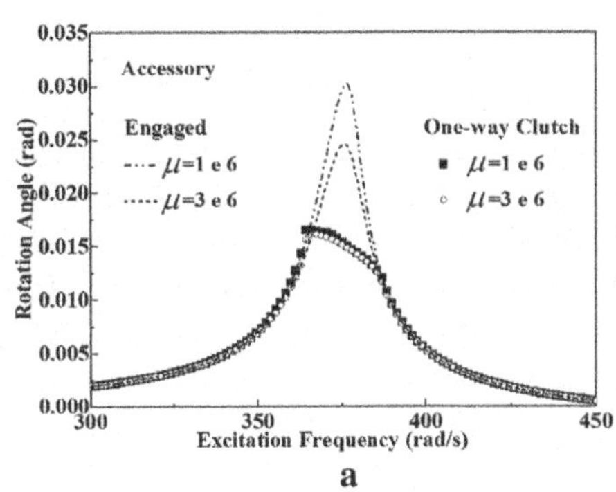

Fig. 5.10 Steady-state response for the different viscous damping coefficient: **a** the accessory, **b** the belt span 1

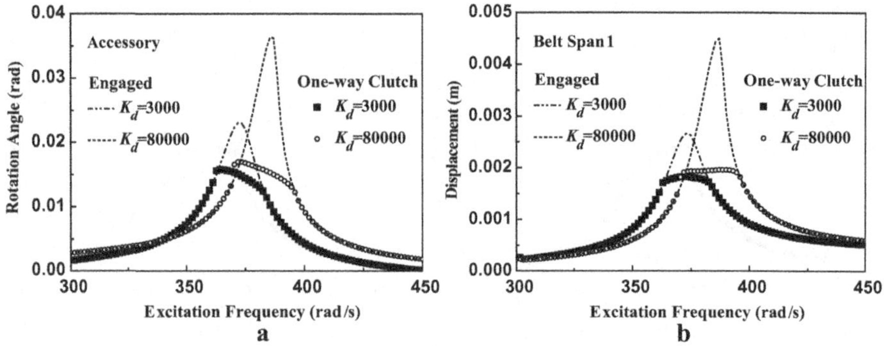

Fig. 5.11 Steady-state response for different stiffnesses of the wrap spring: **a** the accessory, **b** the belt span 1

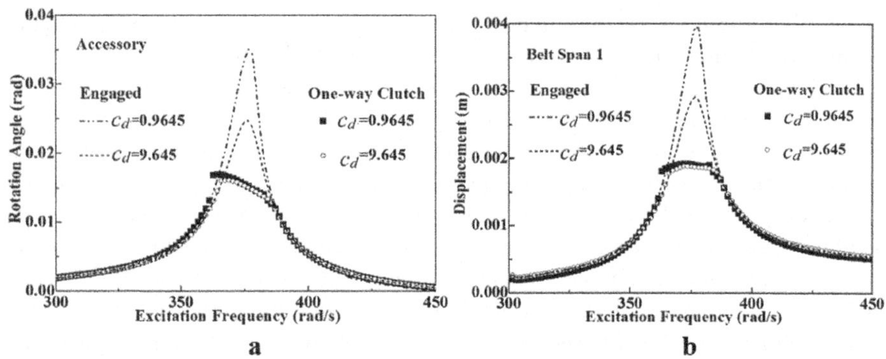

Fig. 5.12 Steady-state response for different damping coefficients of the wrap spring: **a** the accessory, **b** the belt span 1

nonlinearities are consequences of the dynamic system with a one-way clutch. Additionally, the softening behavior results from the clutch disengagement, and thus the wrap spring between the accessory and driven pulley is inactive.

As shown in Fig. 5.11, if the wrap spring's stiffness decreases, the resonance regions of the system will move to a lower frequency. It results in the decrease of the natural frequency of the system with a one-way clutch. Additionally, the stiffness also affects the resonant magnitude. Contrary to the stiffness, the damping coefficient of the wrap spring only affects the resonant magnitude of the pulley–belt system; see Fig. 5.12. For the pulley–belt system without one-way clutch, it is observed that when the stiffness increases and the damping coefficient decreases, the resonance of both the pully's rotation and the belt's lateral vibration becomes stronger. Figure 5.11 reveals that the resonance region of the system is narrowed using a smaller stiffness. Observably, the resonance of the pulley–belt system without one-way clutch is more sensitive to the change of the wrap spring's parameters (i.e., stiffness and damping coefficient) than that of the system with a one-way clutch.

The influence of the rotational inertias of the accessory and driven pulley on the magnitude of the steady-state response is shown in Figs. 5.13 and 5.14. As shown in these figures, the resonance region moves to the right, i.e., a higher excitation frequency, with the decrease of the rotational inertias of the accessory and driven pulley. It follows the increase of the natural frequency of the system. Additionally, the resonant response of the system becomes slightly stronger if the rotational inertia of the driven pulley and accessory decreases. It is concluded that the rotational inertias of the accessory and driven pulley significantly affect the amplitude–frequency relations of the pulley–belt system with and without one-way clutch. Figures 5.13 and 5.14 also show that the accessory's rotations are coupled with the belt's lateral vibration. Additionally, the softening nonlinearities of the resonance become stronger for the larger rotational inertias of the accessory and driven pulley.

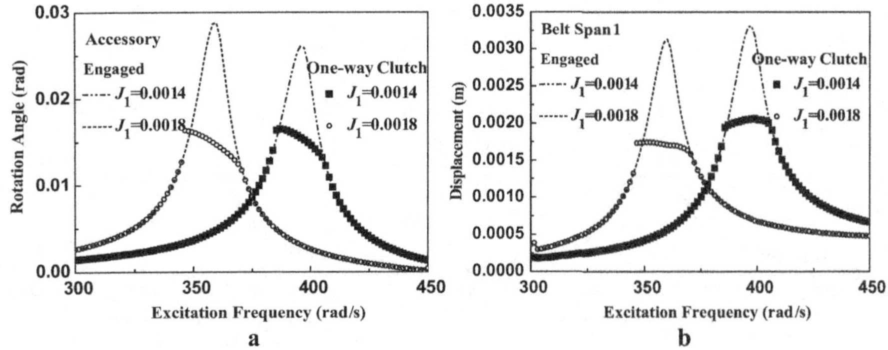

Fig. 5.13 Steady-state response for different rotational inertias of the driven pulley: **a** accessory, **b** belt span 1

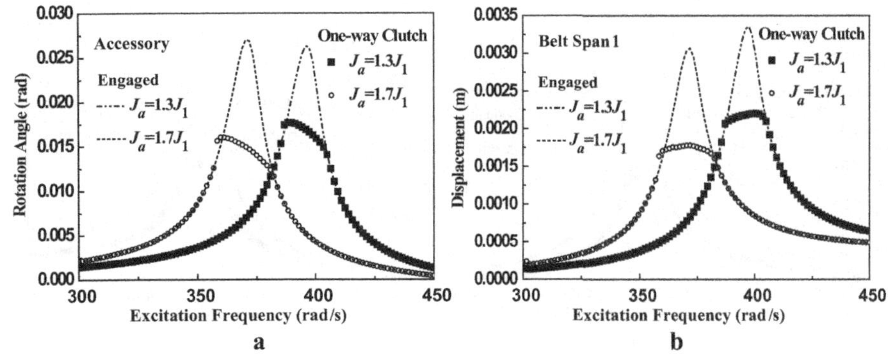

Fig. 5.14 Steady-state response for different rotational inertias of the accessory: **a** accessory, **b** belt span 1

It is noted that the influence of the damping coefficient of the rotating accessory is similar to that of friction between the pulley and belt. Therefore, for simplicity, we set $c_a = c_b$ in the following example. The effect of the rotational damping coefficients on the system response is shown in Fig. 5.15. The resonant magnitude decreases with the increase of the rotation damping coefficients. Additionally, the pulley–belt system without one-way clutch is more sensitive to the change of the rotation damping coefficients than the system with a one-way clutch. Furthermore, as shown in Fig. 5.15, the jumping phenomenon in the response of the pulley–belt system with a one-way clutch can be eliminated by using large rotation damping coefficients.

In Fig. 5.16, the amplitude–frequency relations of the accessory and the belt span are displayed for revealing the influences of the excitation magnitude. Figure 5.16a, b shows that when the excitation magnitude reduces, the magnitude and region of the resonances decrease. The resonant magnitude of the pulley–belt system with a one-way clutch is less sensitive to the change of the excitation magnitude than that

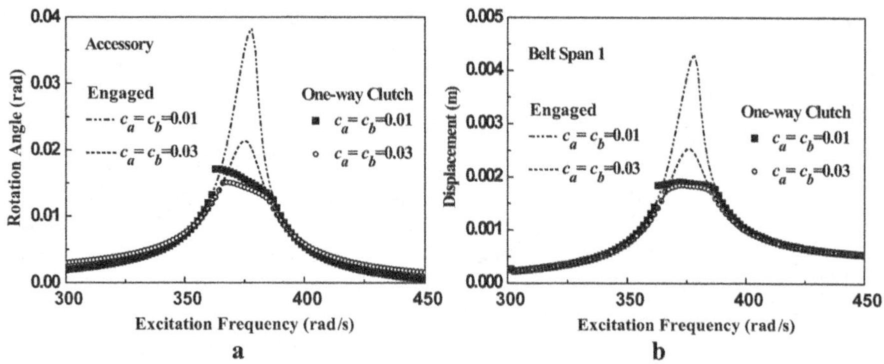

Fig. 5.15 Steady-state response for different friction coefficients of the rotational element: **a** accessory, **b** belt span 1

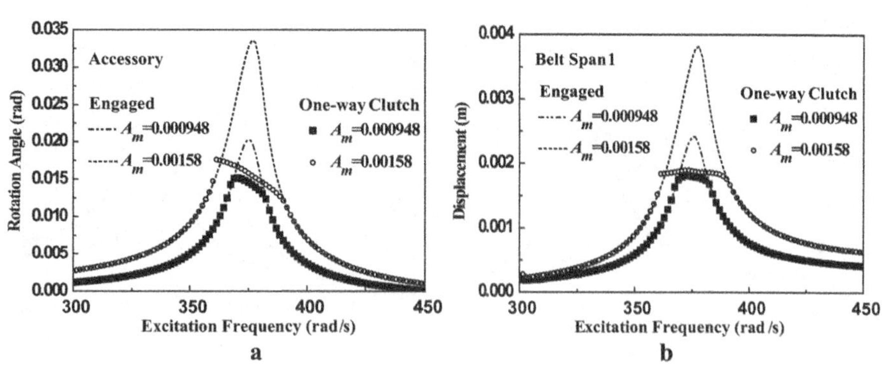

Fig. 5.16 Steady-state response for different amplitudes of excitation: **a** accessory, **b** belt span 1

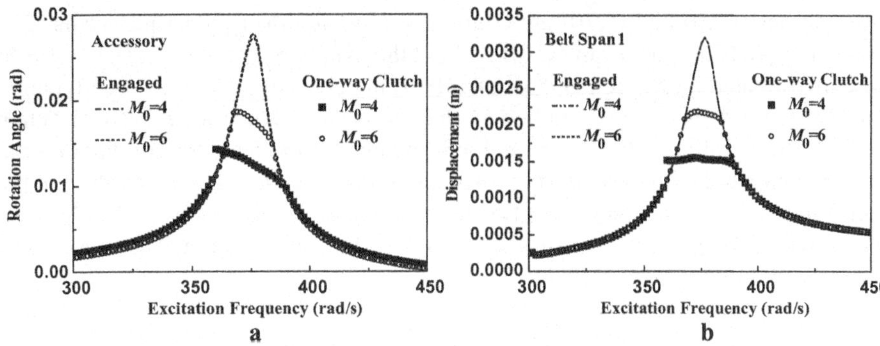

Fig. 5.17 Steady-state response for different preloads of the driven pulley: **a** accessory, **b** belt span 1

of the system without one-way clutch. It is shown that the system with a clutch is more stable.

The influence of the preload of the driven pulley on the amplitude–frequency relations of the periodic response is shown in Fig. 5.17. For the pulley–belt system without one-way clutch, the influence of the preload on the amplitude–frequency curve is insignificant. The amplitude–frequency curve of the system without one-way clutch is almost unchanged when the preload of the driven pulley changes. Contrarily, the resonance of the system with a one-way clutch is significantly affected by the preloads of the driven pulley. When the preload of the driven pulley increases, the resonance regions of the system are narrowed, the resonant magnitude increases, and the softening nonlinearity is strengthened.

5.3 Approximate Model

The axially moving belt involving both longitudinal and lateral vibrations has been modeled as a set of coupled PDEs, which is not convenient to exploit the dynamic behaviors. The Galerkin method is a handy technique to truncate the system into ODEs, which is relatively easier to investigate. Now consider the forced vibration of an axially moving belt. By adding structural damping and neglecting higher nonlinear order other than three, (5.21) and (5.22) lead to the following governing equations.

$$\rho A\left(u_{tt} + 2v_0 u_{xt} + v_0^2 u_{xx}\right) + c_1 u_t - EAu_{xx} - EAw_x w_{xx} = F_1 \cos(\Omega_1 t), \quad (5.69)$$

$$\begin{aligned} &\rho A\left(w_{tt} + 2v_0 w_{xt} + v_0^2 w_{xx}\right) + c_2 w_t - T_0 w_{xx} + EI w_{xxxx} \\ &\quad - EA(u_{xx} w_x + u_x w_{xx}) = F_2 \cos(\Omega_2 t). \end{aligned} \quad (5.70)$$

Equations (5.69) and (5.70) are coupled via the nonlinear quadratic terms. The dynamic models in (5.69) and (5.70) without the axially moving effect (i.e., $v_0 = 0$) were investigated by Nayfeh and Mook (1979) in detail. In their study, the dynamic model of the stationary belt is studied for the cases of $u = O(w^2)$ and $u = O(w)$. When the radius of gyration of the cross section is extremely small, the longitudinal inertia is small compared with the restoring force; therefore, $u = O(w^2)$. The coupling of the lateral and longitudinal vibrations of AMSs, in this case, is also discussed by Ghayesh et al. (2012, 2013). In the case of $u = O(w)$, the radius of gyration is assumed to be insignificant but not extremely small.

The Galerkin truncation method is used to establish the ODE model for an axially moving belt. Accordingly, we can expand the spatiotemporal function in terms of linear undamped natural modes of the corresponding time-independent system by ignoring the excitation, the translating velocity, and the damping effects. The longitudinal and lateral vibrations of an axially moving belt with pinned–pinned end conditions are assumed as follows:

$$u(x, t) = \sum_{n=1}^{\infty} p_n(t) \sin \frac{n\pi x}{l}, \tag{5.71}$$

$$w(x, t) = \sum_{n=1}^{\infty} q_n(t) \sin \frac{n\pi x}{l}. \tag{5.72}$$

Substituting (5.71) and (5.72) into (5.69) and (5.70), multiplying both sides of the resulting equations by $\sin(m\pi x/l)$ and then integrating over the interval $[0, l]$ yields

$$\begin{aligned} \ddot{p}_n + \lambda_n^2 p_n = \Bigg\{ & -k_1 \dot{p}_n - 4v_0 \sum_{m=1}^{\infty} \frac{nm}{m^2 - n^2} \left[(-1)^{m+n} - 1 \right] \dot{p}_m \\ & - k_3 \left[\sum_{m=1}^{\infty} mn(m+n) q_m q_{m+n} + \sum_{m=1}^{n} m^2 (n-m) q_m q_{n-m} \right] \\ & - \frac{(-1)^n}{n} f_1 \cos(\Omega_1 t) \Bigg\}, \end{aligned} \tag{5.73}$$

$$\begin{aligned} \ddot{q}_n + \omega_n^2 q_n = \Bigg\{ & -k_2 \dot{q}_n - 4v_0 \sum_{m=1}^{\infty} \frac{nm}{m^2 - n^2} \left[(-1)^{m+n} - 1 \right] \dot{q}_m \\ & - k_3 \left[\sum_{m=1}^{\infty} mn(m+n)(q_m p_{m+n} + p_m q_{n+m}) \right. \\ & \left. + \sum_{m=1}^{n} m^2 (n-m)(p_m q_{n-m} + q_m p_{n-m}) \right] \\ & - \frac{(-1)^n}{n} f_2 \cos(\Omega_2 t) \Bigg\}, \end{aligned} \tag{5.74}$$

where λ_n and ω_n denote, respectively, the n-th order natural frequency of the static beam in the longitudinal and the lateral directions and

$$k_1 = \frac{c_1}{\rho A l},\ k_2 = \frac{c_2}{\rho A l},\ k_3 = \pi^3 \frac{EA - T_0}{2\rho A l^5},\ f_1 = \frac{4F_1 l}{\rho A \pi},\ f_2 = \frac{4F_2 l}{\rho A \pi}. \tag{5.75}$$

The truncated approximate model (5.73) and (5.74) can be investigated by both analytical and numerical methods. It should be noted that two directions are coupled by nonlinear terms in both equations, and the generalized coordinates in the same direction are coupled by gyroscopic terms.

5.4 Dynamic Analysis of Belt Models

5.4.1 Equilibrium Solutions

If the damping, excitation, and nonlinear coupling terms are neglected, Eqs. (5.73) and (5.74) result in two sets of decoupled equations

$$\ddot{p}_n + \lambda_n^2 p_n + 4v_0 \sum_{m=1}^{\infty} \frac{nm}{m^2 - n^2}\left[(-1)^{m+n} - 1\right]\dot{p}_m = 0, \tag{5.76}$$

$$\ddot{q}_n + \omega_n^2 q_n + 4v_0 \sum_{m=1}^{\infty} \frac{nm}{m^2 - n^2}\left[(-1)^{m+n} - 1\right]\dot{q}_m = 0. \tag{5.77}$$

The longitudinal and the lateral natural frequencies λ_n and ω_n are derived via solving for the roots of the following algebraic equations.

$$\begin{aligned} &\det\left(-\lambda^2 \mathbf{I} + \mathbf{K}_1 + \mathrm{i}\lambda \mathbf{G}\right) = 0, \\ &\det\left(-\omega^2 \mathbf{I} + \mathbf{K}_2 + \mathrm{i}\omega \mathbf{G}\right) = 0 \end{aligned} \tag{5.78}$$

where $\mathbf{I}$ denotes the unit matrix, $\mathbf{K}_1$ and $\mathbf{K}_2$ are diagonal stiffness matrices, and $\mathbf{G}$ denotes the gyroscopic matrix. The elements of $\mathbf{G}$, $\mathbf{K}_1$, and $\mathbf{K}_2$ are

$$\begin{aligned} &\mathrm{G}_{nm} = 4v_0 \frac{nm}{m^2 - n^2}\left[(-1)^{m+n} - 1\right], \\ &\mathrm{K}_{1nn} = \frac{n^2\pi^2}{l^2}\left(\frac{E}{\rho} - v_0^2\right),\ \mathrm{K}_{2nn} = \frac{n^4\pi^4}{l^4}\frac{EI}{\rho A} - \frac{n^2\pi^2}{l^2}v_0^2 + \frac{T_0}{\rho A}. \end{aligned} \tag{5.79}$$

Table 5.3 Values of parameters

ρ	E	A	I	l	P
0.8×10^3 kg/m^3	6.5 MPa	1.0×10^{-3} m^2	8.0×10^{-6} m^4	1 m	500 N

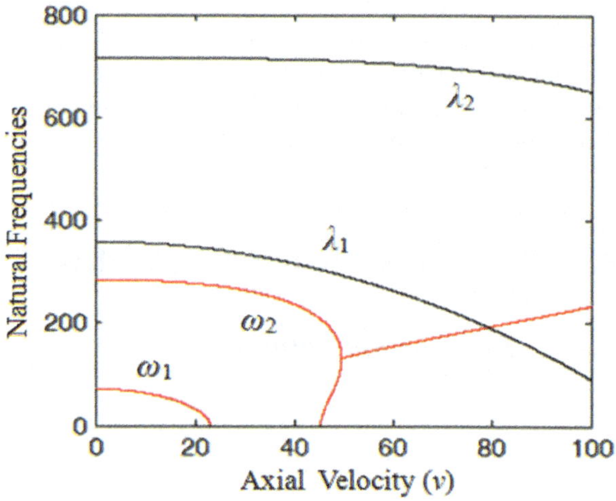

Fig. 5.18 Natural frequencies versus the axially moving velocity

Analysis on the roots of (5.78) reveals the influence of the axial speed on the longitudinal and the lateral natural frequencies. Consider the axially moving system with parameters given in Table 5.3. Figure 5.18 presents the variation of the first two lateral and longitudinal natural frequencies with respect to the translating speed. If the translating speed increases, the natural frequencies will decrease. This is pointed out in many references on the vibration of AMSs. When the translating speed is low, the natural frequencies of the lateral vibration are usually lower than that of the longitudinal vibration. Although all the natural frequencies are reduced with the increasing translating speed, the varying rates of the longitudinal and the lateral natural frequencies are rather different. Hence, in most cases, the lateral vibration rather than the longitudinal vibration has been focused on.

Inspection of (5.73) and (5.74) shows that the trivial solution is the static equilibrium of the system. However, the stability of the trivial equilibrium can be determined by the eigenvalues of (5.78). Figure 5.18 shows the natural frequencies varying with increasing axial speed. Starting from speed zero, all natural frequencies, which are the natural frequencies of the static beam, will decrease. The first natural frequency vanishes at the critical speed $v_{cr} = 24.8$, beyond which the system becomes unstable due to divergence. However, with further increase of the axial speed, the first natural frequency will appear again, and the system will regain stability at $v = 46.2$. Beyond the critical axial speed where the first two natural frequencies coincide, the system loses its stability for the second time due to flutter.

5.4.2 Stability Analysis

Further consider the responses and the stability of an axially moving belt described under an external lateral excitation described by (5.73) and (5.74) (Yang and Zhang 2014). If no internal resonance is considered, the truncated one degree-of-freedom system may be enough to show the contribution of the nonlinear terms and the external excitation parameters. The internal resonance makes the coupling among the multiple degree-of-freedoms strong, and some interesting phenomena may be found.

Some internal resonance points can be located from the frequency varying trend via detuning the translating speed, as shown in Fig. 5.18. For instance, when speed $v = 9.46$, the first natural frequency of the longitudinal vibration $\lambda_1 = 355.78$, the first two lateral natural frequencies are $\omega_1 = 70.77$ and $\omega_2 = 285.01$. In this case, the internal resonance may appear since $\lambda_1 = \omega_1 + \omega_2$.

The method of multiple scales can be used to analyze the governing ODEs (5.73) and (5.74). We assume that the approximate solutions for small but finite amplitude vibrations have in the following form.

$$p_n(t) = p_{n0}(\tau_0, \tau_1) + \varepsilon p_{n1}(\tau_0, \tau_1) + O(\varepsilon^2),$$

$$q_n(t) = q_{n0}(\tau_0, \tau_1) + \varepsilon q_{n1}(\tau_0, \tau_1) + O(\varepsilon^2), \tag{5.80}$$

where $\tau_0 = t$ and $\tau_1 = \varepsilon t$ represent the fast and the slow timescales and ε is the bookkeeping device. Substitute (5.80) and their derivatives into (5.73) and (5.74) yields

$$O(\varepsilon^0): \begin{cases} \mathrm{D}_0^2 p_{n0} + \lambda_n^2 p_{n0} = 0, \\ \mathrm{D}_0^2 q_{n0} + \omega_n^2 q_{n0} = 0, \end{cases} \tag{5.81}$$

$$O\left(\varepsilon^{1}\right):\begin{cases}\mathrm{D}_0^2 p_{n1}+\omega_n^2 p_{n1}=-k_1\mathrm{D}_0 p_{n0}-4v_0\sum_{m=1}^{\infty}\frac{nm}{m^2-n^2}\left[(-1)^{m+n}-1\right]\mathrm{D}_0 p_{m0}\\ \quad -k_3\left[\sum_{m=1}^{\infty}nm(n+m)q_{m0}q_{(m+n)0}+\sum_{m=1}^{n}m^2(n-m)q_{m0}q_{(n-m)0}\right]\\ \quad -2\mathrm{D}_0\mathrm{D}_1 p_{n0}-\frac{(-1)^n}{n}f_1\cos(\Omega_1 t),\\ \mathrm{D}_0^2 q_{n1}+\gamma_n^2 q_{n1}=-k_2\mathrm{D}_0 q_{n0}\\ \quad -4v_0\sum_{m=1}^{\infty}\frac{nm}{m^2-n^2}\left[(-1)^{m+n}-1\right]\mathrm{D}_0 q_{m0}-2\mathrm{D}_0\mathrm{D}_1 q_{n0}\\ \quad -k_3\left[\sum_{m=1}^{\infty}mn(m+n)\left(p_{m0}q_{(n+m)0}+q_{m0}p_{(m+n)0}\right)\right.\\ \quad \left.+\sum_{m=1}^{n}m^2(n-m)\left(p_{m0}q_{(n-m)0}+q_{m0}p_{(n-m)0}\right)\right]\\ \quad -\frac{(-1)^n}{n}\varepsilon f_2\cos(\Omega_2 t),\end{cases} \tag{5.82}$$

In (5.81) and (5.82), the differential operator D_n $(n=0,1)$ stands for the derivative with respect to τ_n.

Equation (5.81) has the following solutions.

$$p_{n0}=A_n(\tau_1)\exp(\mathrm{i}\lambda_n\tau_0)+A_n^*(\tau_1)\exp(-\mathrm{i}\lambda_n\tau_0),$$

$$q_{n0}=B_n(\tau_1)\exp(\mathrm{i}\omega_n\tau_0)+B_n^*(\tau_1)\exp(-\mathrm{i}\omega_n\tau_0). \tag{5.83}$$

Substitution of (5.83) into (5.82) leads to

$$\begin{aligned}\mathrm{D}_0^2 p_{n1}+\lambda_n^2 p_{n1}=&-k_1\mathrm{i}\lambda_n A_n\mathrm{e}^{\mathrm{i}\lambda_n\tau_0}-4v_0\sum_{m=1}^{\infty}\frac{nm}{m^2-n^2}\left[(-1)^{m+n}-1\right]\mathrm{i}\lambda_m A_m\mathrm{e}^{\mathrm{i}\lambda_m\tau_0}\\ &-2\mathrm{D}_1 A_n\mathrm{i}\lambda_n\mathrm{e}^{\mathrm{i}\lambda_n\tau_0}\\ &-k_3\sum_{m=1}^{\infty}mn(n+m)\left\{B_m B_{n+m}\mathrm{e}^{\mathrm{i}(\omega_m+\omega_{n+m})\tau_0}\right.\\ &\left.+B_m^* B_{n+m}\mathrm{e}^{\mathrm{i}(\omega_{n+m}-\omega_m)\tau_0}\right\}\\ &-k_3\sum_{m=1}^{n}m^2(n-m)\left\{B_m B_{n-m}\mathrm{e}^{\mathrm{i}(\omega_m+\omega_{n-m})\tau_0}\right.\\ &\left.+B_m B_{n-m}^*\mathrm{e}^{\mathrm{i}(\omega_m-\omega_{n-m})\tau_0}\right\}\end{aligned}$$

$$- \frac{(-1)^n}{2n} f_1 \exp(\mathrm{i}\Omega_1 \tau_0) + cc, \tag{5.84}$$

$$\begin{aligned}
\mathrm{D}_0^2 q_{n1} + \omega_n^2 q_{n1} = & -k_2 \mathrm{i}\omega_n B_n \mathrm{e}^{\mathrm{i}\omega_n \tau_0} - 4v_0 \sum_{m=1}^{\infty} \frac{nm}{m^2 - n^2} \left[(-1)^{m+n} - 1\right] \mathrm{i}\omega_m B_m \mathrm{e}^{\mathrm{i}\omega_m \tau_0} \\
& - 2i\omega_n \mathrm{D}_1 B_n \mathrm{e}^{\mathrm{i}\omega_n \tau_0} \\
& - k_3 \sum_{m=1}^{\infty} mn(n+m) \big[A_{n+m} B_m \mathrm{e}^{\mathrm{i}(\omega_m + \lambda_{n+m})\tau_0} \\
& + A_{n+m} B_m^* \mathrm{e}^{\mathrm{i}(\lambda_{n+m} - \omega_m)\tau_0} \big] \\
& - k_3 \sum_{m=1}^{\infty} nm(m+n) \big[A_m B_{n+m} \mathrm{e}^{\mathrm{i}(\lambda_m + \omega_{n+m})\tau_0} \\
& + A_m B_{n+m}^* \mathrm{e}^{\mathrm{i}(\lambda_m - \omega_{n+m})\tau_0} \big] \\
& - k_3 \sum_{m=1}^{n} m^2 (n-m) \big[A_m B_{n-m} \mathrm{e}^{\mathrm{i}(\lambda_m + \omega_{n-m})\tau_0} \\
& + A_m B_{n-m}^* \mathrm{e}^{\mathrm{i}(\lambda_m - \omega_{n-m})\tau_0} \big] \\
& - k_3 \sum_{m=1}^{n} m^2 (n-m) \big[A_{n-m} B_m \mathrm{e}^{\mathrm{i}(\omega_m + \lambda_{n-m})\tau_0} \\
& + A_{n-m}^* B_m \mathrm{e}^{\mathrm{i}(\omega_m - \lambda_{n-m})\tau_0} \big] \\
& - \frac{(-1)^n}{n} f_2 \exp(\mathrm{i}\Omega_2 \tau_0) + cc.
\end{aligned} \tag{5.85}$$

In (5.84) and (5.85), cc denotes the complex conjugate of all the terms on the right-hand side of the equation. The study on the possible secular terms of (5.84) leads to the following conclusion: Extra internal resonance links connecting p_{n1} and q_{n1} may occur when $\lambda_n \approx |\omega_m \pm \omega_{n-m}|$ or $\lambda_n \approx |\omega_{n+m} \pm \omega_m|$. By considering the possible secular terms, we can obtain different resonances under those algebraic relations.

Let us study the internal resonance with $\lambda_n \approx \omega_{n+m} + \omega_m$. The external detuning parameter σ_1 and internal detuning parameters σ_2 are introduced as follows:

$$\Omega_1 = \lambda_n + \varepsilon\sigma_1, \quad \lambda_n = \omega_n + \omega_{n+m} + \varepsilon\sigma_2. \tag{5.86}$$

In order to understand the external–internal resonance, we focus on the case of $f_2 = 0$ and $n = m = 1$. The first longitudinal natural frequency is detuned close to the sum of the first two lateral natural frequencies. It is noted that the external excitation in the lateral direction is omitted, whereas the excitation in the longitudinal direction is considered. The solvability conditions of (5.84) and (5.85), in this case, are obtained as follows:

$$
\begin{aligned}
&- \mathrm{i}\lambda_1(2\mathrm{D}_1 A_1 + k_1 A_1) - 2k_3 B_1 B_2 \mathrm{e}^{-\mathrm{i}\sigma_2\tau_1} + \tfrac{1}{2} f_1 \mathrm{e}^{\mathrm{i}\sigma_1\tau_1} = 0,\\
&- \mathrm{i}\omega_1(2\mathrm{D}_1 B_1 + k_2 B_1) - 2k_3 A_1 \overline{B}_2 \mathrm{e}^{\mathrm{i}\sigma_2\tau_1} = 0,\\
&- \mathrm{i}\omega_2(2\mathrm{D}_1 B_2 + k_2 B_2) - 2k_3 A_1 \overline{B}_1 \mathrm{e}^{\mathrm{i}\sigma_2\tau_1} = 0.
\end{aligned} \tag{5.87}
$$

The polar form for the amplitudes is introduced as follows:

$$
A_n = \frac{1}{2} a_n \mathrm{e}^{\mathrm{i}\alpha_n}, \quad B_n = \frac{1}{2} b_n \mathrm{e}^{\mathrm{i}\beta_n} \tag{5.88}
$$

where α_n, β_n, a_n, and b_n are real functions of τ_1.

Substituting (5.88) into (5.87) and separating imaginary and real parts of the resulting equations yield the following differential equations

$$
\begin{aligned}
\dot{a}_1 &= -\frac{1}{2} k_1 a_1 + \frac{k_3}{\lambda_1} b_1 b_2 \sin \xi_2 + \frac{f_1}{4\lambda_1} \sin \xi_1,\\
\dot{b}_1 &= -\frac{1}{2} k_2 b_1 - \frac{k_3}{\omega_1} a_1 b_2 \sin \xi_2,\\
\dot{b}_2 &= -\frac{1}{2} k_2 b_2 - \frac{k_3}{\omega_2} a_1 b_1 \sin \xi_2,\\
\dot{\xi}_1 &= \sigma_1 - \frac{k_3}{\lambda_1} \frac{b_1 b_2}{a_1} \cos \xi_2 + \frac{f_1}{4\lambda_1 a_1} \cos \xi_1,\\
\dot{\xi}_2 &= \sigma_2 + \frac{k_3}{\lambda_1} \frac{b_1 b_2}{a_1} \cos \xi_2 - \frac{f_1}{4\lambda_1 a_1} \cos \xi_1 - \frac{k_3}{\omega_1} \frac{a_1 b_2}{b_1} \cos \xi_2 - \frac{k_3}{\omega_2} \frac{a_1 b_1}{b_2} \cos \xi_2
\end{aligned} \tag{5.89}
$$

where the dot over a variable stands for the derivative of the variable with respect to τ_1. In (5.89), two new variables are introduced

$$
\xi_1 = \sigma_1 \tau_1 - \alpha_1, \quad \xi_2 = \sigma_2 \tau_1 + \alpha_1 - \beta_1 - \beta_2. \tag{5.90}
$$

To determine the steady-state response, we set time-dependent variables to constants and time derivatives to zero in (5.89) and (5.90). By letting the right-hand side to zero, we can obtain the steady-state solutions via a set of algebraic equations. The set of algebraic equations yield the magnitude response curve. The following two possible steady-state solutions are found: (i) $a_1 \neq 0$ and $b_1 = b_2 = 0$ and (ii) all a_1, b_1, and b_2 are nonzero. For the first case, $b_1 = b_2 = 0$, the analytical solution of a_1 is derived as follows.

$$
a_1 = \frac{f_1}{2\lambda_1 \left(k_1^2 + 4\sigma_1^2\right)^{1/2}}. \tag{5.91}
$$

For the case that a_1, b_1, and b_2 are nonzero, the solutions are obtained as follows:

$$a_1 = \frac{\left[(\sigma_1+\sigma_2)^2 + \frac{1}{4}(k_1+k_2)^2\right]^{1/2}}{\left(2\sqrt{\frac{k_3^2}{\omega_1\omega_2}}\right)} = \frac{1}{2k_3}\sqrt{\omega_1\omega_2\left[(\sigma_1+\sigma_2)^2 + \frac{1}{4}(k_1+k_2)^2\right]}, \tag{5.92}$$

$$b_1 = \left[\zeta\left(\omega_2/\omega_1\right)^{1/2}\right]^{1/2}, \quad b_2 = \left[\zeta\left(\omega_1/\omega_2\right)^{1/2}\right]^{1/2}. \tag{5.93}$$

In (5.93), the value of ζ is obtained via the roots of the following quadratic equation.

$$\frac{k_3^2}{\lambda_1^2}\zeta^2 + a_1\frac{k_3}{\lambda_1}(k_1\sin\xi_2 - 2\sigma_1\cos\xi_2)\zeta + \frac{1}{4}k_1^2a_1^2 + a_1^2\sigma_1^2 - \left(\frac{f_1}{4\lambda_1}\right)^2 = 0, \tag{5.94}$$

where

$$\begin{aligned}\sin\xi_2 &= -k_2\Big/\left[(\sigma_1+\sigma_2)^2 + k_2^2\right]^{1/2},\\ \cos\xi_2 &= (\sigma_1+\sigma_2)\Big/\left[(\sigma_1+\sigma_2)^2 + k_2^2\right]^{1/2}.\end{aligned} \tag{5.95}$$

According to the eigenvalues of the Jacobian matrix at the corresponding steady-state solution, the stability of steady-state solutions in (5.91)–(5.93) can be determined. The solution is unstable if the real part of any eigenvalue of the matrix is positive. Contrarily, the corresponding steady-state solution is stable if the real part of each eigenvalue is negative.

Based on the steady-state solutions, nonlinear phenomena found in the system are shown in Figs. 5.19, 5.20, 5.21, and 5.22. In Fig. 5.19, the first-order longitudinal

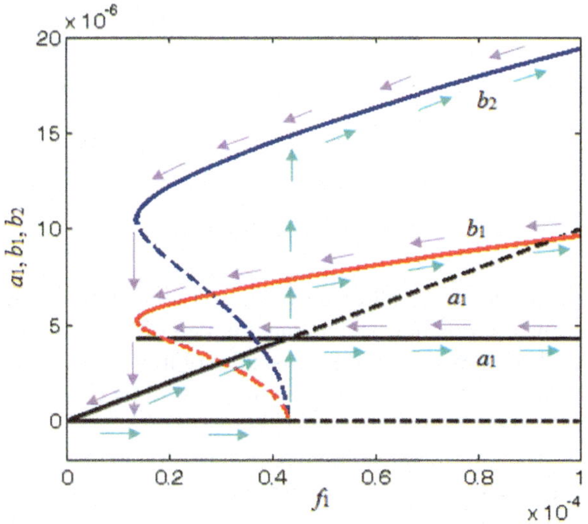

Fig. 5.19 Magnitudes of the responses as functions of excitation magnitudes ($\sigma_1 = 0.005$ and $\sigma_2 = 0$)

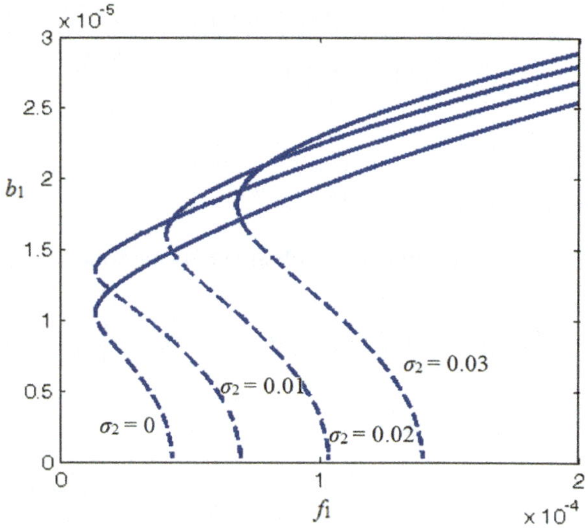

Fig. 5.20 Magnitudes of response b_1 as functions of the excitation amplitude ($\sigma_1 = 0.005$)

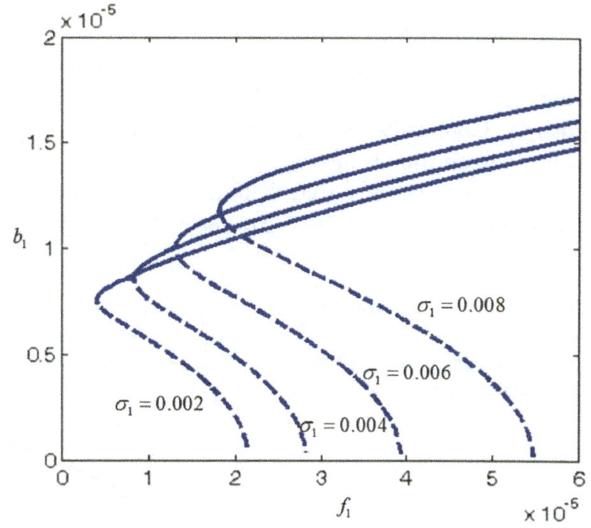

Fig. 5.21 Magnitudes of response b_1 as functions of the excitation's amplitude ($\sigma_2 = 0.01$)

magnitude a_1 and the first-order and second-order lateral magnitudes, b_1 and b_2, are shown as functions of the magnitude f_1 of the longitudinal excitation. In this figure, dashed lines stand for the unstable solutions, whereas the solid lines denote the stable solutions. The external resonance is slightly detuned as $\sigma_1 = 0.005$, and the internal resonance is precisely tuned $\sigma_2 = 0$.

The saturation phenomenon can be found in Fig. 5.19. The first-order longitudinal magnitude response a_1 is indicated in the solid black line as the excitation magnitude f_1 increases from zero. This natural relationship can be obtained based on the linear

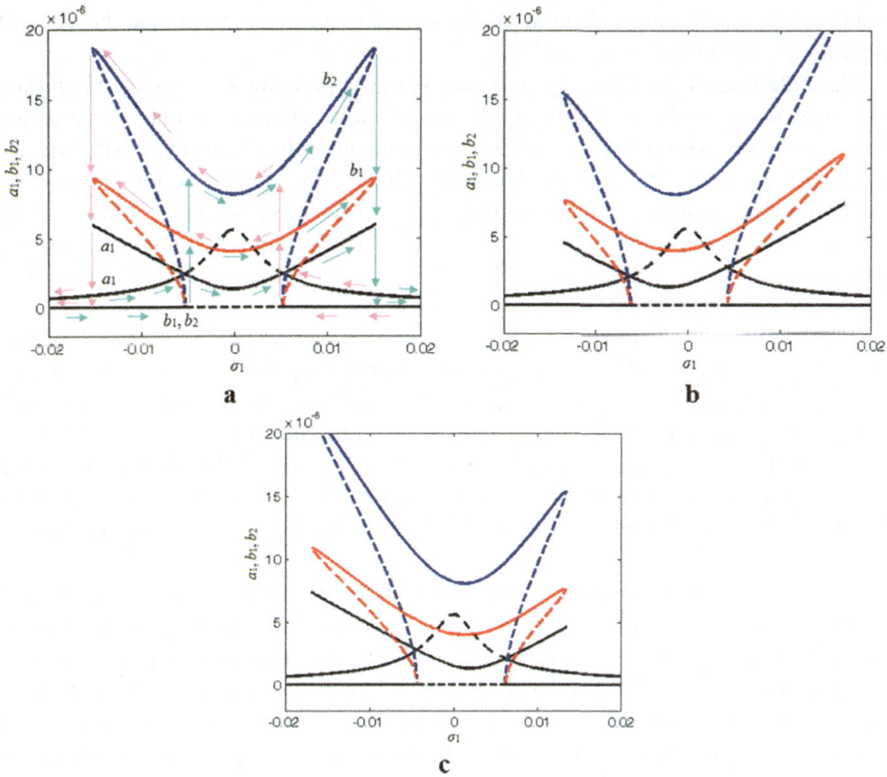

Fig. 5.22 Frequency–response curves: **a** $\sigma_2 = 0$, **b** $\sigma_2 = 0.02$, **c** $\sigma_2 = -0.02$

equation in (5.91), which coincides the solution of the corresponding linear problem. If the magnitude f_1 exceeds the critical value, the solution of (5.91) will become unstable (i.e., dashed black line), and another branch of the solution obtained via (5.92) will dominate. It is noted that the first-order longitudinal response a_1 of (5.92) does not depend on external longitudinal excitation magnitude f_1. Therefore, a_1 keeps a maximum value even if the excitation magnitude f_1 increases further. The saturation phenomenon occurs in the longitudinal vibration mode.

In Fig. 5.19, the magnitudes of the first-order and second-order lateral vibrations, b_1 and b_2, are denoted by the red and blue lines, respectively. It is observed that these magnitudes jump from zero to a finite value when the excitation magnitude f_1 exceeds the saturation point. These magnitudes of the lateral vibrations keep increasing with the further increase of the excitation magnitude f_1. In the saturation case, the energy due to the excitation in the longitudinal direction is transferred to the lateral motion.

The jumping phenomenon due to the varying excitation magnitude may appear as multiple stable solutions exist. This phenomenon can be observed via the magenta and cyan arrows in Fig. 5.19. There are two ways, i.e., from the zero excitation magnitude to the higher excitation magnitude and vice versa, to pursue the tread of magnitude responses a_1, b_1, and b_2.

The influence of the internal resonance detuning parameter σ_2 on the magnitude response curves is shown in Fig. 5.20. The relation between the magnitude of the first-order lateral vibrations b_1 and the excitation magnitude f_1 is demonstrated with different values of σ_2. It is observed that the critical value of the excitation magnitude f_1, wherein the saturation phenomenon occurs, may be postponed when the detuning parameter σ_2 increases. In other words, the increasing detuning parameter σ_2 delays the saturation phenomenon. However, the response magnitude becomes higher after the excitation magnitude f_1 exceeds the critical value.

The saturation has been investigated in some engineering fields (i.e., see Eftekhari et al. (2013) and Zhao and Xu (2012) for detailed investigations). One of the characteristics of saturation is the energy transfer. Accordingly, Oueini and Nayfeh (1997) presented a vibration suppression technique based on this method.

The influence of the external detuning parameter σ_1 on the magnitude of the first-order lateral vibration b_1 is presented in Fig. 5.21. Observably, the effect of the external detuning parameter is similar to the influence of the internal detuning parameter σ_2.

The magnitude responses of a_1, b_1, and b_2 described as functions of the internal detuning parameter σ_1 are shown in Fig. 5.22. Figure 5.22a reveals the magnitude responses with $\sigma_2 = 0$, whereas Fig. 5.22b, c depicts the magnitude responses with $\sigma_2 = 0.02$ and $\sigma_2 = -0.02$, respectively. In Fig. 5.22a, the jumping phenomenon due to the varying external excitation frequency can be observed via the magenta and cyan arrows. In the case of $\sigma_2 = 0$ (i.e., the internal resonance is precisely detuned), the frequency–response curve has a symmetrical configuration (Fig. 5.22a). Contrarily, the nonsymmetric behaviors of the frequency–response curve, in the case of the nonzero value of σ_2, are depicted in Fig. 5.22b, c.

5.4.3 Bifurcation and Chaotic Dynamics

Bifurcation and chaos of axially translating belts have been investigated in Ghayesh (2012) and Ghayesh et al. (2012). The axially moving speed is assumed to pulsate around a mean value. Then, the governing Eqs. (5.28) and (5.29) without external damping become

$$\begin{aligned}&\rho A\left(u_{tt} + 2vu_{xt} + v^2 u_{xx} + \dot{v}(1 + u_x)\right) - EAu_{xx} - EAw_x w_{xx}\\&\quad - \mu A u_{xxt} - \mu A w_{xt} w_{xx} - \mu A w_x w_{xxt} = 0,\end{aligned} \tag{5.96}$$

$$\begin{aligned}&\rho A\left(w_{tt} + 2vw_{xt} + \dot{v}w_x + v^2 w_{xx}\right) - T_0 w_{xx} + EIw_{xxxx} - EA(u_{xx}w_x + u_x w_{xx})\\&\quad + \mu I w_{xxxxt} - \mu A(u_{xxt}w_x + u_{xx}w_{xt} + u_{xt}w_{xx} + u_x w_{xxt}) = 0.\end{aligned} \tag{5.97}$$

Introduce the following dimensionless quantities:

$$\overline{x} = \frac{x}{l}, \overline{w} = \frac{w}{l}, \overline{u} = \frac{u}{l}, \overline{t} = t\sqrt{\frac{T_0}{\rho A l^2}}, \overline{v} = v\sqrt{\frac{\rho A}{T_0}}, k_1 = \sqrt{\frac{EA}{T_0}},$$

$$\alpha_2 = \frac{I\mu}{l^3\sqrt{\rho A T_0}}, \alpha_1 = \frac{A\mu}{l\sqrt{\rho A T_0}}, k_f = \sqrt{\frac{EI}{T_0 l^2}}. \tag{5.98}$$

Then, the governing equations in the dimensionless form are rewritten as follows:

$$\begin{aligned}&\overline{u}_{\overline{t}\overline{t}} + 2\overline{v}\overline{u}_{\overline{x}\overline{t}} + \overline{v}^2\overline{u}_{\overline{x}\overline{x}} + \dot{\overline{v}}(1+\overline{u}_{\overline{x}}) - k_1\overline{u}_{\overline{x}\overline{x}} - k_1\overline{w}_{\overline{x}}\overline{w}_{\overline{x}\overline{x}}\\ &- \alpha_1\overline{u}_{\overline{x}\overline{x}\overline{t}} - \alpha_1\overline{w}_{\overline{x}\overline{t}}\overline{w}_{\overline{x}\overline{x}} - \alpha_1\overline{w}_{\overline{x}}\overline{w}_{\overline{x}\overline{x}\overline{t}} = 0,\end{aligned} \tag{5.99}$$

$$\begin{aligned}&\overline{w}_{\overline{t}\overline{t}} + 2\overline{v}\overline{w}_{\overline{x}\overline{t}} + \dot{\overline{v}}\overline{w}_{\overline{x}} + \left(\overline{v}^2 - 1\right)\overline{w}_{\overline{x}\overline{x}} + k_2\overline{w}_{\overline{x}\overline{x}\overline{x}\overline{x}} - k_1\left(\overline{u}_{\overline{x}\overline{x}}\overline{w}_{\overline{x}} + \overline{u}_{\overline{x}}\overline{w}_{\overline{x}\overline{x}}\right)\\ &+ \alpha_2\overline{w}_{\overline{x}\overline{x}\overline{x}\overline{x}\overline{t}} - \alpha_1\left(\overline{u}_{\overline{x}\overline{x}\overline{t}}\overline{w}_{\overline{x}} + \overline{u}_{\overline{x}\overline{x}}\overline{w}_{\overline{x}\overline{t}} + \overline{u}_{\overline{x}\overline{t}}\overline{w}_{\overline{x}\overline{x}} + \overline{u}_{\overline{x}}\overline{w}_{\overline{x}\overline{x}\overline{t}}\right) = 0.\end{aligned} \tag{5.100}$$

Based on the Galerkin truncation method, the time-spatial response functions for the conditions of the pinned–pinned end are assumed as follows:

$$\overline{u}(\overline{x}, \overline{t}) = \sum_{n=1}^{N} p_n(\overline{t}) \sin\frac{n\pi\overline{x}}{l}, \tag{5.101}$$

$$\overline{w}(\overline{x}, \overline{t}) = \sum_{n=1}^{M} q_n(\overline{t}) \sin\frac{n\pi\overline{x}}{l}. \tag{5.102}$$

A set of $M+N$ second-order ODEs with time-dependent coefficients can be obtained. The moving speed pulsates around a mean value as follows:

$$\overline{v}(\overline{t}) = \gamma_0 + \gamma_1 \sin\Omega\overline{t} \tag{5.103}$$

where γ_0 and γ_1 are positive constants.

By varying the magnitude of the speed variations or the mean axial speed, we can investigate the global nonlinear dynamics of the system. Dynamic analysis is performed by applying the variable step size Runge–Kutta method for conducting direct time integration. The system's responses in the nondimensional time interval of [0 2000] are considered. It is noted that the first 70% of the data are excluded to remove the transient effects. The phase space is sectioned in every period of the speed variations, and the bifurcation diagrams of the Poincaré map are plotted.

We consider the system with the following parameters: $k_1 = 33.526$, $k_f = 0.173$, and $\mu = 0.04$. In Figs. 5.23, 5.24, 5.25, 5.26, 5.27, and 5.28, the mean axial speed γ_0 is treated as a bifurcation parameter. These figures show the bifurcation diagrams associating with the mean axial speed. Additionally, we consider different values of

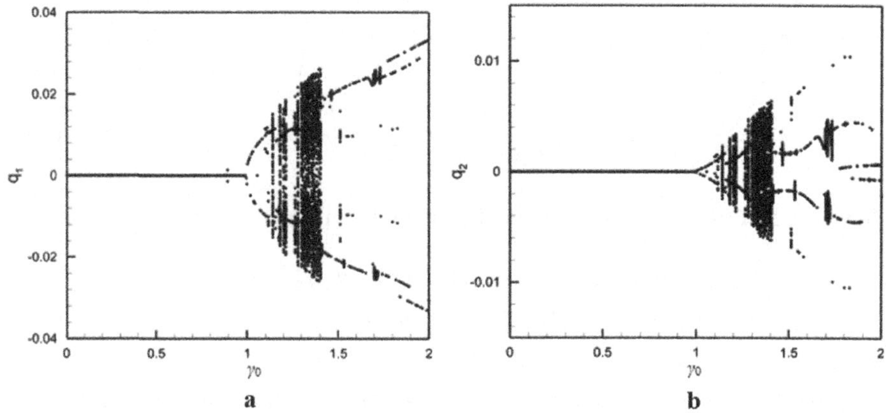

Fig. 5.23 Bifurcation diagrams of Poincaré points by increasing the mean axial speed ($\gamma_1 = 0.12$): **a** first generalized coordinate, **b** second generalized coordinate

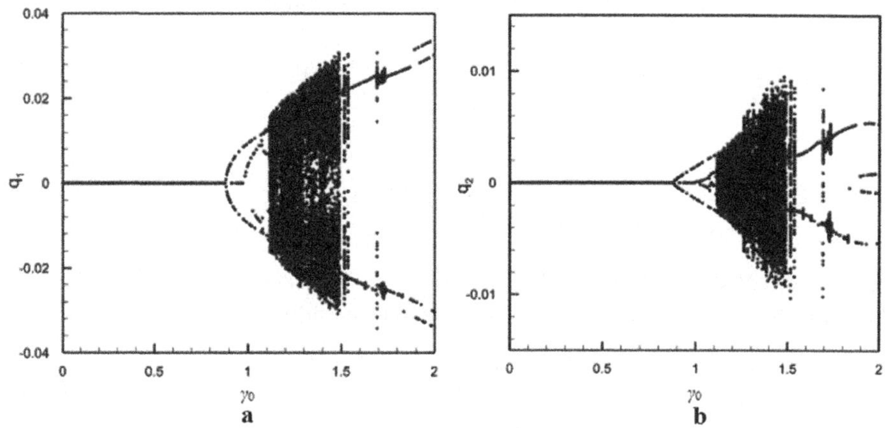

Fig. 5.24 Bifurcation diagrams of Poincaré points by increasing the mean axial speed ($\gamma_1 = 0.16$): **a** first generalized coordinate, **b** second generalized coordinate

the magnitude of the speed variations γ_1 (i.e., $\gamma_1 = 0.12$, 0.16, 0.20, 0.24, 0.40, and 0.60).

Figure 5.23a shows the first generalized coordinate of the lateral vibration (i.e., q_1) with $\gamma_1 = 0.12$. With the increasing mean translating speed γ_0, a zero response bifurcates to a period-2 motion at $\gamma_0 = 0.896$. In the mean translating speed range of [0.996 1.096], there are two coexisting periodic attractors, where the response jumps from one to another repeatedly. A typical periodic motion is demonstrated for $\gamma_0 = 1.012$ in Fig. 5.29. Figure 5.29a, b shows the time histories of the q_1 and q_2 motions,

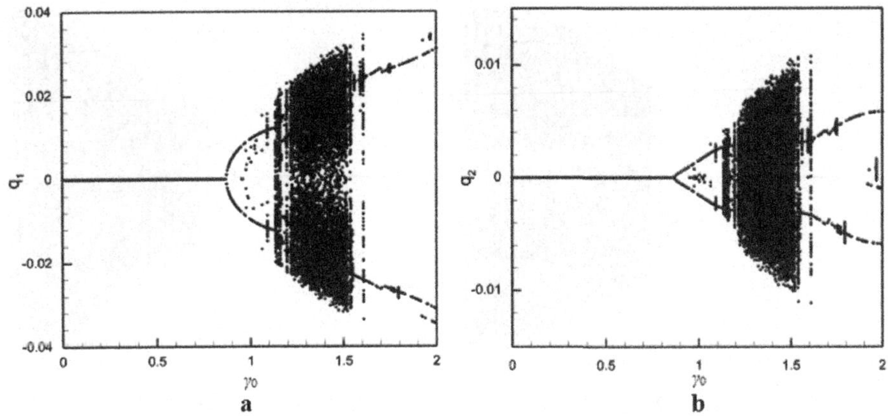

Fig. 5.25 Bifurcation diagrams of Poincaré points by increasing the mean axial speed ($\gamma_1 = 0.20$): **a** first generalized coordinate, **b** second generalized coordinate

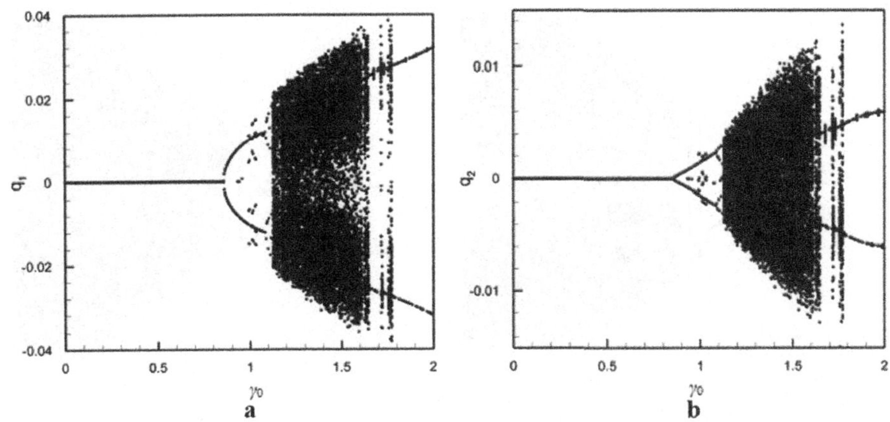

Fig. 5.26 Bifurcation diagrams of Poincaré points by increasing the mean axial speed ($\gamma_1 = 0.24$): **a** first generalized coordinate, **b** second generalized coordinate

whereas Fig. 5.29c, d reveals the phase-plane portraits of the q_1 and q_2 motions and Fig. 5.29a, b shows the fast Fourier transforms of the q_1 and q_2 motions, respectively. With a considerable mean translating speed, dynamic behaviors, including period-2, period-3, periodic, chaotic, and quasiperiodic motions, occur. Figure 5.30 shows the characteristics of a period-2 motion at $\gamma_0 = 1.164$. In the interval of [1.300 1.416], the chaotic motion is dominant. The periodic and period-2 motion are dominant when the mean translating speed is in the range of [1.420 1.692]. If the mean translating speed increases further, the quasiperiodic motion will be dominant (i.e., in the range of [1.700 1.732]). After that, the periodic motion is dominant until $\gamma_0 = 2.00$.

In Figs. 5.31, 5.32, 5.33, and 5.34, the magnitude of the translating speed variations

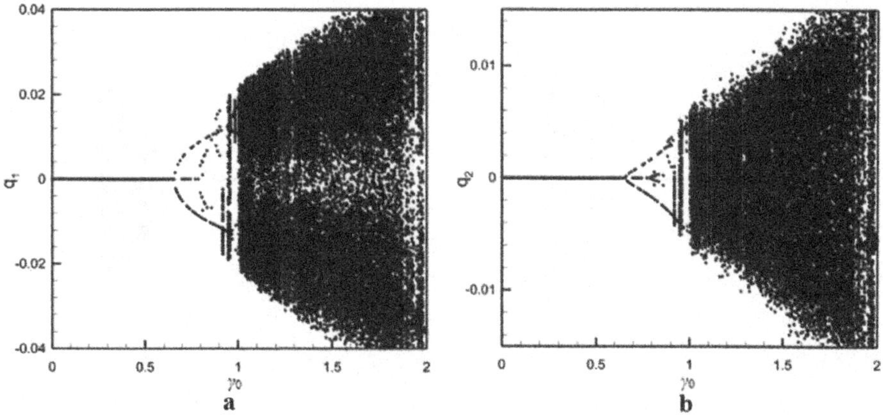

Fig. 5.27 Bifurcation diagrams of Poincaré points by increasing the mean axial speed ($\gamma_1 = 0.40$): **a** first generalized coordinate, **b** second generalized coordinate

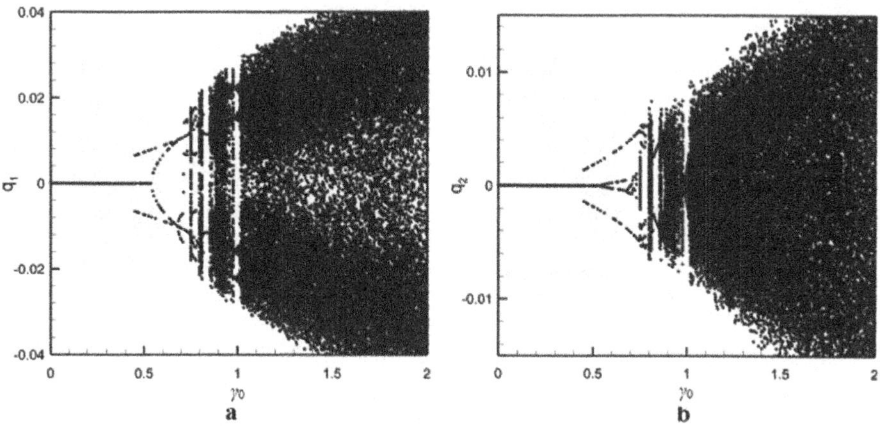

Fig. 5.28 Bifurcation diagrams of Poincaré points by increasing the mean axial speed ($\gamma_1 = 0.60$): **a** first generalized coordinate, **b** second generalized coordinate

is treated as a bifurcation parameter. In the numerical example shown in Fig. 5.31, we consider the system with the following parameters: $k_1 = 33.526$, $k_f = 0.173$, $\mu = 0.05$, and $\gamma_0 = 1.13$. It is noted that the mean translating speed, $\gamma_0 = 1.13$, is close but smaller than the critical value of the translating speed. Figure 5.31 shows the resonant responses of the first and second modes of the lateral vibration. Observably, in the vicinity of $\gamma_1 = 0.018$, the chaotic motion appears for the first time. The system experiences period-2 and chaotic oscillations when the magnitude of speed variations further increases.

In Figs. 5.32, 5.33, and 5.34, we set γ_0 to 1.0, 1.13, and 1.5, respectively. For all the examples, we set the excitation frequency to 1.2 and choose the other system

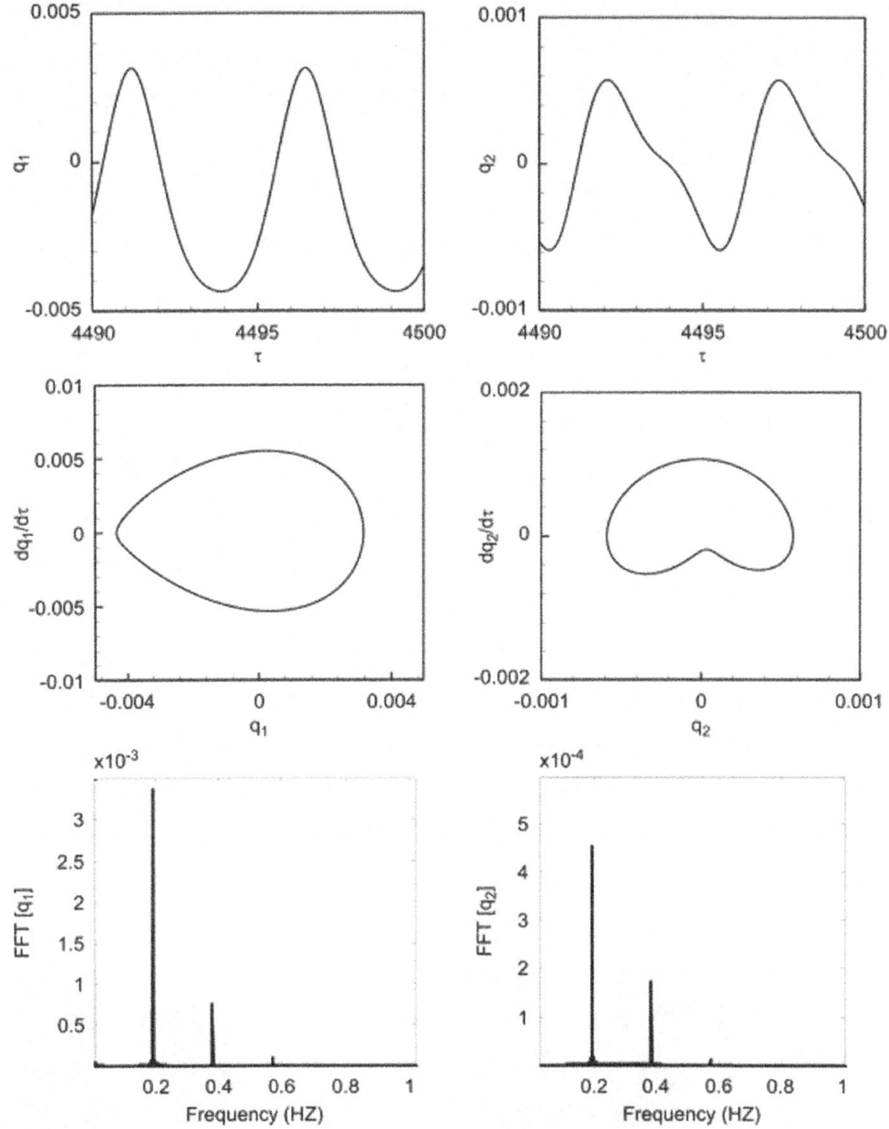

Fig. 5.29 Periodic oscillation for the system with $\gamma_0 = 1.012$

parameters as follows: $\mu = 0.05$, $k_f = 0.173$, and $k_1 = 33.526$. As shown in Fig. 5.32a, the system shows a zero response when the magnitude of the speed variations γ_1 is in the range of 0, 0.1. At $\gamma_1 = 0.1$, the zero response bifurcates. The further increase of the magnitude of the speed variations results in the period-doubling bifurcation and the chaotic motion. Figures 5.32, 5.33, and 5.34 show that the number of strange attractors, coexisting attractors, and higher-order bifurcations is increased when we

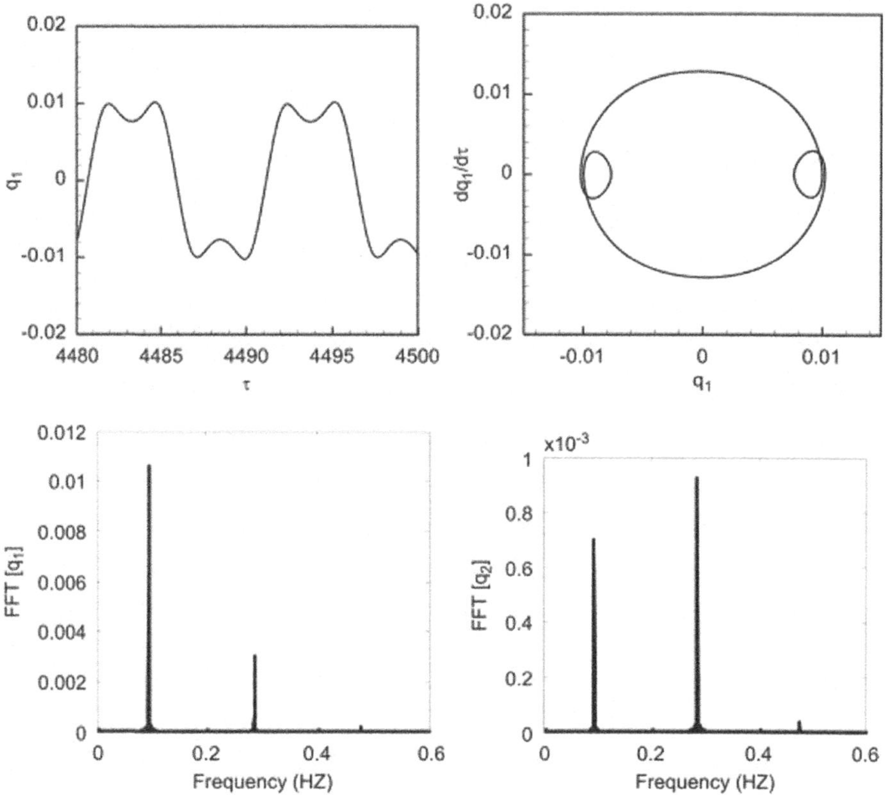

Fig. 5.30 Period-2 oscillation for the system with $\gamma_0 = 1.164$

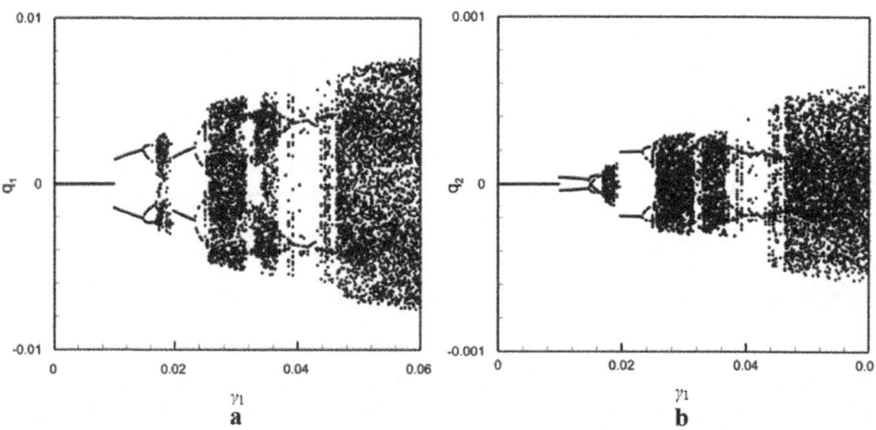

Fig. 5.31 Bifurcation diagrams of Poincaré points by increasing the excitation amplitude ($\gamma_0 = 1.13$), while the frequency of the speed variations is set to the first linear natural frequency of the lateral vibrations: **a** first generalized coordinate, **b** second generalized coordinate

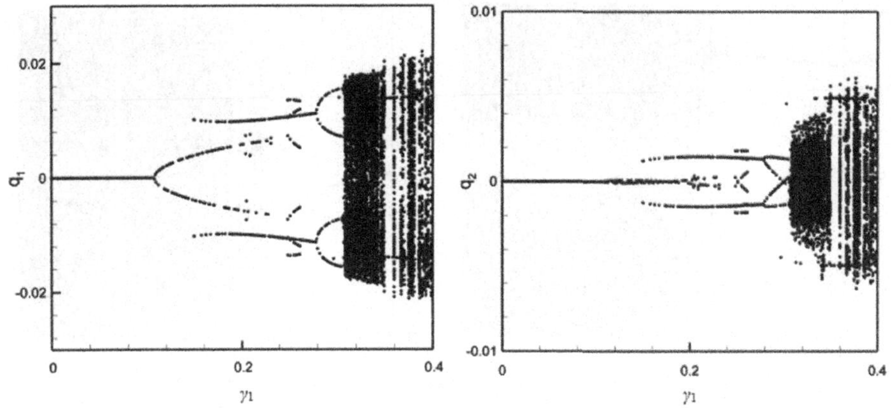

a b

Fig. 5.32 Bifurcation diagrams of Poincaré points by increasing γ_1 (with $\gamma_0 = 1.0$): **a** first generalized coordinate, **b** second generalized coordinate

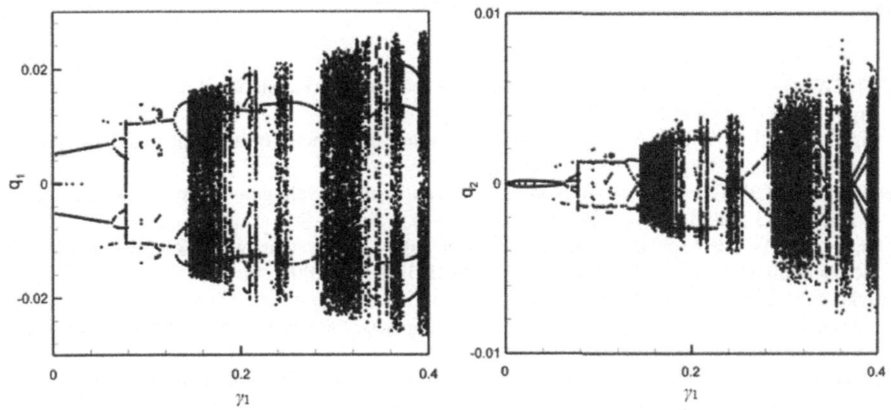

a b

Fig. 5.33 Bifurcation diagrams of Poincaré points by increasing γ_1 (with $\gamma_0 = 1.13$): **a** first generalized coordinate, **b** second generalized coordinate

set the mean translating speed to a more significant value and vary the magnitude of the speed variations. In these cases, the system displays chaotic oscillations at small excitation magnitudes.

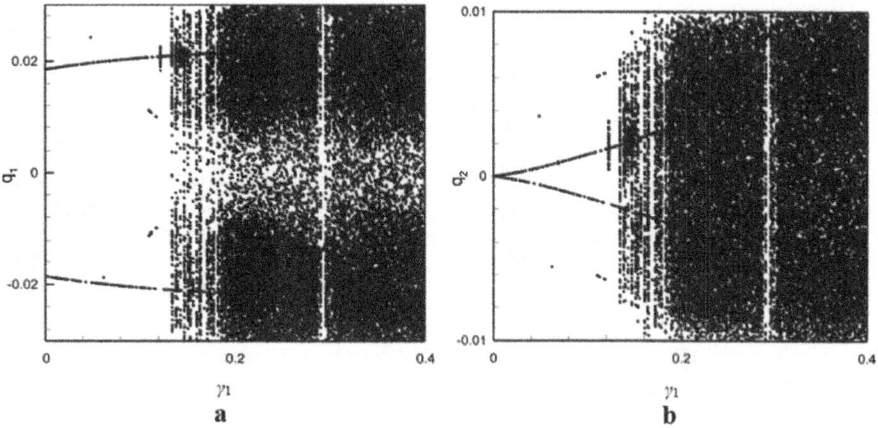

Fig. 5.34 Bifurcation diagrams of Poincaré points by increasing γ_1 (with $\gamma_0 = 1.5$): **a** first generalized coordinate, **b** second generalized coordinate

References

Bert CW, Malik M (1996) Differential quadrature method in computational mechanics: a review. Appl Mech Rev 49(1):1–28

Ding H, Chen LQ (2010) Galerkin methods for natural frequencies of high-speed axially moving beams. J Sound Vib 329(17):3484–3494

Ding H, Lim CW, Chen LQ (2018) Nonlinear vibration of a traveling belt with non-homogeneous boundaries. J Sound Vib 424:78–93

Ding H, Zu JW (2014) Steady-state responses of pulley-belt systems with a one-way clutch and belt bending stiffness. J Vib Acoust-Trans ASME 136(4):041006

Eftekhari M, Ziaei-Rad S, Mahzoon M (2013) Vibration suppression of a symmetrically cantilever composite beam using internal resonance under chordwise base excitation. Int J Non-Linear Mech 48:86–100

Farokhi H, Ghayesh MH, Hussain S (2016) Three-dimensional nonlinear global dynamics of axially moving viscoelastic beams. J Vib Acoust-Trans ASME 138(1):011007

Ghayesh M, Kazemirad S, Amabili M (2012) Coupled longitudinal-transverse dynamics of an axially moving beam with an internal resonance. Mech Mach Theory 52:18–34

Ghayesh MH (2012) Coupled longitudinal-transverse dynamics of an axially accelerating beam. J Sound Vib 331(23):5107–5124

Ghayesh MH, Amabili M (2013) Nonlinear dynamics of an axially moving Timoshenko beam with an internal resonance. Nonlinear Dyn 73(1–2):39–52

Ghayesh MH, Amabili M (2013) Nonlinear vibrations and stability of an axially moving Timoshenko beam with an intermediate spring support. Mech Mach Theory 67:1–16

Ghayesh MH, Amabili M, Farokhi H (2013) Coupled global dynamics of an axially moving viscoelastic beam. Int J Non-Linear Mech 51:54–74

Ghayesh MH, Farokhi H (2016) Nonlinear dynamical behavior of axially accelerating beams: Three-dimensional analysis. J Comput Nonlin Dyn 11(1):011007

Kong L, Parker RG (2003) Equilibrium and belt-pulley vibration coupling in serpentine belt drives. J Appl Mech-Trans ASME 70(5):739–750

Kong L, Parker RG (2004) Coupled belt-pulley vibration in serpentine drives with belt bending stiffness. J Appl Mech-Trans ASME 71(1):109–119

Mockensturm EM, Balaji R (2005) Piece-wise linear dynamic systems with one-way clutches. J Vib Acoust-Trans ASME 127(5):475–482

Nayfeh AH, Mook DT (1979) Nonlinear Oscillations. Wiley-Interscience, New York

Oueini SS, Nayfeh AH (1997) A theoretical and experimental implementation of a control method based on saturation. Nonlinear Dyn 13(2):189–202

Riedel CH, Tan CA (2002) Coupled, forced response of an axially moving strip with internal resonance. Int J Nonlinear Mech 37(1):101–116

Thurman AL, Mote CD (1969) Free, periodic, nonlinear oscillation of an axially moving strip. J Appl Mech-Trans ASME 36(1):83–91

Wang KW, Mote CD (1986) Vibration coupling analysis of band/wheel mechanical systems. J Sound Vib 109(2):237–258

Wickert JA (1992) Nonlinear vibration of a traveling tensioned beam. Int J Nonlinear Mech 27(3):503–517

Yang XD, Zhang W (2014) Nonlinear dynamics of axially moving beam with coupled longitudinal-transversal vibrations. Nonlinear Dyn 78(4):2547–2556

Zhao YY, Xu J (2012) Using the delayed feedback control and saturation control to suppress the vibration of the dynamical system. Nonlinear Dyn 67(1):735–754

Zhu F, Parker RG (2005) Nonlinear dynamics of a one-way clutch in belt-pulley systems. J Sound Vib 279(1–2):285–308

Zhu F, Parker RG (2006) Perturbation analysis of a clearance-type nonlinear system. J Sound Vib 292(3–5):969–979

Zhu F, Parker RG (2008) Influence of tensioner dry friction on the vibration of belt drives with belt bending stiffness. J Appl Mech-Trans ASME 130(1):011002

Chapter 6
Control of Axially Moving Belts

Abstract This chapter discusses the lateral vibration control of axially moving belts via a boundary control. The performance of the belt system under the proposed control law is shown via numerical results. The control schematic for simultaneous longitudinal and lateral vibration suppression based on the Lyapunov method is presented.

Keywords Boundary control · Lyapunov method · Belt · Quasi-static stretching · Lateral vibration control

In most practical axially moving systems, the control problem of lateral vibrations usually receives significantly more attention than longitudinal vibration suppression. Therefore, this chapter discusses suppressing the lateral vibration of the axially moving belt via a boundary control scheme.

6.1 Boundary Control Based on the Lyapunov Method

Although an axially moving belt is characterized by the dynamic relationship between the two vibrations (i.e., lateral and longitudinal vibrations), it is possible to suppress the lateral vibration by using only one actuator. In this case, the IPDE model that describes only lateral vibration based on the quasi-static stretch assumption [i.e., (5.27)] is used as the "control-oriented" model. This approach was proposed by Hong et al. (2004).

Let us consider an axially moving belt with one actuator, as shown in Fig. 6.1. The equations of motion of the belt with insufficient bending stiffness are given as

$$\rho A\left(u_{tt} + 2v_0 u_{xt} + v_0^2 u_{xx}\right) - EA\left(u_x + \frac{1}{2}w_x^2\right)_x = 0, \tag{6.1}$$

$$\rho A\left(w_{tt} + 2v_0 w_{xt} + v_0^2 w_{xx}\right) - T_0 w_{xx} - EA\left(\left(u_x + \frac{1}{2}w_x^2\right)w_x\right)_x = 0, \tag{6.2}$$

© The Author(s), under exclusive license to Springer Nature Singapore Pte Ltd. 2022

K.-S. Hong et al., *Control of Axially Moving Systems*,
https://doi.org/10.1007/978-981-16-2915-0_6

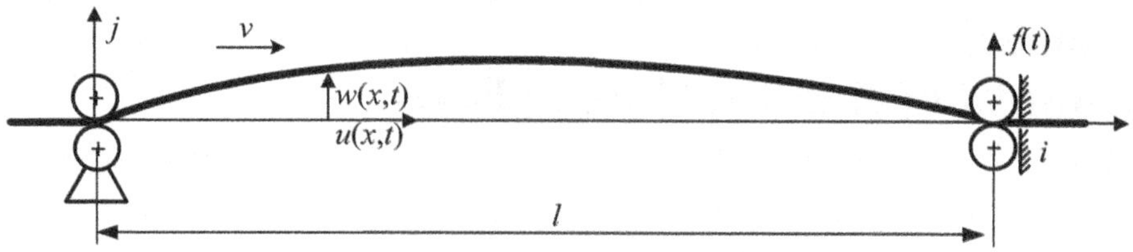

Fig. 6.1 Axially moving belt with a boundary control force

with the boundary conditions:

$$u(0, t) = 0,\ u(l, t) = 0,\ w(0, t) = 0, \tag{6.3}$$

$$T_0 w_x(l, t) + EA\left(u_x(l, t) + \frac{w_x^2(l, t)}{2}\right) w_x(l, t) = f(t) \tag{6.4}$$

where $f(t)$ is the control input force. The "control-oriented" model for the dynamic model given by (6.1)–(6.4) is derived as follows:

$$\rho A\left(w_{tt} + 2v_0 w_{xt} + v_0^2 w_{xx}\right) - T_0 w_{xx} - \frac{EA}{2l} w_{xx} \int_0^l w_x^2(x, t)\mathrm{d}x = 0, \tag{6.5}$$

$$u_x + \frac{1}{2} w_x^2 = \frac{1}{2l} \int_0^l w_\xi^2(\xi, t)\mathrm{d}\xi, \tag{6.6}$$

and the boundary conditions are

$$u(0, t) = 0,\ u(l, t) = 0,\ w(0, t) = 0, \tag{6.7}$$

$$\left(T_0 + \frac{EA}{2l} \int_0^l w_\xi^2(\xi, t)\mathrm{d}\xi\right) w_x(l, t) = f(t). \tag{6.8}$$

Our control objective is to suppress the belt's lateral vibration via right boundary control and further guarantee that the closed-loop system is exponentially stable. The closed-loop system's stability proof is performed by using the Lyapunov second method, wherein the "control-oriented" model is used. Consider the following Lyapunov function candidate:

$$V(t) = \frac{1}{2}\rho A \int_0^l (v_0 + u_t + v_0 u_x)^2 \mathrm{d}x + \frac{1}{2}\rho A \int_0^l (w_t + v_0 w_x)^2 \mathrm{d}x$$

$$+ T_0 \int_0^l \left(u_x + \frac{1}{2}w_x^2\right)\mathrm{d}x + \frac{1}{2}EA \int_0^l \left(u_x + \frac{1}{2}w_x^2\right)^2 \mathrm{d}x. \tag{6.9}$$

The time derivative of $V(t)$ is derived as follows:

$$\begin{aligned}
\frac{\mathrm{d}V(t)}{\mathrm{d}t} &= \rho A \int_0^l (v_0 + u_t + v_0 u_x)(u_{tt} + 2v_0 u_{xt} + v_0^2 u_{xx})\mathrm{d}x \\
&\quad + \rho A \int_0^l (w_t + v_0 w_x)(w_{tt} + 2v_0 w_{xt} + v_0^2 w_{xx})\mathrm{d}x \\
&\quad + T_0 \int_0^l (u_{xt} + v_0 u_{xx} + w_x(w_{tx} + v_0 w_{xx}))\mathrm{d}x \\
&\quad + EA \int_0^l \left(u_x + \frac{1}{2}w_x^2\right)(u_{xt} + v_0 u_{xx} + w_x(w_{tx} + v_0 w_{xx}))\mathrm{d}x.
\end{aligned} \tag{6.10}$$

According to the quasi-static stretch assumption, the axial strain is a time function (i.e., $\varepsilon(t) = (1/2l)\int_0^l w_x^2(x,t)\mathrm{d}x \approx u_x + w_x^2/2$ and $\varepsilon_x = u_{xx} + w_x w_{xx} \approx 0$) and $u_t = u_{tt} = u_{xt} \approx 0$. Hence,

$$\begin{aligned}
\frac{\mathrm{d}V(t)}{\mathrm{d}t} &= \rho A v_0^3 \int_0^l (1 + u_x)u_{xx}\mathrm{d}x + \rho A \int_0^l (w_t + v_0 w_x)(w_{tt} + 2v_0 w_{xt} + v_0^2 w_{xx})\mathrm{d}x \\
&\quad + T_0 \int_0^l \left(w_x w_{tx} + v_0\left(u_x + \frac{1}{2}w_x^2\right)_x\right)\mathrm{d}x \\
&\quad + EA \int_0^l \varepsilon(t)\left(w_x w_{tx} + v_0\left(u_x + \frac{1}{2}w_x^2\right)_x\right)\mathrm{d}x \\
&= \rho A v_0^3 \int_0^l (1 + u_x)u_{xx}\mathrm{d}x + (T_0 + EA\varepsilon(t)) \int_0^l (w_t w_{xx} + v_0 w_x w_{xx})\mathrm{d}x \\
&\quad + (T_0 + EA\varepsilon(t)) \int_0^l w_x w_{tx}\mathrm{d}x \\
&= -\rho A v_0^3 \int_0^l (1 + \varepsilon(t) - \frac{1}{2}w_x^2)w_x w_{xx}\mathrm{d}x + (T_0 + EA\varepsilon(t)) \int_0^l v_0 w_x w_{xx}\mathrm{d}x
\end{aligned}$$

$$
\begin{aligned}
&+ (T_0 + EA\varepsilon(t)) \int_0^l (w_x w_{tx} + w_t w_{xx})\mathrm{d}x \\
&= -\rho A(1+\varepsilon(t))v_0^3 \int_0^l w_x w_{xx}\mathrm{d}x + \frac{1}{2}\rho A v_0^3 \int_0^l w_x^3 w_{xx}\mathrm{d}x \\
&+ (T_0 + EA\varepsilon(t)) \int_0^l v_0 w_x w_{xx}\mathrm{d}x \\
&+ (T_0 + EA\varepsilon(t)) \int_0^l (w_x w_{tx} + w_t w_{xx})\mathrm{d}x \\
&= \frac{v_0}{2}\left((T_0 + EA\varepsilon(t)) - \rho A v_0^2(1+\varepsilon(t))\right) w_x^2\Big|_{x=0}^{x=l} \\
&+ \frac{1}{8}\rho A v_0^3 w_x^4\Big|_{x=0}^{x=l} \\
&+ (T_0 + EA\varepsilon(t))\, w_t w_x\Big|_{x=0}^{x=l}.
\end{aligned} \tag{6.11}
$$

By using the boundary conditions given by (6.7) and (6.8), (6.11) is further simplified:

$$
\begin{aligned}
\frac{\mathrm{d}V(t)}{\mathrm{d}t} &= -\Gamma w_x^2(0,t) - \frac{1}{8}\rho A v_0^3 w_x^4(0,t) \\
&+ w_x(l,t)\left\{\Gamma w_x(l,t) + \frac{1}{8}\rho A v_0^3 w_x^3(l,t)\right. \\
&+ \left.(T_0 + EA\varepsilon(t))w_t(l,t)\right\},
\end{aligned} \tag{6.12}
$$

where $\Gamma = v_0\left[(T_0 - \rho A v_0^2) + (EA - \rho A v_0^2)\varepsilon(t)\right]/2$. Γ is positive in the subcritical regime (i.e., $v_0 \leq \sqrt{T_0/\rho A}$). It follows that the first two terms of (6.12) are always negative. Therefore, we aim to design the control force $f(t)$ such that the remainder of (6.12) is also negative. The feedback control law is proposed as follows.

$$
w_t(l,t) = -\alpha_1(T_0 + EA\varepsilon(t))w_x(l,t) - \alpha_2((T_0 + EA\varepsilon(t))w_x(l,t))^3 \tag{6.13}
$$

where α_1 and α_2 are positive constants. Substituting (6.13) into (6.12) yields

$$
\begin{aligned}
\frac{\mathrm{d}V(t)}{\mathrm{d}t} &= -\Gamma w_x^2(l,t) - \frac{1}{8}\rho A v_0^3 w_x^4(0,t) \\
&- (T_0 + EA\varepsilon(t))^2\left[\alpha_1 - \frac{\Gamma}{(T_0 + EA\varepsilon(t))^2}\right] w_x^2(l,t) \\
&- (T_0 + EA\varepsilon(t))^4\left[\alpha_2 - \frac{1}{8}\frac{\rho A v_0^3}{(T_0 + EA\varepsilon(t))^4}\right] w_x^4(l,t).
\end{aligned} \tag{6.14}
$$

Equation (6.14) shows that if $\alpha_1 > 2\Gamma/(T_0 + EA\varepsilon(t))^2$ and $\alpha_2 > \rho A v_0^3/4(T_0 + EA\varepsilon(t))^4$, then $\mathrm{d}V(t)/\mathrm{d}t \leq 0$. We can assure the existence of α_1 and α_2 because two terms $2\Gamma/(T_0 + EA\varepsilon(t))^2$ and $\rho A v_0^3/4(T_0 + EA\varepsilon(t))^4$ are upper bounded. The boundedness of these terms is proven by the following inequalities:

$$\frac{2\Gamma}{(T_0 + EA\varepsilon(t))^2} = v_0\left(\frac{1}{(T_0 + EA\varepsilon(t))} - \frac{\rho A v_0^2(1+\varepsilon(t))}{(T_0 + EA\varepsilon(t))^2}\right) < v_0 < \alpha_1, \quad (6.15)$$

$$\frac{1}{4}\frac{\rho A v_0^3}{(T_0 + EA\varepsilon(t))^4} < \frac{1}{4} < \alpha_2. \quad (6.16)$$

Now, we determine the control force $f(t)$ from the feedback control law in (6.13). Notably, $f(t) = (T_0 + EA\varepsilon(t))w_x(l,t)$; therefore, (6.13) can be rewritten as follows:

$$w_t(l,t) = -\alpha_1 f(t) - \alpha_2 f(t)^3. \quad (6.17)$$

Based on (6.17), the control force can be solved in terms of $w_t(l,t)$. The control force, which corresponds to the unique real root of (6.17), is

$$f(t) = \sqrt[3]{-\frac{w_t(l,t)}{2\alpha_2} + \sqrt{\frac{\alpha_1^3}{27\alpha_2^3} + \frac{w_t^2(l,t)}{4\alpha_2^2}}} + \sqrt[3]{-\frac{w_t(l,t)}{2\alpha_2} - \sqrt{\frac{\alpha_1^3}{27\alpha_2^3} + \frac{w_t^2(l,t)}{4\alpha_2^2}}}. \quad (6.18)$$

The control force can be generated by using only the time derivative of the lateral vibration and the control gains.

Furthermore, we investigate the exponential stability of the lateral vibration of the belt under the control law in (6.13). For stability analysis, we define the following state-space X:

$$X := \left\{(w, w_t)^{\mathrm{T}} \middle| w \in H_0^1, w_t \in L^2\right\} \quad (6.19)$$

where L^2 and H_0^k are the spaces defined in (4.105) and (4.106). The state-space form of (6.5) is derived as follows:

$$\dot{\mathbf{z}} = \mathbf{A}\mathbf{z}, \ \mathbf{z}(0) = \mathbf{z}_0 \in X \quad (6.20)$$

where $\mathbf{z} = [w \ \mathrm{D}w/\mathrm{D}t]^{\mathrm{T}} \in X$ and the operator $\mathbf{A} : X \to X$ is defined as

$$\mathbf{Az} = \begin{bmatrix} w_t + v_0 w_x \\ \frac{T_0}{\rho A} w_{xx} + \frac{EA}{2l\rho A} \int_0^l w_x^2(x,t)\mathrm{d}x \, w_{xx} \end{bmatrix}. \quad (6.21)$$

The domain $D(\mathbf{A})$ of operator $\mathbf{A}$ for the closed-loop system with (6.13) is

$$D(\mathbf{A}) := \{[w\ \mathrm{D}w/\mathrm{D}t]^{\mathrm{T}} | w \in H_0^2, \mathrm{D}w/\mathrm{D}t \in H_0^1,$$
$$w_t(l,t) + \alpha_1(T_0 + EA\varepsilon(t))w_x(l,t) + \alpha_2((T_0 + EA\varepsilon(t))w_x(l,t))^3 = 0\}. \quad (6.22)$$

The lateral energy of the belt is introduced as

$$E_{\mathrm{bt}}(t) = \|\mathbf{z}\|^2 \triangleq \langle \mathbf{z}, \mathbf{z} \rangle := \frac{1}{2}\rho A \int_0^l (w_t + v_0 w_x)^2 \mathrm{d}x + \frac{T_0}{2} \int_0^l w_x^2 \mathrm{d}x + \frac{EA}{8} \left(\int_0^l w_x^2 \mathrm{d}x \right)^2. \quad (6.23)$$

By using (6.6) and the fact $\int_0^l w_x^2 \mathrm{d}x = (1/l) \int_0^l \left(\int_0^l w_x^2 \mathrm{d}x \right) \mathrm{d}x = 2 \int_0^l \left(u_x + \frac{1}{2} w_x^2 \right) \mathrm{d}x$, (6.23) can be rewritten as follows:

$$E_{\mathrm{bt}}(t) = \frac{1}{2}\rho A \int_0^l (w_t + v_0 w_x)^2 \mathrm{d}x + T_0 \int_0^l \left(u_x + \frac{1}{2} w_x^2 \right) \mathrm{d}x + \frac{EA}{2} \int_0^l \left(u_x + \frac{1}{2} w_x^2 \right)^2 \mathrm{d}x. \quad (6.24)$$

Lemma 6.1 *Operator* $\mathbf{A}$ *in* (6.21) *generates a semigroup of contractions on* X.

Proof We can prove that the time derivative of the lateral energy of the belt under the control law in (6.13) is negative semi-definite. Thus, the lateral dynamics of the closed-loop system are dissipative. According to Komura (1967), operator $\mathbf{A}$ generates a semigroup $S(t)$ of contraction on X, which implies that (6.20) has a unique solution that can be represented by using the semigroup $S(t)$, namely $\mathbf{z} = S(t)\mathbf{z}_0$, $t \geq 0$, where $\mathbf{z}_0 \in D(A)$.

Theorem 6.1 *Consider the axially moving belt described by* (6.5) *under the feedback control law in* (6.18)*, wherein the control gain satisfies* (6.15) *and* (6.16)*. The lateral energy of the belt is upper bounded as*

$$E_{\mathrm{bt}}(t) \leq \lambda_1 \mathrm{e}^{-\lambda_2}, \quad t \geq 0. \quad (6.25)$$

It follows that the closed-loop system is exponentially stable.

Proof Theorem 6.1 can be proven by using a similar method, which is presented in Sect. 4.2.3. Let us consider the following function.

$$\overline{E}_{\mathrm{bt}}(t) = tE_{\mathrm{bt}}(t) + 2\rho A \int_0^l x(w_t + v_0 w_x) w_x \mathrm{d}x \quad (6.26)$$

where $E_{\mathrm{bt}}(t)$ is the lateral energy of the belt, as defined in (6.23). Using Lemma 4.2 yields

$$2\rho A\int_0^l x(w_t+v_0w_x)w_x\mathrm{d}x \le \rho Al\int_0^l (w_t+v_0w_x)^2\mathrm{d}x+\rho Al\int_0^l w_x^2\mathrm{d}x \le \lambda_3 E_{\mathrm{bt}}(t) \tag{6.27}$$

where $\lambda_3 = \max\{2l, 8\rho Al^3/(4T_0l+EA)\}$ is a positive constant. The following inequalities hold:

$$0 \le (t-\lambda_3)E_{\mathrm{bt}}(t) \le \overline{E}_{\mathrm{bt}}(t) \le (t+\lambda_3)E_{\mathrm{bt}}(t), \quad \text{for } t > \lambda_3. \tag{6.28}$$

The time derivative of $\overline{E}_{\mathrm{bt}}(t)$ is derived as

$$\begin{aligned}
\frac{\mathrm{d}\overline{E}_{\mathrm{bt}}(t)}{\mathrm{d}t} &= E_{\mathrm{bt}}(t)+t\frac{\mathrm{d}E_{\mathrm{bt}}(t)}{\mathrm{d}t}+2\rho A\int_0^l x(w_{tt}+2v_0w_{tx}+v_0^2w_{xx})w_x\mathrm{d}x \\
&\quad +2\rho A\int_0^l x(w_t+v_0w_x)(w_{xt}+v_0w_{xx})\mathrm{d}x \\
&= E_{\mathrm{bt}}(t)+t\frac{\mathrm{d}E_{\mathrm{bt}}(t)}{\mathrm{d}t}+\Psi(t)lw_x^2(l,t) \\
&\quad -\Psi(t)\int_0^l w_x^2\mathrm{d}x+\rho Al(w_t(l,t)+v_0w_x(l,t))^2 \\
&\quad -\rho A\int_0^l (w_t+v_0w_x)^2\mathrm{d}x
\end{aligned} \tag{6.29}$$

where $\Psi(t) = T_0 + EA\varepsilon(t)$. Under the control law in (6.18), (6.29) can be rewritten as follows:

$$\begin{aligned}
\frac{\mathrm{d}\overline{E}_{\mathrm{bt}}(t)}{\mathrm{d}t} &= E_{\mathrm{bt}}(t)+t\frac{\mathrm{d}E_{\mathrm{bt}}(t)}{\mathrm{d}t}+\rho Al\left((\alpha_1\Psi(t)-v_0)w_x(l,t)+\alpha_2\Psi(t)^3w_x^3(l,t)\right)^2 \\
&\quad +\Psi(t)lw_x^2(l,t)-\rho A\int_0^l (w_t+v_0w_x)^2\mathrm{d}x-\Psi(t)\int_0^l w_x^2\mathrm{d}x
\end{aligned} \tag{6.30}$$

where $\mathrm{d}E_{\mathrm{bt}}(t)/\mathrm{d}t \le 0$. Since $E_{\mathrm{bt}}(t)$ and $w_x(l,t)$ are bounded, $\mathrm{d}\overline{E}_{\mathrm{bt}}(t)/\mathrm{d}t$ is negative for a sufficiently large time, $t = t_\Omega$. Furthermore, when $t > t_\Omega$, (6.28) also holds:

$$E_{\mathrm{bt}}(t) \le \frac{\overline{E}_{\mathrm{bt}}(t)}{(t-\lambda_3)}. \tag{6.31}$$

Based on the semigroup property of the solution, the following inequality is derived:

$$\int_0^\infty E_{\mathrm{bt}}^2(t)\mathrm{d}t = \int_0^{t_\Omega} E_{\mathrm{bt}}^2(t)\mathrm{d}t + \int_{t_\Omega}^\infty E_{\mathrm{bt}}^2(t)\mathrm{d}t = \int_0^{t_\Omega} \|S(t)\mathbf{z}_0\|_X^4 \mathrm{d}t + \int_{t_\Omega}^\infty \frac{\overline{E}_{\mathrm{bt}}^2}{(t-\lambda_3)^2}\mathrm{d}t < \infty. \tag{6.32}$$

This inequality implies that.

$$\int_0^\infty \|S(t)\mathbf{z}_0\|_X^4 \mathrm{d}t < \infty. \tag{6.33}$$

Applying the semigroup theorem to (6.33) yields

$$\|\mathbf{z}\|_X \le \beta_1 \|\mathbf{z}_0\|_X \mathrm{e}^{-\beta_2 t}. \tag{6.34}$$

where β_1 and β_2 are positive constants. Consequently, the following inequality holds.

$$E(t) = \|\mathbf{z}\|_X^2 \le \beta_1^2 \|\mathbf{z}_0\|_X^2 \mathrm{e}^{-2\beta_2 t} \text{ or } E(t) \le \lambda_1 e^{-\lambda_1 t}. \tag{6.35}$$

where $\lambda_1 = \beta_1^2 \|\mathbf{z}_0\|_X^2$ and $\lambda_2 = 2\beta_2$. Theorem 6.1 is proven. ■

In addition to the control problem of only lateral vibrations of axially moving belts, simultaneous suppression of longitudinal and lateral vibration based on the Lyapunov method was investigated by Nguyen and Hong (2012). In their work, an axially moving belt that represents a roll-to-roll system is considered (i.e., Fig. 6.2). The lateral vibrations, longitudinal vibrations, and axial velocity are controlled using a control force and two control torques applied via a hydraulic actuator and rollers. A coupled PDE-ODE model that represents the longitudinal vibration, the lateral vibration, and the dynamics at the boundaries are derived as follows:

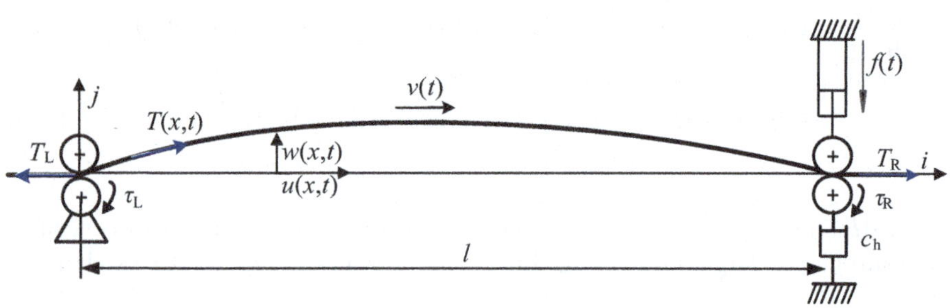

Fig. 6.2 Axially moving belt with a boundary control force and two control torques at the boundaries

$$\rho A\big(u_{tt} + \dot{v}(t) + \dot{v}(t)u_x + 2v(t)u_{xt} + v(t)^2 u_{xx}\big) - EA(u_{xx} + w_{xx}w_x) = 0, \tag{6.36}$$

$$\rho A\big(w_{tt} + \dot{v}(t)w_x + 2v(t)w_{xt} + v(t)^2 w_{xx}\big) + c_e(w_t + v(t)w_x) - T_0 w_{xx} - EA\left(w_{xx}u_x + w_x u_{xx} + \frac{3}{2}w_x^2 w_{xx}\right) = 0, \tag{6.37}$$

and the boundary conditions are given by

$$\frac{J}{r^2}u_{tt}(0,t) - EA\left(u_x(0,t) + \frac{w_x^2(0,t)}{2}\right) = \frac{\tau_L(t)}{r} - T_L + T_0, \tag{6.38}$$

$$\frac{J}{r^2}u_{tt}(l,t) + EA\left(u_x(l,t) + \frac{w_x^2(l,t)}{2}\right) = \frac{\tau_R(t)}{r} + T_R - T_0, \tag{6.39}$$

$$w(0,t) = 0, \tag{6.40}$$

$$m_a w_{tt}(l,t) + \left[T_0 + EA\left(u_x(l,t) + \frac{w_x^2(l,t)}{2}\right)\right]w_x(l,t) + (c_h - \rho A v(t))w_t(l,t) - \rho A v(t)^2 w_x(l,t) = f(t) \tag{6.41}$$

where c_e is the damping coefficient of the material; J and r indicate the rotation inertia and the radius of the two rollers, respectively; T_L and T_R are constants that are related to the material tensions in the respective adjacent spans; and c_h denotes the damping coefficient of the actuator. $f(t)$, $\tau_L(t)$, and $\tau_R(t)$ are the control force and control torques. Notably, the control torques not only affect the longitudinal vibrations but also change the axial velocity. The relationship between the axial velocity and the control torques is given as follows:

$$\left(\rho A r + \frac{2J}{r}\right)\dot{v}(t) = \tau_L(t) + \tau_R(t) + (T_R - T_L)r. \tag{6.42}$$

This relationship implies that vibration suppression and velocity control are coupled. The following control torques and control force are proposed to suppress both longitudinal oscillation and lateral oscillation and control the axial velocity to track the desired profile $v_d(t)$.

$$f(t) = -\frac{\alpha_1\alpha_2 l T_0}{2(\alpha_1 v(t) + 2\alpha_2 l)^2}w_t(l,t) - \frac{\alpha_2 l T_0}{2(\alpha_1 v(t) + 2\alpha_2 l)}w_x(l,t) - \rho A v(t)^2 w_x(l,t) + (c_h - \rho A v(t))w_t(l,t) - (v(t) + 2\alpha_2 l/\alpha_1)m_a w_{xt}(l,t), \tag{6.43}$$

$$\begin{aligned}
\tau_{\mathrm{L}}(t) = \frac{EAr}{\alpha_1 v(t)} \Bigg\{ & \frac{\alpha_1 v(t)(T_{\mathrm{L}} - T_0)}{EA} + \alpha_2 l EAu_x(l,t) w_x(l,t)^2 \\
& - \frac{\alpha_2 EA}{2l}\left(1 - \frac{1}{2\beta_1}\right) u^2(l,t) - \alpha_1 \left(v(t) + u_t + v(t)u_x\right)\left(u_x + w_x^2/2\right)\big|_{x=l} \\
& + \alpha_1 \left(u_t + v(t)u_x\right)\left(u_x + w_x^2/2\right)\big|_{x=0} \\
& - \frac{EA\alpha_1}{2l} u(l,t)u_t(l,t) \\
& + \frac{J\alpha_1}{r^2}\left[\left(\frac{\alpha_1 v(t)}{EA} - u_t(0,t)\right)u_{tt}(0,t) - u_t(l,t)u_{tt}(l,t)\right]\Bigg\} \\
& - \alpha_2 EAlu_x^2(l,t) + \alpha_1 v(t) u_x(0,t) + \alpha_1 m_{\mathrm{a}} w_x(l,t) w_{xt}(l,t) \\
& + \alpha_2 \rho Al(w_t(l,t) + v(t)w_x(l,t))^2 - \left(u_t^2(0,t) + u_t^2(l,t)\right)/\beta_1 \\
& - \alpha_2 \rho Al(v(t) + u_t(l,t) + v(t)u_x(l,t))^2,
\end{aligned} \tag{6.44}$$

$$\tau_{\mathrm{R}}(t) = \left(\rho AlR + 2J/r\right)\dot{v}_d(t) - \beta_2(v(t) - v_{\mathrm{d}}(t)) - \tau_{\mathrm{L}}(t) + (T_{\mathrm{L}} - T_{\mathrm{R}})r, \tag{6.45}$$

where $\alpha_1, \alpha_2, \beta_1$, and β_2 are the control gains. The asymptotic stability of the system under the designed control law can be proven by using a similar approach presented in the previous section, wherein the Lyapunov function candidate is defined based on the modified total mechanical energy of the system as follows:

$$V(t) = \alpha_1 V_1(t) + \alpha_1 V_2(t) + \alpha_2 V_3(t) \tag{6.46}$$

where

$$\begin{aligned}
V_1(t) = & \frac{1}{2}\rho A \int_0^l (v(t) + u_t + v(t)u_x)^2 \mathrm{d}x + \frac{1}{2}\rho A \int_0^l (w_t + v(t)w_x)^2 \mathrm{d}x \\
& + \frac{1}{2}T_0 \int_0^l w_x^2 \mathrm{d}x + \frac{1}{2}EA \int_0^l u_x^2 \mathrm{d}x + \frac{1}{2}EA \int_0^l u_x w_x^2 \mathrm{d}x + \frac{1}{8}EA \int_0^l w_x^4 \mathrm{d}x,
\end{aligned} \tag{6.47}$$

$$\begin{aligned}
V_2(t) = & \frac{J}{2r^2}\left(u_t^2(0,t) + u_t^2(l,t)\right) + \frac{EA}{4l}u(l,t)^2 + \frac{m_{\mathrm{a}} w_x(l,t)^2}{2} \\
& + \frac{1}{2}m_{\mathrm{a}}\left[\left(w_t(l,t) + \left(v(t) + \frac{2\alpha_2 l}{\alpha_1}\right)w_x(l,t)\right)^2 + w_x^2(l,t)\right] \\
& + \frac{1}{2}(v(t)(t) - v_{\mathrm{d}}(t))^2,
\end{aligned} \tag{6.48}$$

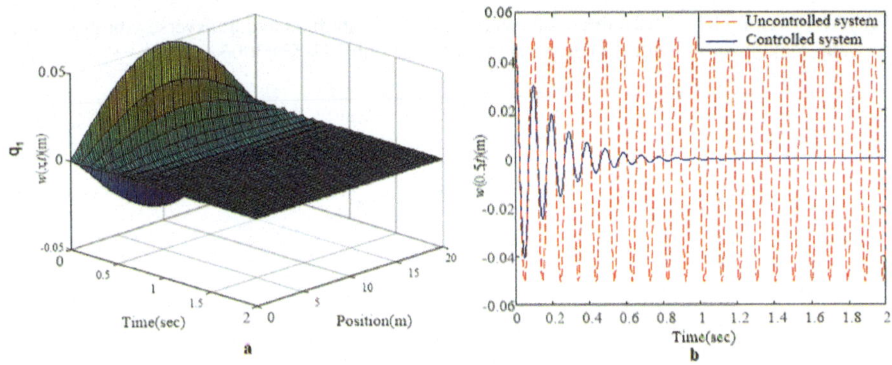

Fig. 6.3 Lateral vibration of the belt using the boundary control law based on the Lyapunov method: **a** $w(x, t)$, **b** $w(0.5, t)$

$$V_3(t) = 2\rho A \int_0^l x(u_x(v(t) + u_t + v(t)u_x) + w_x(w_t + v(t)w_x))\mathrm{d}x. \tag{6.49}$$

The proposed control scheme for simultaneous suppression of longitudinal and lateral vibrations of axially moving belts with velocity tracking is applicable to roll-to-roll systems, especially roll-to-roll printing processes. However, the control law in (6.43)–(6.45) has some limitations: The longitudinal vibration and its derivatives must be measured. Contrary to lateral vibration, the installation of sensors to measure these values and avoid interfering with the system's operation is somewhat involved.

Example 6.1 In this example, we present simulation results of controlling the lateral vibrations of a moving steel strip (i.e., Fig. 6.1). The strip is modeled by an axially moving belt given by (6.1)–(6.4), wherein $E = 2 \times 10^{11}$ N/m^2, $A = 7.5 \times 10^{-3}$ m^2, $T_0 = 10^4$ kN, $\rho = 7850$ kg/m^2, $L = 20$ m, and $v_0 = 2$ m/s. The lateral vibration of the strip is suppressed by using the control law in (6.18). The initial conditions are given as $w(x, 0) = 0.05\sin(\pi x/l)$ and $u(x, 0) = 0.01\sin(\pi x/l)$. The PDEs are solved by using a finite difference scheme, wherein the small segments that correspond to spatial and temporal variables are set to $\Delta x = 0.2$ and $\Delta t = 10^{-5}$. Figure 6.3 shows that the lateral vibration of the coupled model in (6.1)–(6.4) can be attenuated by using only one control input (i.e., 6.18).

References

Hong KS, Kim CW, Hong KT (2004) Boundary control of an axially moving belt system in a thin-metal production line. Int J Control Autom 2(1):55–67

Komura Y (1967) Nonlinear semi-groups in Hilbert space. J Math Soc Japan 19(4):493–507

Nguyen QC, Hong KS (2012) Simultaneous control of longitudinal and transverse vibrations of an axially moving string with velocity tracking. J Sound Vib 331(13):3006–3019

Chapter 7
Plate Model

Abstract This chapter provides various mathematical models of axially moving orthotropic, isotropic, viscoelastic, and composite plates. Two approximate models are discussed. The dynamic behavior analyses using PDE models are presented.

Keywords Axially moving plate · Membrane · Two-dimensional model · Viscoelastic · Laminate composite plate · Approximate models · Vibration analysis · Bifurcations

7.1 Introduction

In the string/beam/belt models, only one independent spatial variable, x, is considered. Therefore, these models are known as the one-dimensional model. In numerous situations, the use of these models produces satisfactory results. Nevertheless, in several cases wherein the width material is significant, e.g., a metal layer, the one-dimensional models cannot be implemented. The dynamic of these systems should be analyzed based on two-dimensional models that involve two independent spatial variables x and z, such as the plate model or membrane model (Marynowski 2012). Ulsoy and Mote (1982) are recognized as one of the pioneering studies on axially moving plates. In their study, a dynamic model of a blade of a band saw was established via the extended Hamilton principle. Investigating the dynamics of axially moving plates/membranes has received an increasing amount of attention since the 1990s. Lin and Mote (1995) applied the von Karman strain theory to analyze the lateral oscillation of a translating plate. Later, a nonlinear model of a translating orthotropic plate, wherein the lateral, longitudinal, and transverse vibrations are coupled, was developed by Marynowski and Kolakowski (1999). These works are the premise for many subsequent studies on axially moving plates. Investigations on viscoelastic plates—based on the Kelvin viscoelastic model and standard linear solid model—were also published by Zhou and Wang (2007), Abedi et al. (2014), and Marynowski (2006). Furthermore, the studies of Hatami et al. (2007), Zhang et al. (2014) and Esmaeilzadeh and Kadkhodayan (2019) were devoted to the dynamics of axially moving laminated composite plates.

© The Author(s), under exclusive license to Springer Nature Singapore Pte Ltd. 2022
K.-S. Hong et al., *Control of Axially Moving Systems*,
https://doi.org/10.1007/978-981-16-2915-0_7

As previously mentioned, the plate and membrane models are appropriate for analyzing axially moving materials with considerable width. The membrane model can be used for systems that withstand only tensile stresses, whereas the plate model should be utilized to consider systems that withstand both tensile stresses and bending. These models can describe the vibration behaviors of many systems, e.g., metal layers in nanoscale metal printing and metal sheets in coil-coating processes and thin-metal production lines.

7.2 Dynamic Models of Axially Moving Plates

7.2.1 *Elastic Plate*

In this section, we establish the dynamic model of an axially moving orthotropic plate. Figure 7.1 demonstrates an axially moving orthotropic plate. The length of the plate, width of the plate, and thickness of the plate are denoted by l, b, and h, respectively, while the lateral vibration, longitudinal vibration, and transverse vibration are indicated by $w(x, z, t)$, $u(x, z, t)$, and $\eta(x, z, t)$, respectively. The plate is simply supported at both ends, as shown in Fig. 7.1.

The equation of motion of the axially moving plate is developed based on the Hamilton principle and the von Karman strain theory. The displacement and velocity vectors of a unit element of the plate at (x, z) are given as follows:

$$\mathbf{r}_{\mathrm{d}}(x, z, t) = (x + u)\mathbf{i} + w\mathbf{j} + (z + \eta)\mathbf{k}, \tag{7.1}$$

$$\mathbf{r}_{\mathrm{v}}(x, z, t) = (v_0 + u_t + v_0 u_x)\mathbf{i} + (w_t + v_0 w_x)\mathbf{j} + (\eta_t + v_0 \eta_x)\mathbf{k}. \tag{7.2}$$

The kinetic energy of the system is computed as

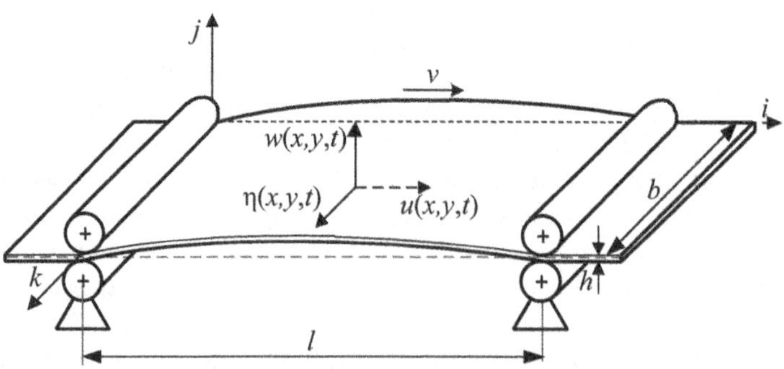

Fig. 7.1 Axially moving plate

$$K = \frac{1}{2}\rho h \int\limits_S \mathbf{r}_v \cdot \mathbf{r}_v \mathrm{d}A = \frac{1}{2}\rho h \int\limits_S \big((w_t + v_0 w_x)^2 + (v_0 + u_t + v_0 u_x)^2 + (\eta_t + v_0 \eta_x)^2\big)\mathrm{d}S \tag{7.3}$$

where S is the plate area. The potential energy is expressed as follows:

$$U = \int\limits_S (N_X \varepsilon_X^0 + N_Z \varepsilon_Z^0 + 2N_{XZ}\varepsilon_{XZ}^0)\mathrm{d}S + \int\limits_S (M_X \kappa_X + M_Z \kappa_Z + 2M_{XZ}\kappa_{XZ})\mathrm{d}S \tag{7.4}$$

where N_X and N_Z are the sectional plate forces along the i-axis and k-axis, respectively; N_{XZ} denotes the sectional shear force; M_X and M_Z are the sectional bending moments with respect to the i-axis and k-axis, respectively, and M_{XZ} indicates the sectional torsion moment. ε_X^0, ε_Z^0, and ε_{XZ}^0 are the strain components for the mid-surface, whereas κ_X, κ_Z, and κ_{XZ} represent the curvature modifications and torsions of the mid-surface of the plate. These terms are given as follows:

$$\begin{aligned} &\varepsilon_X^0 = u_x + \frac{1}{2}w_x^2,\ \varepsilon_Z^0 = \eta_z + \frac{1}{2}w_z^2, \\ &\varepsilon_{XZ}^0 = \frac{1}{2}(u_z + \eta_x + w_x w_z), \\ &\kappa_X = -w_{xx},\ \kappa_Z = -w_{zz},\ \kappa_{XZ} = -w_{xz}. \end{aligned} \tag{7.5}$$

In an orthotropic plate, the relationship between Young's modulus and Poisson's number is given by

$$E_X \upsilon_{ZX} = E_Z \upsilon_{XZ}, \tag{7.6}$$

where E_X and E_Z are Young's moduli of the plate along the i-axis and the k-axis, respectively, and υ_{ZX} and υ_{XZ} are Poisson's numbers with respect to the lateral direction and longitudinal direction, respectively. By letting $E_X = E$, $\upsilon_{XZ} = \upsilon$, and $\chi = E_Z/E_X$, the strain–strain relationship of an orthotropic elastic plate is given as follows:

$$\sigma_X = \frac{E}{1-\chi\upsilon^2}(\varepsilon_X + \chi\upsilon\varepsilon_Z),\ \sigma_Z = \frac{\chi E}{1-\chi\upsilon^2}(\varepsilon_Z + \upsilon\varepsilon_X),\ \sigma_{XZ} = 2Gh\varepsilon_{XZ}, \tag{7.7}$$

where χ is the orthotropic factor, G denotes the shear modulus of the plate, and ε_X, ε_Z, and ε_{XZ} denote the strain components in the x and z coordinates, that is,

$$\varepsilon_X = \varepsilon_X^0 + y\kappa_X,\ \varepsilon_Z = \varepsilon_Z^0 + y\kappa_Z,\ \varepsilon_{XZ} = \varepsilon_{XZ}^0 + 2y\kappa_{XZ}. \tag{7.8}$$

The sectional plate forces and shear force—N_X, N_Z, and N_{XZ}—are obtained by integrating the stresses over the plate thickness, namely

$$N_X = \int_{-h/2}^{h/2} \sigma_X \mathrm{d}y = \frac{Eh}{1-\chi\upsilon^2}\left(\varepsilon_X^0 + \chi\upsilon\varepsilon_Z^0\right),$$

$$N_Z = \int_{-h/2}^{h/2} \sigma_Z \mathrm{d}y = \frac{\chi Eh}{1-\chi\upsilon^2}\left(\varepsilon_Z^0 + \upsilon\varepsilon_X^0\right),$$

$$N_{XZ} = \int_{-h/2}^{h/2} \sigma_{XZ} \mathrm{d}y = 2G\varepsilon_{XZ}^0. \tag{7.9}$$

The sectional bending moments and torsion moment—M_X, M_Z, and M_{XZ}—are computed as follows:

$$M_X = \int_{-h/2}^{h/2} y\sigma_X \mathrm{d}y = -D(w_{xx} + \chi\upsilon w_{zz}),$$

$$M_Z = \int_{-h/2}^{h/2} y\sigma_Z \mathrm{d}y = -D\chi(w_{zz} + \upsilon w_{xx}),$$

$$M_{XZ} = \int_{-h/2}^{h/2} y\sigma_{XZ} \mathrm{d}y = -\frac{Gh^3}{6} w_{xz} \tag{7.10}$$

where $D = Eh^3/12(1-\chi\upsilon^2)$ is the "flexural rigidity" of the plate. Substituting (7.9) and (7.10) into (7.4) and then applying the Hamilton principle yields

$$0 = \int_{t_1}^{t_2}\int_S (\delta L_{\mathrm{p1}} + \delta L_{\mathrm{p2}})\mathrm{d}S\mathrm{d}t \tag{7.11}$$

where

$$\begin{aligned} L_{\mathrm{p1}} = {} & \frac{1}{2}\rho h\left((w_t + v_0 w_x)^2 + (v_0 + u_t + v_0 u_x)^2 + (\eta_t + v_0\eta_x)^2\right) \\ & - N_X\left(u_x + \frac{1}{2}w_x^2\right) \\ & - N_Z\left(\eta_z + \frac{1}{2}w_z^2\right) - N_{XZ}(u_z + \eta_x + w_x w_z), \end{aligned}$$

$$L_{p2} = (M_X w_{xx} + M_Z w_{zz} + 2M_{XZ} w_{xz}). \tag{7.12}$$

Applying the procedure for integration by parts to (7.11) yields the following results:

$$\begin{aligned}
0 = &\int_{t_1}^{t_2}\int_S \Bigg(\frac{\partial L_{p1}}{\partial w} - \frac{\partial}{\partial x}\frac{\partial L_{p1}}{\partial w_x} - \frac{\partial}{\partial z}\frac{\partial L_{p1}}{\partial w_z} - \frac{\partial}{\partial t}\frac{\partial L_{p1}}{\partial w_t} \\
&+ \frac{\partial^2}{\partial x^2}\frac{\partial L_{p2}}{\partial w_{xx}} + \frac{\partial^2}{\partial z^2}\frac{\partial L_{p2}}{\partial w_{zz}} + \frac{\partial^2}{\partial x \partial z}\frac{\partial L_{p2}}{\partial w_{xz}}\Bigg)\delta w \mathrm{d}S\mathrm{d}t \\
&+ \int_{t_1}^{t_2}\int_S \left(\frac{\partial L_{p1}}{\partial u} - \frac{\partial}{\partial x}\frac{\partial L_{p1}}{\partial u_x} - \frac{\partial}{\partial z}\frac{\partial L_{p1}}{\partial u_z} - \frac{\partial}{\partial t}\frac{\partial L_{p1}}{\partial u_t}\right)\delta u \mathrm{d}S\mathrm{d}t \\
&+ \int_{t_1}^{t_2}\int_S \left(\frac{\partial L_{p1}}{\partial \eta} - \frac{\partial}{\partial x}\frac{\partial L_{p1}}{\partial \eta_x} - \frac{\partial}{\partial z}\frac{\partial L_{p1}}{\partial \eta_z} - \frac{\partial}{\partial t}\frac{\partial L_{p1}}{\partial \eta_t}\right)\delta \eta \mathrm{d}S\mathrm{d}t \\
&+ \int_{t_1}^{t_2}\int_0^b \left(\frac{\partial L_{p1}}{\partial w_x} - \frac{\partial}{\partial x}\frac{\partial L_{p2}}{\partial w_{xx}} - \frac{\partial}{\partial z}\frac{\partial L_{p2}}{\partial w_{xz}}\right)\delta w\Bigg|_{x=0}^{x=l} \mathrm{d}z\mathrm{d}t \\
&+ \int_{t_1}^{t_2}\int_0^b \frac{\partial L_{p1}}{\partial u_x}\delta u\Bigg|_{x=0}^{x=l} \mathrm{d}z\mathrm{d}t + \int_{t_1}^{t_2}\int_0^b \frac{\partial L_{p1}}{\partial \eta_x}\delta \eta\Bigg|_{x=0}^{x=l} \mathrm{d}z\mathrm{d}t \\
&+ \int_{t_1}^{t_2}\int_0^l \left(\frac{\partial L_{p1}}{\partial w_z} - \frac{\partial}{\partial z}\frac{\partial L_{p2}}{\partial w_{zz}} - \frac{\partial}{\partial x}\frac{\partial L_{p2}}{\partial w_{xz}}\right)\delta w\Bigg|_{z=0}^{z=b} \mathrm{d}x\mathrm{d}t \\
&+ \int_{t_1}^{t_2}\int_0^l \frac{\partial L_{p1}}{\partial u_z}\delta u\Bigg|_{z=0}^{z=b} \mathrm{d}x\mathrm{d}t + \int_{t_1}^{t_2}\int_0^l \frac{\partial L_{p1}}{\partial \eta_z}\delta \eta\Bigg|_{z=0}^{z=b} \mathrm{d}x\mathrm{d}t \\
&+ \int_S \frac{\partial L_{p1}}{\partial w_t}\delta w\Bigg|_{t=t_1}^{t=t_2} \mathrm{d}S + \int_S \frac{\partial L_{p1}}{\partial u_t}\delta u\Bigg|_{t=t_1}^{t=t_2} \mathrm{d}S + \int_S \frac{\partial L_{p1}}{\partial \eta_t}\delta \eta\Bigg|_{t=t_1}^{t=t_2} \mathrm{d}S \\
&+ \int_{t_1}^{t_2}\int_0^b \frac{\partial L_{p2}}{\partial w_{xx}}\delta w_x\Bigg|_{x=0}^{x=l} \mathrm{d}z\mathrm{d}t + \int_{t_1}^{t_2}\int_0^l \frac{\partial L_{p2}}{\partial w_{zz}}\delta w_z\Bigg|_{z=0}^{z=l} \mathrm{d}x\mathrm{d}t + \int_{t_1}^{t_2}\left(\frac{\partial L_{p2}}{\partial w_{xz}}\delta w\Bigg|_{x=0}^{x=l}\right)\Bigg|_{z=0}^{z=b} \mathrm{d}t.
\end{aligned} \tag{7.13}$$

The expansion of (7.13) is shown in Appendix D. Accordingly, the dynamic model of the axially moving orthotropic plate can be derived as follows:

$$\begin{aligned}
&\rho h\left(w_{tt} + 2v_0 w_{xt} + v_0^2 w_{xx}\right) - M_{Xxx} - 2M_{XZxz} - M_{Zzz} - (N_X w_x)_x \\
&- (N_Z w_z)_z - (N_{XZ} w_x)_z - (N_{XZ} w_z)_x = 0,
\end{aligned} \tag{7.14}$$

$$\rho h\left(u_{tt}+2v_0u_{xt}+v_0^2u_{xx}\right)-N_{Xx}-N_{XZz}=0, \tag{7.15}$$

$$\rho h\left(\eta_{tt}+2v_0\eta_{xt}+v_0^2\eta_{xx}\right)-N_{XZx}-N_{Zz}=0. \tag{7.16}$$

As shown in Fig. 7.1, the considered plate has two simply supported edges (i.e., $x = 0$ and $x = l$) and two free edges (i.e., $z = 0$ and $z = b$). Therefore, the boundary conditions are given as follows:

For two simply supported edges

$$\begin{aligned} &w(0,z,t)=w(l,z,t)=0,\\ &M_Z(0,z,t)=M_Z(l,z,t)=0. \end{aligned} \tag{7.17}$$

For two freely supported edges

$$\begin{aligned} &M_Z(x,0,t)=M_Z(x,b,t)=0,\\ &N_Z(x,0,t)w_z(x,0,t)-N_{XZ}(x,0,t)w_z(x,0,t)\\ &\quad+M_{Zz}(x,0,t)+2M_{XZx}(x,0,t)=0,\\ &N_Z(x,b,t)w_z(x,b,t)-N_{XZ}(x,b,t)w_z(x,b,t)\\ &\quad+M_{Zz}(x,b,t)+2M_{XZx}(x,b,t)=0. \end{aligned} \tag{7.18}$$

Equations (7.14)–(7.18) are the mathematical model for the coupling dynamics of the lateral, longitudinal, and transverse vibrations of a moving orthotropic plate.

Remark 7.1 In the case of the isotropic material, $E_X = E_Y = E$ and $\upsilon_{ZX} = \upsilon_{XZ} = \upsilon$, where E and υ denote Young's modulus and Poisson's number of the isotropic material, respectively. Furthermore, if we assume that the displacement is small enough, the sectional plate forces and shear force can be approximated as follows:

$$N_X=\frac{Eh}{1-\upsilon^2}(u_x+\upsilon\eta_z);\quad N_Z=\frac{Eh}{1-\upsilon^2}(\eta_z+\upsilon u_x);\quad N_{XZ}=2G(u_z+\upsilon\eta_x) \tag{7.19}$$

where the shear modulus is given by $G = E/2(1+\upsilon)$. With this assumption, the equations of motion of the in-plane displacements (i.e., longitudinal and transverse displacements) can be decoupled from the dynamic model in (7.14)–(7.16).

Remark 7.2 In the case that the propagation of the elastic wave in the *i-k* plane is disregarded, the lateral vibration of the plate can be described via the following coupled nonlinear differential equation:

$$\rho h\left(w_{tt}+2v_0w_{xt}+v_0^2w_{xx}\right)-M_{Xxx}-2M_{XZxz}-M_{Zzz}$$

$$- \Phi_{zz} w_{xx} + 2\Phi_{xz} w_{xz} - \Phi_{xx} w_{zz} = 0, \tag{7.20}$$

$$\frac{1}{Eh\chi}\left[\Phi_{zzzz} + \left(\frac{E\chi}{G} - 2\chi\upsilon\right)\Phi_{xxzz} + \chi\Phi_{xxxx}\right] = w_{xz}^2 - w_{xx} w_{zz}, \tag{7.21}$$

where Φ is the Airy stress function of the cross-sectional force, which is related to the sectional plate forces and sectional shear force by

$$N_{\mathrm{X}} = \Phi_{zz}, N_{\mathrm{Z}} = \Phi_{xx}, N_{\mathrm{XZ}} = -\Phi_{xz}. \tag{7.22}$$

Remark 7.3 If the bending moment is neglected, (7.14)–(7.16) become the dynamic model of an axially moving membrane, that is,

$$\rho h\left(w_{tt} + 2v_0 w_{xt} + v_0^2 w_{xx}\right) - (N_{\mathrm{X}} w_x)_x - (N_{\mathrm{Z}} w_z)_z - (N_{\mathrm{XZ}} w_x)_z - (N_{\mathrm{XZ}} w_z)_x = 0, \tag{7.23}$$

$$\rho h\left(u_{tt} + 2v_0 u_{xt} + v_0^2 u_{xx}\right) - N_{\mathrm{X}x} - N_{\mathrm{XZ}z} = 0, \tag{7.24}$$

$$\rho h\left(\eta_{tt} + 2v_0 \eta_{xt} + v_0^2 \eta_{xx}\right) - N_{\mathrm{XZ}x} - N_{\mathrm{Z}z} = 0. \tag{7.25}$$

7.2.2 *Viscoelastic Plate*

The two-dimensional standard linear solid model (Zener model) for axially moving viscoelastic plates was presented by Marynowski (2004). According to this model, the stress–strain relation is given as follows:

$$\begin{gathered} \sigma_{\mathrm{XZ}} = \int_0^t \psi_{\mathrm{XZ}}(t-\tau)\varepsilon_{\mathrm{XZ}\tau}\mathrm{d}\tau, \\ \sigma_{\mathrm{X}} = \Gamma^{-1}\Delta_1\varepsilon_{\mathrm{X}}, \sigma_{\mathrm{Z}} = \Gamma^{-1}\Delta_2\varepsilon_{\mathrm{Z}}, \end{gathered} \tag{7.26}$$

where ψ_{XZ} is the relaxation function of the Zener-type viscoelastic element, which is determined based on the relaxation time constant t_G and the shear modulus of the material G_1 and G_2, that is,

$$\psi_{\mathrm{XZ}} = G_1 + G_2 \mathrm{e}^{-t/t_G}. \tag{7.27}$$

In (7.26), Γ, Δ_1, and Δ_2 are differential operators, which are defined as follows:

$$\Gamma = \frac{1}{Q_{11}^2 Q_{22}^2}\left(q_6\frac{d^6}{dt^6} + q_5(t)\frac{d^5}{dt^5} + q_4(t)\frac{d^4}{dt^4} + q_3(t)\frac{d^3}{dt^3} + q_2(t)\frac{d^2}{dt^2}\right), \quad (7.28)$$

$$\Delta_1 = \frac{1}{Q_{11}^2 Q_{22}^2}\left(\frac{\mu_Z}{E_{1Z}^2}\frac{d^5}{dt^5} + \left(E_{1Z}^{-1} + \phi_Z(t)\right)\frac{d^2}{dt^2} + \frac{d\phi_Z}{dt}\frac{d}{dt}\right), \quad (7.29)$$

$$\Delta_2 = \frac{1}{Q_{11}^2 Q_{22}^2}\left(\frac{\mu_X}{E_{1X}^2}\frac{d^5}{dt^5} + \left(E_{1X}^{-1} + \phi_X(t)\right)\frac{d^2}{dt^2} + \frac{d\phi_X}{dt}\frac{d}{dt}\right), \quad (7.30)$$

where Q_{11} and Q_{22} are the stiffness elements of the elastic compliance matrix of the material,

$$\begin{aligned}
q_2(t) &= \frac{d\phi_Z}{dt}\frac{d\phi_X}{dt},\\
q_3(t) &= \frac{d\phi_X}{dt}\left(E_{1X}^{-1} + \phi_X(t)\right) + \frac{d\phi_Z}{dt}\left(E_{1Z}^{-1} + \phi_Z(t)\right),\\
q_4(t) &= \frac{\mu_X}{E_{1X}^2}\frac{d\phi_Z}{dt} + \frac{\mu_Z}{E_{1Z}^2}\frac{d\phi_X}{dt} + \left(E_{1X}^{-1} + \phi_X(t)\right)\left(E_{1Z}^{-1} + \phi_Z(t)\right),\\
q_5(t) &= \frac{\mu_Z}{E_{1Z}^2}\left(E_{1X}^{-1} + \phi_X(t)\right) + \frac{\mu_X}{E_{1X}^2}\left(E_{1Z}^{-1} + \phi_Z(t)\right),\\
q_6 &= \frac{\mu_X\mu_Z}{E_{1X}^2 E_{1Z}^2}.
\end{aligned} \quad (7.31)$$

Here, the creep functions ϕ_X and ϕ_Z are derived as follows:

$$\begin{aligned}
\phi_X(t) &= \frac{1}{E_{1X}} - \frac{E_{2X}}{E_{1X}(E_{1X} + E_{2X})}e^{-\frac{E_{1X}E_{2X}}{\mu_X(E_{1X}+E_{2X})}},\\
\phi_Z(t) &= \frac{1}{E_{1Z}} - \frac{E_{2Z}}{E_{1Z}(E_{1Z} + E_{2Z})}e^{-\frac{E_{1Z}E_{2Z}}{\mu_Z(E_{1Z}+E_{2Z})}}
\end{aligned} \quad (7.32)$$

where E_{1X}, E_{2X}, E_{1Z}, and E_{2Z} denote Young's moduli of the material and μ_X and μ_Z indicate the internal viscous damping coefficients.

If only a small displacement is considered, the sectional plate forces, shear force, sectional bending moments, and torsion moment can be computed based on the strain–stress relation given in (7.26), that is,

$$N_X = \Gamma^{-1}\Delta_1 u_x h, \quad N_Z = \Gamma^{-1}\Delta_2 \eta_x h,$$

$$N_{XZ} = \int_0^t \psi_{XZ}(t-\tau)(u_z + \eta_x)_\tau d\tau,$$

$$M_X = -\Gamma^{-1}\Delta_1 w_{xx}\frac{h^3}{12}, \quad M_Z = -\Gamma^{-1}\Delta_2 w_{zz}\frac{h^3}{12},$$

$$M_{XZ} = -\frac{h^3}{6}\int_0^t \psi_{XZ}(t-\tau)w_{xy\tau}\mathrm{d}\tau. \tag{7.33}$$

The dynamic model of the lateral vibration of an axially moving viscoelastic plate can be derived by using (7.33).

7.2.3 *Laminated Composite Plate*

In Chap. 3, we investigated axially moving laminated composite beams via classical laminate theory. In this section, the dynamic model of moving laminated composite plates with m layers is developed using a similar approach. Accordingly, the stress–strain relationship of the n-th layer is given as follows:

$$\begin{aligned}\begin{bmatrix}\sigma_X^n\\ \sigma_Z^n\\ \sigma_{XZ}^n\end{bmatrix} &= \begin{bmatrix}\overline{Q}_{11}^n & \overline{Q}_{12}^n & \overline{Q}_{16}^n\\ \overline{Q}_{12}^n & \overline{Q}_{22}^n & \overline{Q}_{26}^n\\ \overline{Q}_{16}^n & \overline{Q}_{26}^n & \overline{Q}_{66}^n\end{bmatrix}\begin{bmatrix}\varepsilon_X^0\\ \varepsilon_Z^0\\ 2\varepsilon_{XZ}^0\end{bmatrix}\\ &\quad + y\begin{bmatrix}\overline{Q}_{11}^n & \overline{Q}_{12}^n & \overline{Q}_{16}^n\\ \overline{Q}_{12}^n & \overline{Q}_{22}^n & \overline{Q}_{26}^n\\ \overline{Q}_{16}^n & \overline{Q}_{26}^n & \overline{Q}_{66}^n\end{bmatrix}\begin{bmatrix}\kappa_X\\ \kappa_Z\\ \kappa_{XZ}\end{bmatrix},\end{aligned} \tag{7.34}$$

where

$$\begin{bmatrix}\overline{Q}_{11}^n & \overline{Q}_{12}^n & \overline{Q}_{16}^n\\ \overline{Q}_{12}^n & \overline{Q}_{22}^n & \overline{Q}_{26}^n\\ \overline{Q}_{16}^n & \overline{Q}_{26}^n & \overline{Q}_{66}^n\end{bmatrix} = \mathbf{T}_L^{-1}\begin{bmatrix}Q_{11}^n & Q_{12}^n & Q_{16}^n\\ Q_{12}^n & Q_{22}^n & Q_{26}^n\\ Q_{16}^n & Q_{26}^n & Q_{66}^n\end{bmatrix}\mathbf{R}\mathbf{T}_L\mathbf{R}^{-1}, \tag{7.35}$$

where $\mathbf{T}_L$ and $\mathbf{R}$ are the transformation matrix and the Reuter matrix, respectively, which are defined in (3.113); Q_{ij}^n are the stiffness elements of the material used in the n-th layer, which are given in (3.115). The forces and moments in the laminated composite plate are computed as follows:

$$\begin{aligned}\begin{bmatrix}N_X\\ N_Z\\ N_{XZ}\end{bmatrix} &= \int_{-h/2}^{h/2}\begin{bmatrix}\sigma_X\\ \sigma_Z\\ \sigma_{XZ}\end{bmatrix}\mathrm{d}z\\ &= \sum_{n=1}^{m}\int_{h_{n-1}}^{h_n}\begin{bmatrix}\sigma_X^n\\ \sigma_Z^n\\ \sigma_{XZ}^n\end{bmatrix}\mathrm{d}z\end{aligned}$$

$$
\begin{aligned}
&= \sum_{n=1}^{m} \int_{h_{n-1}}^{h_n} \begin{bmatrix} \overline{Q}_{11}^n & \overline{Q}_{12}^n & \overline{Q}_{16}^n \\ \overline{Q}_{12}^n & \overline{Q}_{22}^n & \overline{Q}_{26}^n \\ \overline{Q}_{16}^n & \overline{Q}_{26}^n & \overline{Q}_{66}^n \end{bmatrix} \begin{bmatrix} \varepsilon_{\mathrm{X}}^0 \\ \varepsilon_{\mathrm{Z}}^{0n} \\ 2\varepsilon_{\mathrm{XZ}}^{0n} \end{bmatrix} \mathrm{d}z \\
&+ \sum_{n=1}^{m} \int_{h_{n-1}}^{h_n} y \begin{bmatrix} \overline{Q}_{11}^n & \overline{Q}_{12}^n & \overline{Q}_{16}^n \\ \overline{Q}_{12}^n & \overline{Q}_{22}^n & \overline{Q}_{26}^n \\ \overline{Q}_{16}^n & \overline{Q}_{26}^n & \overline{Q}_{66}^n \end{bmatrix} \begin{bmatrix} \kappa_{\mathrm{X}} \\ \kappa_{\mathrm{Z}} \\ \kappa_{\mathrm{XZ}} \end{bmatrix} \mathrm{d}z \\
&= \begin{bmatrix} A_{11} & A_{12} & A_{16} \\ A_{12} & A_{22} & A_{26} \\ A_{16} & A_{26} & A_{66} \end{bmatrix} \begin{bmatrix} \varepsilon_{\mathrm{X}}^0 \\ \varepsilon_{\mathrm{Z}}^0 \\ 2\varepsilon_{\mathrm{XZ}}^0 \end{bmatrix} \\
&+ \begin{bmatrix} B_{11} & B_{12} & B_{16} \\ B_{12} & B_{22} & B_{26} \\ B_{16} & B_{26} & B_{66} \end{bmatrix} \begin{bmatrix} \kappa_{\mathrm{X}} \\ \kappa_{\mathrm{Z}} \\ \kappa_{\mathrm{XZ}} \end{bmatrix},
\end{aligned} \tag{7.36}
$$

$$
\begin{aligned}
\begin{bmatrix} M_{\mathrm{X}} \\ M_{\mathrm{Z}} \\ M_{\mathrm{XZ}} \end{bmatrix} &= \int_{-h/2}^{h/2} y \begin{bmatrix} \sigma_{\mathrm{X}} \\ \sigma_{\mathrm{Z}} \\ \sigma_{\mathrm{XZ}} \end{bmatrix} \mathrm{d}z \\
&= \sum_{n=1}^{m} \int_{h_{n-1}}^{h_n} y \begin{bmatrix} \sigma_{\mathrm{X}}^n \\ \sigma_{\mathrm{Z}}^n \\ \sigma_{\mathrm{XZ}}^n \end{bmatrix} \mathrm{d}z \\
&= \sum_{n=1}^{m} \int_{h_{n-1}}^{h_n} y \begin{bmatrix} \overline{Q}_{11}^n & \overline{Q}_{12}^n & \overline{Q}_{16}^n \\ \overline{Q}_{12}^n & \overline{Q}_{22}^n & \overline{Q}_{26}^n \\ \overline{Q}_{16}^n & \overline{Q}_{26}^n & \overline{Q}_{66}^n \end{bmatrix} \begin{bmatrix} \varepsilon_{\mathrm{X}}^0 \\ \varepsilon_{\mathrm{Z}}^{0n} \\ 2\varepsilon_{\mathrm{XZ}}^{0n} \end{bmatrix} \mathrm{d}z \\
&+ \sum_{n=1}^{m} \int_{h_{n-1}}^{h_n} y^2 \begin{bmatrix} \overline{Q}_{11}^n & \overline{Q}_{12}^n & \overline{Q}_{16}^n \\ \overline{Q}_{12}^n & \overline{Q}_{22}^n & \overline{Q}_{26}^n \\ \overline{Q}_{16}^n & \overline{Q}_{26}^n & \overline{Q}_{66}^n \end{bmatrix} \begin{bmatrix} \kappa_{\mathrm{X}} \\ \kappa_{\mathrm{Z}} \\ \kappa_{\mathrm{XZ}} \end{bmatrix} \mathrm{d}z \\
&= \begin{bmatrix} B_{11} & B_{12} & B_{16} \\ B_{12} & B_{22} & B_{26} \\ B_{16} & B_{26} & B_{66} \end{bmatrix} \begin{bmatrix} \varepsilon_{\mathrm{X}}^0 \\ \varepsilon_{\mathrm{Z}}^0 \\ 2\varepsilon_{\mathrm{XZ}}^0 \end{bmatrix} \\
&+ \begin{bmatrix} D_{11} & D_{12} & D_{16} \\ D_{12} & D_{22} & D_{26} \\ D_{16} & D_{26} & D_{66} \end{bmatrix} \begin{bmatrix} \kappa_{\mathrm{X}} \\ \kappa_{\mathrm{Z}} \\ \kappa_{\mathrm{XZ}} \end{bmatrix},
\end{aligned} \tag{7.37}
$$

where

$$
A_{ij} = \sum_{n=1}^{m} \overline{Q}_{ij}^n (h_n - h_{n-1}), \quad i = 1, 2, 6; \quad j = 1, 2, 6,
$$

$$
B_{ij} = \frac{1}{2} \sum_{n=1}^{m} \overline{Q}_{ij}^n (h_n^2 - h_{n-1}^2), \quad i = 1, 2, 6; \quad j = 1, 2, 6,
$$

$$D_{ij} = \frac{1}{3}\sum_{n=1}^{m} \overline{Q}_{ij}^{n}(h_n^3 - h_{n-1}^3), \quad i = 1, 2, 6; \quad j = 1, 2, 6, \tag{7.38}$$

Combining (7.36) and (7.37) yields the following equations:

$$\begin{bmatrix} N_{\mathrm{X}} \\ N_{\mathrm{Z}} \\ N_{\mathrm{XZ}} \\ M_{\mathrm{X}} \\ M_{\mathrm{Z}} \\ M_{\mathrm{XZ}} \end{bmatrix} = \begin{bmatrix} A_{11} & A_{12} & A_{16} & B_{11} & B_{12} & B_{16} \\ A_{12} & A_{22} & A_{26} & B_{12} & B_{22} & B_{26} \\ A_{16} & A_{26} & A_{66} & B_{16} & B_{26} & B_{66} \\ B_{11} & B_{12} & B_{16} & D_{11} & D_{12} & D_{16} \\ B_{12} & B_{22} & B_{26} & D_{12} & D_{22} & D_{26} \\ B_{16} & B_{26} & B_{66} & D_{16} & D_{26} & D_{66} \end{bmatrix} \begin{bmatrix} \varepsilon_{\mathrm{X}}^0 \\ \varepsilon_{\mathrm{Z}}^0 \\ 2\varepsilon_{\mathrm{XZ}}^0 \\ \kappa_{\mathrm{X}} \\ \kappa_{\mathrm{Z}} \\ \kappa_{\mathrm{XZ}} \end{bmatrix}$$

$$= \begin{bmatrix} \mathbf{A} & \mathbf{B} \\ \mathbf{B} & \mathbf{D} \end{bmatrix} \begin{bmatrix} \varepsilon_{\mathrm{X}}^0 \\ \varepsilon_{\mathrm{Z}}^0 \\ 2\varepsilon_{\mathrm{XZ}}^0 \\ \kappa_{\mathrm{X}} \\ \kappa_{\mathrm{Z}} \\ \kappa_{\mathrm{XZ}} \end{bmatrix}. \tag{7.39}$$

In (7.39), **A** is the extensional stiffness matrix that relates the in-plate forces and the strain of the plate, whereas the bending stiffness matrix **D** is related to the plate's bending moments. Additionally, **B** is the coupling stiffness matrix that couples the force and moment terms to the strain and curvature components in the mid-surface. By substituting the strain–displacement relationship in Sect. 7.2.1 into (7.39) and then substituting the resulting equations into the dynamic model in (7.14)–(7.18), we obtain the governing equations describing the dynamics of a laminated composite plate. Furthermore, if ρ_n is the mass density of the n-th layer, the dynamic model of the laminated composite plate is shown as follows:

$$\begin{aligned} &\left(w_{tt} + 2v_0 w_{xt} + v_0^2 w_{xx}\right)\sum_{n=1}^{m} \rho_n(h_n - h_{n-1}) \\ &- M_{\mathrm{X}xx} - 2M_{\mathrm{XZ}xz} - M_{\mathrm{Z}zz} - (N_{\mathrm{X}} w_x)_x \\ &- (N_{\mathrm{Z}} w_z)_z - (N_{\mathrm{XZ}} w_x)_z - (N_{\mathrm{XZ}} w_z)_x = 0, \end{aligned} \tag{7.40}$$

$$\left(u_{tt} + 2v_0 u_{xt} + v_0^2 u_{xx}\right)\sum_{n=1}^{m} \rho_n(h_n - h_{n-1}) - N_{\mathrm{X}x} - N_{\mathrm{Z}z} = 0, \tag{7.41}$$

$$\left(\eta_{tt} + 2v_0 \eta_{xt} + v_0^2 \eta_{xx}\right)\sum_{n=1}^{m} \rho_n(h_n - h_{n-1}) - N_{\mathrm{X}x} - N_{\mathrm{Z}z} = 0. \tag{7.42}$$

7.3 Approximate Models

In establishing the approximate model for the axially moving plate, the Galerkin method is extended to discretize the equations of motion and boundary conditions of two-dimensional models. Similar to one-dimensional continuous systems, the vibration of the plate can be represented by a series of admissible functions, namely

$$w(x, z, t) = \sum_{i=1}^{K} q_i(t)\varphi_i(x, z) \tag{7.43}$$

where $\varphi_i(x, z)$ is the basis function. The extended Galerkin method requires the basis functions to satisfy the geometric boundary conditions while not necessarily satisfying the dynamic boundary conditions, that is $\varphi_i(x, z)$ satisfies the geometric boundary conditions that relate the displacements and slopes at the boundaries. Furthermore, the basis function $\varphi_i(x, z)$ can be chosen as the product of the basis functions of two one-dimensional continuous systems. The expression of $w(x, z, t)$ can be rewritten as

$$w(x, z, t) = \sum_{i=1}^{N}\sum_{j=1}^{M} q_{ij}(t)X_i(x)Z_j(z) \tag{7.44}$$

where $X_i(x)$ and $Z_j(z)$ are the eigenfunctions of the linear static beams in the i-direction and k-direction, respectively. It should be noted that $X_i(x)$ and $Z_j(z)$ must satisfy the geometric boundary conditions of the corresponding beams. For example, in the case shown in Fig. 7.1, the basis functions $X_i(x)$ can be chosen similar to the case of a simply supported beam, whereas $Z_j(z)$ possesses the same form as the case of a free–free beam. By using (7.44), a set of $N \times M$ ODEs, wherein $q_{ij}(t)$ is the state variable, is established based on the Galerkin method. Since the plate's vibration involves two independent spatial variables, the process of deriving the approximate model of the axially moving plate is more complicated mathematically than the string, beam, and belt.

Now, we establish an approximate model for the following axially moving isotropic plate.

$$\begin{aligned} &\rho h\left(w_{tt} + 2v_0 w_{xt} + v_0^2 w_{xx}\right) + D(w_{xxxx} + 2w_{xxzz} + w_{zzzz}) \\ &\quad - \Phi_{zz}w_{xx} + 2\Phi_{xz}w_{xz} - \Phi_{xx}w_{zz} = 0, \end{aligned} \tag{7.45}$$

$$\frac{1}{Eh}\left[\Phi_{zzzz} + \left(\frac{E}{G} - 2\upsilon\right)\Phi_{xxzz} + \Phi_{xxxx}\right] = w_{xz}^2 - w_{xx}w_{zz}. \tag{7.46}$$

If we assume that the plate is loaded along the i-axis only by an axial tension T_0 and the edges of the plate are supported simply, the boundary conditions are given as follows:

$$\begin{aligned}&\text{At } x=0,l:\ w=w_{xx}=0;\quad \Phi_{zz}=T_0/h;\quad \Phi_{xz}=0,\\&\text{At } z=0,b:\ w=w_{xx}=0;\quad \Phi_{xx}=\Phi_{xz}=0.\end{aligned}\tag{7.47}$$

With the assumed boundary conditions, the sine function can be used as the basis function. The approximations of $w(x, z, t)$ and $\Phi(x, z, t)$ are expressed as follows:

$$w(x,z,t)=\sum_{i=1}^{N}\sum_{j=1}^{M}q_{ij}(t)\sin\left(\frac{i\pi x}{l}\right)\sin\left(\frac{j\pi z}{b}\right),\tag{7.48}$$

$$\Phi(x,z,t)=\frac{z^2}{2hl}+\sum_{i=1}^{N}\sum_{j=1}^{M}\phi_{ij}(t)\sin^2\left(\frac{i\pi x}{l}\right)\sin^2\left(\frac{j\pi z}{b}\right),\tag{7.49}$$

where $q_{ij}(t)$ and $\phi_{ij}(t)$ denote the sets of generalized displacements and stress variables concerning the plate. First, the stress variables are determined by using the Galerkin method for (7.46). Accordingly, we obtain $\phi_{ij}(t)$ that depends on the generalized displacement $q_{ij}(t)$. Only the first- and second-order vibrations are considered (i.e., $N = 2$ and $M = 2$). According to the Galerkin method, we derive

$$\begin{aligned}0=&\int_0^l\int_0^b\left\{\left[\frac{\Phi_{zzzz}}{Eh}+\left(\frac{1}{Gh}-2\frac{\upsilon}{Eh}\right)\Phi_{xxzz}+\frac{\Phi_{xxxx}}{Eh}-w_{xz}^2+w_{xx}w_{zz}\right]\right.\\&\left.\sin^2\left(\frac{m\pi x}{l}\right)\sin^2\left(\frac{n\pi z}{b}\right)\right\}\mathrm{d}x\mathrm{d}z;\\&m=1,2;\ n=1,2.\end{aligned}\tag{7.50}$$

Expanding (7.50) yields the algebraic relationship between the stress variables $\phi_{ij}(t)$ and the displacement variables $q_{ij}(t)$, that is,

$$\begin{bmatrix}\alpha_1 & -4Gl^4 & -4Gb^4 & 0\\ -6Gl^4 & \alpha_2 & 0 & -128Gb^4\\ -8Gb^4 & 0 & \alpha_2 & -128Gb^4\\ 0 & -32Gb^4 & -32Gl^4 & 8\alpha_1\end{bmatrix}\begin{bmatrix}\phi_{11}\\ \phi_{21}\\ \phi_{12}\\ \phi_{22}\end{bmatrix}=EGb^2l^2h\begin{bmatrix}1&2&2&0\\1&8&0&16\\1&0&8&16\\0&1&1&8\end{bmatrix}\begin{bmatrix}q_{11}^2\\ q_{21}^2\\ q_{12}^2\\ q_{22}^2\end{bmatrix}\tag{7.51}$$

where

$$\begin{aligned}\alpha_1&=4Gb^2l^2\upsilon-2Eb^2l^2-6Gb^2-6Gl^4,\\ \alpha_2&=32Gb^2l^2\upsilon-192Gb^2-12Gl^4-16Eb^2l^2.\end{aligned}\tag{7.52}$$

The stress variables in the algebraic equations (7.51) can be solved in terms of q_{ij}^2. Let $\phi_{ij} = f_{ij}(q_{ij}^2), i = 1, 2; j = 1, 2$. Now, the Airy stress function $\Phi(x, z, t)$ has been determined. We can substitute this Airy stress function into (7.45) and then apply the Galerkin method for the resulting equation. The Galerkin integration of (7.45) is

$$0 = \int_0^l \int_0^b \{[\rho h(w_{tt} + 2v_0 w_{xt} + v_0^2 w_{xx}) + D(w_{xxxx} + 2w_{xxzz} + w_{zzzz}) - \Phi_{zz} w_{xx} \\ + 2\Phi_{xz} w_{xz} - \Phi_{xx} w_{zz}] \sin\left(\frac{m\pi x}{l}\right) \sin\left(\frac{n\pi z}{b}\right)\} \mathrm{d}x\mathrm{d}z; \quad m = 1, 2; \quad n = 1, 2. \tag{7.53}$$

Accordingly, we obtain a set of four coupled nonlinear ODEs, which is shown in the following matrix form:

$$\mathbf{M}\ddot{\mathbf{q}} + \mathbf{C}\dot{\mathbf{q}} + (\mathbf{K} + \mathbf{N}(\mathbf{q}^2))\mathbf{q} = \mathbf{0} \tag{7.54}$$

where $\mathbf{q} = [q_{11}\ q_{21}\ q_{12}\ q_{22}]^{\mathrm{T}}$ is the state vector. **M**, **C**, and **K** refer to the global matrices of the mass, damping coefficient, and stiffness, respectively, and **N** is the matrix associated with the nonlinear terms. These matrices are defined as follows:

$$\mathbf{M} = \begin{bmatrix} 1\,0\,0\,0 \\ 0\,1\,0\,0 \\ 0\,0\,1\,0 \\ 0\,0\,0\,1 \end{bmatrix}, \mathbf{C} = \begin{bmatrix} 0 & -16v_0/3l & 0 & 0 \\ 16v_0/3l & 0 & 0 & 0 \\ 0 & 0 & 0 & -16v_0/3l \\ 0 & 0 & 16v_0/3l & 0 \end{bmatrix},$$
$$\mathbf{K} = \begin{bmatrix} K_{11} & 0 & 0 & 0 \\ 0 & K_{22} & 0 & 0 \\ 0 & 0 & K_{33} & 0 \\ 0 & 0 & 0 & K_{44} \end{bmatrix}, \mathbf{N} = \begin{bmatrix} N_{11} & 0 & 0 & 0 \\ 0 & N_{22} & 0 & 0 \\ 0 & 0 & N_{33} & 0 \\ 0 & 0 & 0 & N_{44} \end{bmatrix}, \tag{7.55}$$

where

$$K_{11} = \frac{\pi^4}{\rho h l^4 b^4}(Db^4 + 2Dl^2b^2 + Dl^4 + Fl^2b^4 - \rho h v_0^2 l^2 b^4),$$
$$K_{22} = \frac{\pi^4}{\rho h l^4 b^4}(16Db^4 + 8Dl^2b^2 + Dl^4 + 4Fl^2b^4 - 4\rho h v_0^2 l^2 b^4),$$
$$K_{33} = \frac{\pi^4}{\rho h l^4 b^4}(Db^4 + 8Dl^2b^2 + 16Dl^4 + Fl^2b^4 - \rho h v_0^2 l^2 b^4),$$
$$K_{44} = \frac{\pi^4}{\rho h l^4 b^4}(16Db^4 + 32Dl^2b^2 + 16Dl^4 + 4Fl^2b^4 - 4\rho h v_0^2 l^2 b^4),$$
$$N_{11} = -\frac{\pi^4}{2\rho h l^2 b^2}(2f_{11} + f_{21} + f_{12}),$$

$$N_{22} = -\frac{\pi^4}{\rho h l^2 b^2}(2f_{11} + 4f_{21} + 2f_{22}),$$
$$N_{33} = -\frac{\pi^4}{\rho h l^2 b^2}(2f_{11} + 4f_{12} + 2f_{22}),$$
$$N_{44} = -\frac{\pi^4}{\rho h l^2 b^2}(8f_{21} + 8f_{12} + 16f_{22}). \tag{7.56}$$

Another kind of approximate model for two-dimensional continuum was presented by Shin et al. (2006). We consider a moving membrane described by the following dynamic model:

$$\rho h\left(w_{tt} + \dot{v}w_x + 2vw_{xt} + v^2 w_{xx}\right) - (N_X w_x + N_{XZ} w_z)_x - (N_Z w_z + N_{XZ} w_x)_z = 0, \tag{7.57}$$

$$\rho h\left(u_{tt} + \dot{v}u_x + 2vu_{xt} + v_0^2 u_{xx}\right) - N_{Xx} - N_{XZz} = -\rho h\dot{v}, \tag{7.58}$$

$$\rho h\left(\eta_{tt} + \dot{v}\eta_x + 2v\eta_{xt} + v^2\eta_{xx}\right) - N_{XZx} - N_{Zz} = 0, \tag{7.59}$$

where the membrane forces and torque are given in (7.19). The geometric boundary conditions are given as follows:

$$\begin{aligned}
&\text{At } x = 0 : w = \eta = 0;\ \ N_X = T_0;\\
&\text{At } x = l : w = \eta = 0;\ \ N_X = T_l;\\
&\text{At } z = 0, b : N_Z = N_{XZ} = N_Z w_z + N_{XZ} w_{xz} = 0
\end{aligned} \tag{7.60}$$

where T_0 is the tension per unit width on the right end of the membrane and $T_l = T_0 + \rho h l \dot{v}$ is the tension per unit width on the left end of the membrane. The dynamic model is transformed into two-week forms: one form for the motions in the *i*-*k* plane [i.e., Eqs. (7.24) and (7.25)] and one form for the lateral motion [i.e., Eq. (7.23)]. Let $\overline{w}$, $\overline{u}$, and $\overline{\eta}$ be the weight functions of w, u, and η, respectively. The two-week forms relating to the lateral motion and the motions in the *i*-*k* plane are shown as follows:

$$\int_0^l \int_0^b \left[\rho h\overline{w}\left(w_{tt} + \dot{v}w_x + 2vw_{xt} + v^2 w_{xx}\right) + \overline{\boldsymbol{\alpha}}^{\mathrm{T}}\mathbf{N}\boldsymbol{\alpha}\right]\mathrm{d}x\mathrm{d}z = 0, \tag{7.61}$$

$$\begin{aligned}
&\int_0^l \int_0^b \left[\rho h\overline{\boldsymbol{u}}^{\mathrm{T}}\left(\mathbf{u}_{tt} + \dot{v}\mathbf{u}_x + 2v\mathbf{u}_{xt} + v^2\mathbf{u}_{xx}\right) + \overline{\boldsymbol{\beta}}^{\mathrm{T}}\mathbf{D}\beta\right]\mathrm{d}x\mathrm{d}z\\
&= \int_0^z (N_X\bar{u})|_{x=0}^{x=l}\mathrm{d}z - \rho h\dot{v}\int_0^l \int_0^b \bar{u}\,\mathrm{d}x\mathrm{d}z
\end{aligned} \tag{7.62}$$

where

$$\boldsymbol{\alpha}=\begin{bmatrix} w_x \\ w_z \end{bmatrix}, \overline{\boldsymbol{\alpha}}=\begin{bmatrix} \overline{w}_x \\ \overline{w}_z \end{bmatrix}, \mathbf{N}=\begin{bmatrix} N_X & N_{XZ} \\ N_{XZ} & N_Z \end{bmatrix}, \mathbf{u}=\begin{bmatrix} u \\ \eta \end{bmatrix}, \overline{\mathbf{u}}=\begin{bmatrix} \overline{u} \\ \overline{\eta} \end{bmatrix},$$
$$\boldsymbol{\beta}=\begin{bmatrix} u_x \\ \eta_z \\ u_z+\eta_x \end{bmatrix}, \overline{\boldsymbol{\beta}}=\begin{bmatrix} \overline{u}_x \\ \overline{\eta}_z \\ \overline{u}_z+\overline{\eta}_x \end{bmatrix}, \mathbf{D}=\frac{Eh}{1-\upsilon^2}\begin{bmatrix} 1 & \upsilon & 0 \\ \upsilon & 1 & 0 \\ 0 & 0 & (1-\upsilon)/2 \end{bmatrix}. \quad (7.63)$$

Note that the geometric boundary conditions have been included in the week forms. Therefore, the membrane vibrations can be approximated by using admissible functions as the basis function, that is,

$$w(x,z,t)=\sum_{i=0}^{N}\sum_{j=0}^{N} q_{ij}^{w}(t)\varphi_i(x)Z_j(z), \quad (7.64)$$

$$u(x,z,t)=\sum_{i=0}^{N}\sum_{j=0}^{N} q_{ij}^{u}(t)X_i(x)Z_j(z), \quad (7.65)$$

$$\eta(x,z,t)=\sum_{i=0}^{N}\sum_{j=0}^{N} q_{ij}^{\eta}(t)\varphi_i(x)Z_j(z), \quad (7.66)$$

where $\varphi_i(x)=\sin((i+1)\pi x/l)$ and $X_i(x)$ and $Z_j(z)$ are the Legendre polynomials, which are defined as

$$X_i(x)=\sum_{k=0}^{P_i}\frac{(-1)^k(2i-2k)!}{2^i k!(i-k)!(i-2k)!}\left(\frac{2x}{l}-1\right)^{i-2k},$$
$$Z_j(z)=\sum_{k=0}^{P_j}\frac{(-1)^k(2j-2k)!}{2^j k!(j-k)!(j-2k)!}\left(\frac{2z}{b}-1\right)^{j-2k},$$
$$P_i=\begin{cases} i/2, & \text{if } i \text{ is even}, \\ (i-1)/2, & \text{if } i \text{ is odd}. \end{cases} \quad (7.67)$$

The weight functions $\overline{w}$, $\overline{u}$, and $\overline{\eta}$ are expressed in the same forms as (7.64)–(7.66), where the time-dependent functions $q_{ij}^{w}(t)$, $q_{ij}^{u}(t)$, and $q_{ij}^{\eta}(t)$ are replaced with the arbitrary functions $\overline{q}_{ij}^{w}(t)$, $\overline{q}_{ij}^{u}(t)$, and $\overline{q}_{ij}^{\eta}(t)$. By using these weight functions and (7.64)–(7.66), two matrix–vector equations are established based on the two-week forms, namely

$$\mathbf{M}_{\mathrm{w}}\ddot{\mathbf{q}}_{\mathrm{w}}+\mathbf{C}_{\mathrm{w}}\ddot{\mathbf{q}}_{\mathrm{w}}+\mathbf{K}_{\mathrm{w}}\left(\mathbf{q}_{\mathrm{u}\eta}\right)\mathbf{q}_{\mathrm{w}}=0, \quad (7.68)$$

$$\mathbf{M}_{u\eta}\ddot{\mathbf{q}}_{u\eta} + \mathbf{C}_{u\eta}\dot{\mathbf{q}}_{u\eta} + \mathbf{K}_{u\eta}\mathbf{q}_{u\eta} = \mathbf{G}_{u\eta} \tag{7.69}$$

where $\mathbf{q}_{u\eta}$ denotes the state vector consisting of the time-dependent functions $q_{ij}^{u}(t)$ and $q_{ij}^{\eta}(t)$; $\mathbf{q}_{w}$ is the vector of $q_{ij}^{w}(t)$; and $\mathbf{M}_{u\eta}$, $\mathbf{C}_{u\eta}$, and $\mathbf{K}_{u\eta}$ denote the global matrices for the mass, damping component, and stiffness, respectively, in the in-plane deflections. The corresponding notations for the lateral motion are $\mathbf{M}_{w}$, $\mathbf{C}_{w}$, and $\mathbf{K}_{w}$. The elements of these matrices are defined in Appendix E. Notably, $\mathbf{K}_{w}$ is a function of $\mathbf{q}_{u\eta}$. The vector $\mathbf{q}_{u\eta}$ that contains the information of the longitudinal and transverse displacements, $u(x, z, t)$ and $\eta(x, z, t)$, can be obtained independently without considering the lateral displacement $w(x, z, t)$ by solving (7.69). Once $\mathbf{q}_{u\eta}$ is determined, the matrix–vector equation in (7.68) can be solved to determine the lateral displacement.

7.4 Dynamic Analysis of Plate Models

The dynamic behaviors of axially moving plates have been investigated in the literature. Similar to one-dimensional models of AMSs, the influence of the lateral vibration on the dynamic behaviors of the plates is more impactful than other vibrations (i.e., longitudinal and transverse vibrations). Therefore, most of the studies on the dynamics of axially moving plates consider only lateral vibration. The boundaries of the instability regions of the traveling plate are related to its axial velocity: The axially moving plate experiences divergence instability at a sufficiently high velocity, which is known as the critical speed. The critical speed can be determined based on two approaches: Static analysis can be performed to determine the existence of the nontrivial equilibrium configuration, and the dynamic method can be used to examine the eigenvalues of a discretized system. In an early study on the vibration characteristics of axially moving isotropic plates published by Lin (1997), the critical speed was derived via these two approaches.

Based on the finite difference method, the natural frequencies of free vibration of a translating viscoelastic plate were investigated by Yang et al. (2012). Furthermore, the authors analyzed the bifurcation and chaotic dynamics of the forced vibration. In this study, bifurcation diagrams reveal the influence of the translating speed, damping viscosity, and excitation magnitude on the system's dynamics. The equilibrium loses its stability and bifurcates into periodic motions with the increase of these parameters. If these parameters further increase, the chaotic motion will exist after a sequence of period-doubling bifurcations. Zhou and Wang (2019) addressed the divergence and flutter instabilities of an axially moving viscoelastic plate, wherein different boundary conditions were considered. The authors used the power series method to develop complex eigenvalue equations. Based on the numerical results, they also presented the influences of the geometrical ratio of the plate, boundary conditions, and viscosity coefficient on the system instability.

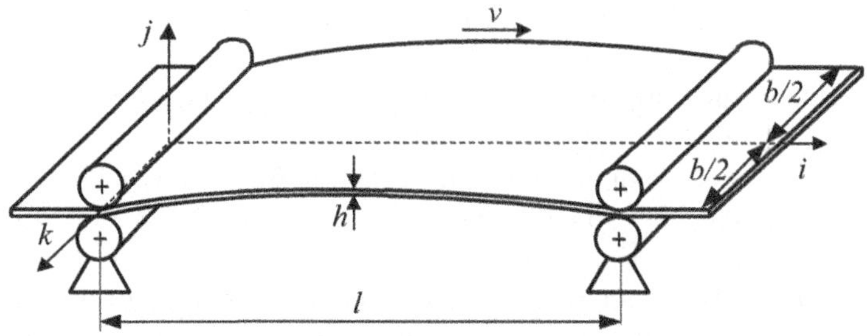

Fig. 7.2 Axially moving elastic plate with free sides at $z = \pm b/2$

Concerning axially moving nanoscale plates, Liu et al. (2017) addressed the stability analysis of a translating nanoscale plate in the sub- and supercritical regimes. Later, Duan et al. (2019) considered an axially moving nanoscale plate subjected to a viscoelastic foundation. The stability of the considered system was investigated in the study.

The dynamics of axially moving composite materials were also studied in the literature. In Yang et al. (2011), a dynamic model of a translating rectangular antisymmetric cross-ply composite plate was developed. Subsequently, the authors analyzed the complex natural frequencies; and accordingly, they studied the flutter and divergence instabilities. Furthermore, Wang et al. (2014), Wang and Zu (2017), Wang and Zu (2017) and Li et al. (2017) also discussed the stability of axially moving plates in diverse environments, such as thermos-electro-mechanical fields, magnetic fields, and fluids.

The determination of the critical velocity and equilibrium configuration of the axially moving plate is similar to the buckling problem of compressed stationary plates presented in the book *Theory of Elastic Stability* by Timoshenko and Gere (2009). The equilibrium problem of moving elastic plates has been introduced in the literature (Lin 1997; Banichuk et al. 2020). We focus on investigating the dynamics of axially moving plates, wherein only the lateral vibration is considered. We further assume that the plate has two simply supported edges and two free edges (i.e., Fig. 7.2), and only the constant longitudinal tension T_0 is considered. In this section, we derive the boundaries of instability regions of axially moving isotropic and orthotropic plates by determining the existence of the nontrivial equilibrium configuration using static analysis.

7.4.1 *Equilibrium Solutions and Stability of an Isotropic Plate*

For an axially moving isotropic plate, the governing equation of the lateral vibration of the considered plate is given as follows:

$$\rho h\left(w_{tt}+2v_0w_{xt}+v_0^2w_{xx}\right)-T_0w_{xx}+D(w_{xxxx}+2w_{xxzz}+w_{zzzz})=0. \quad (7.70)$$

The boundary conditions are shown as follows:

$$\begin{aligned} &w(0,z,t)=w(l,z,t)=0,\\ &w_{zz}(0,z,t)+\upsilon w_{xx}(0,z,t)\\ &\quad=w_{zz}(l,z,t)+\upsilon w_{xx}(l,z,t)=0, \end{aligned} \quad (7.71)$$

$$\begin{aligned} &w_{zz}(x,-b,t)+\upsilon w_{xx}(x,-b,t)\\ &\quad=w_{zz}(x,b,t)+\upsilon w_{xx}(x,b,t)=0,\\ &w_{zzz}(x,-b,t)+(2-\upsilon)w_{xxz}(x,-b,t)\\ &\quad=w_{zzz}(x,b,t)+(2-\upsilon)w_{xxz}(x,b,t)=0. \end{aligned} \quad (7.72)$$

The equilibrium solution $w_\infty(x,z)$ of the plate is obtained by solving the following equation:

$$\left(\frac{\partial^4 w_\infty}{\partial x^4}+2\frac{\partial^4 w_\infty}{\partial x^2\partial z^2}+\chi\frac{\partial^4 w_\infty}{\partial z^4}\right)+\frac{(\rho h v_0^2-T_0)}{D}\frac{\partial^2 w_\infty}{\partial x^2}=0. \quad (7.73)$$

It is apparent that $w_\infty(x,z)=0$ is always a trivial solution of (7.73), which is also the trivial equilibrium configuration of the plate. By comparing (7.73) with the equation of a stationary plate under in-plane compression in Timoshenko and Gere (2009), it is seen that (7.73) has only a trivial equilibrium configuration if $(\rho h v_0^2-T_0)/D<0$; equivalently, $v_0<\sqrt{T_0/\rho h}$. In this situation, the plate is stable. The nontrivial equilibrium configurations, which bifurcate from the trivial equilibrium configuration, exist when the axial velocity reaches the critical speed v_{cr}. With the existence of the nontrivial equilibrium configuration, the plate can become unstable. We will derive the critical speed in the range $v_0\geq\sqrt{T_0/\rho h}$. The following representation is used:

$$w_\infty(x,z)=g(\overline{z})\sin\frac{\pi x}{l} \quad (7.74)$$

where $\overline{z}=\pi z/l$ and $g(\overline{z})$ is an unknown function that is to be subsequently determined. This representation satisfies the boundary conditions in (7.71). By using the representation of (7.73) and the boundary conditions in (7.72), it can be rewritten as follows:

$$\frac{d^4}{d\overline{z}^4}g(\overline{z})-2\frac{d^2}{d\overline{z}^2}g(\overline{z})+\left(1-\gamma^2\right)g(\overline{z})=0, \quad (7.75)$$

$$\frac{\mathrm{d}^2}{\mathrm{d}\bar{z}^2}g(\bar{z}) - \upsilon g(\bar{z}) = 0, \quad \text{for } \bar{z} = \pm\frac{\zeta\pi}{2}, \tag{7.76}$$

$$\frac{\mathrm{d}^3}{\mathrm{d}\bar{z}^3}g(\bar{z}) - (2-\upsilon)\frac{\mathrm{d}}{\mathrm{d}\bar{z}}g(\bar{z}) = 0, \quad \text{for } \bar{z} = \pm\frac{\zeta\pi}{2} \tag{7.77}$$

where $\zeta = b/l$ is the slenderness ratio of the plate and $\gamma = \sqrt{l^2(\rho h v_0^2 - T_0)/\pi^2 D}$. We consider the case in which $\gamma < 1$. The general form of the solution of (7.75) can be represented in the following form:

$$g(\bar{z}) = A_1\cosh(\alpha\bar{z}) + A_2\sinh(\alpha\bar{z}) + A_3\cosh(\beta\bar{z}) + A_4\sinh(\beta\bar{z}) \tag{7.78}$$

where $\alpha = \sqrt{1+\gamma}$ and $\beta = \sqrt{1-\gamma}$; A_1, A_2, A_3, and A_4 are constants. By using the boundary conditions in (7.76) and (7.77), we obtained the following equations:

$$\begin{bmatrix} a_1\cosh\left(\frac{\alpha\zeta\pi}{2}\right) & -a_1\sinh\left(\frac{\alpha\zeta\pi}{2}\right) & a_2\cosh\left(\frac{\beta\zeta\pi}{2}\right) & -a_2\sinh\left(\frac{\beta\zeta\pi}{2}\right) \\ -a_3\sinh\left(\frac{\alpha\zeta\pi}{2}\right) & a_3\cosh\left(\frac{\alpha\zeta\pi}{2}\right) & -a_4\sinh\left(\frac{\beta\zeta\pi}{2}\right) & a_4\cosh\left(\frac{\beta\zeta\pi}{2}\right) \\ a_1\cosh\left(\frac{\alpha\zeta\pi}{2}\right) & a_1\sinh\left(\frac{\alpha\zeta\pi}{2}\right) & a_2\cosh\left(\frac{\beta\zeta\pi}{2}\right) & a_2\sinh\left(\frac{\beta\zeta\pi}{2}\right) \\ a_3\sinh\left(\frac{\alpha\zeta\pi}{2}\right) & a_3\cosh\left(\frac{\alpha\zeta\pi}{2}\right) & a_4\sinh\left(\frac{\beta\zeta\pi}{2}\right) & a_4\cosh\left(\frac{\beta\zeta\pi}{2}\right) \end{bmatrix} \begin{bmatrix} A_1 \\ A_2 \\ A_3 \\ A_4 \end{bmatrix} = 0, \tag{7.79}$$

where

$$a_1 = \alpha^2 - \upsilon,\ a_2 = \beta^2 - \upsilon,\ a_3 = \alpha^3 - (2-\upsilon)\alpha,\ a_4 = \beta^3 - (2-\upsilon)\beta. \tag{7.80}$$

Equations (7.79) can be transformed into the following equations:

$$\mathbf{P}_\mathrm{s}\left[A_1\ A_3\right]^\mathrm{T} = 0, \tag{7.81}$$

$$\mathbf{P}_\mathrm{a}\left[A_2\ A_4\right]^\mathrm{T} = 0, \tag{7.82}$$

where

$$\mathbf{P}_\mathrm{s} = \begin{bmatrix} a_1\cosh\left(\frac{\alpha\zeta\pi}{2}\right) & a_2\cosh\left(\frac{\beta\zeta\pi}{2}\right) \\ a_3\sinh\left(\frac{\alpha\zeta\pi}{2}\right) & a_4\sinh\left(\frac{\beta\zeta\pi}{2}\right) \end{bmatrix}, \tag{7.83}$$

$$\mathbf{P}_\mathrm{a} = \begin{bmatrix} a_1\sinh\left(\frac{\alpha\zeta\pi}{2}\right) & a_2\sinh\left(\frac{\beta\zeta\pi}{2}\right) \\ a_3\cosh\left(\frac{\alpha\zeta\pi}{2}\right) & a_4\cosh\left(\frac{\beta\zeta\pi}{2}\right) \end{bmatrix}. \tag{7.84}$$

The constants A_1, A_2, A_3, and A_4 can be determined by solving (7.81) and (7.82). $\{A_1, A_2, A_3, A_4\} = \{0, 0, 0, 0\}$ is the trivial solution of the previously mentioned system. The existence of the nontrivial solution that satisfies (7.81) and (7.82) can be assured if and only if $\mathbf{P}_\mathrm{s}$ or $\mathbf{P}_\mathrm{a}$ is singular; equivalently, $\det(\mathbf{P}_\mathrm{s}) = 0$ or $\det(\mathbf{P}_\mathrm{a}) = 0$. These zero determinant conditions yield the following equations:

$$
\begin{aligned}
\det(\mathbf{P}_\mathrm{s}) &= a_1 a_4 \sinh\left(\frac{\beta\zeta\pi}{2}\right)\cosh\left(\frac{\alpha\zeta\pi}{2}\right) \\
&\quad - a_2 a_3 \sinh\left(\frac{\alpha\zeta\pi}{2}\right)\cosh\left(\frac{\beta\zeta\pi}{2}\right) = 0 \\
&\Leftrightarrow \frac{a_2 a_3}{a_1 a_4} = \frac{\sinh\left(\frac{\beta\zeta\pi}{2}\right)\cosh\left(\frac{\alpha\zeta\pi}{2}\right)}{\cosh\left(\frac{\beta\zeta\pi}{2}\right)\sinh\left(\frac{\alpha\zeta\pi}{2}\right)} \\
&= \tanh\left(\frac{\beta\zeta\pi}{2}\right)\coth\left(\frac{\alpha\zeta\pi}{2}\right),
\end{aligned}
\tag{7.85}
$$

$$
\begin{aligned}
\det(\mathbf{P}_\mathrm{a}) &= a_1 a_4 \cosh\left(\frac{\beta\zeta\pi}{2}\right)\sinh\left(\frac{\alpha\zeta\pi}{2}\right) \\
&\quad - a_2 a_3 \cosh\left(\frac{\alpha\zeta\pi}{2}\right)\sinh\left(\frac{\beta\zeta\pi}{2}\right) = 0 \\
&\Leftrightarrow \frac{a_1 a_4}{a_2 a_3} = \frac{\sinh\left(\frac{\beta\zeta\pi}{2}\right)\cosh\left(\frac{\alpha\zeta\pi}{2}\right)}{\cosh\left(\frac{\beta\zeta\pi}{2}\right)\sinh\left(\frac{\alpha\zeta\pi}{2}\right)} \\
&= \tanh\left(\frac{\beta\zeta\pi}{2}\right)\coth\left(\frac{\alpha\zeta\pi}{2}\right).
\end{aligned}
\tag{7.86}
$$

Letting

$$
\begin{aligned}
\Delta(\gamma, \zeta) &= \tanh\left(\frac{\beta\zeta\pi}{2}\right)\coth\left(\frac{\alpha\zeta\pi}{2}\right) \\
&= \tanh\left(\frac{\sqrt{1-\gamma}}{2}\zeta\pi\right)\coth\left(\frac{\sqrt{1+\gamma}}{2}\zeta\pi\right), \\
\Psi(\gamma, \upsilon) &= \frac{a_2 a_3}{a_1 a_4} = \frac{\sqrt{1+\gamma}(1-\upsilon-\gamma)^2}{\sqrt{1-\gamma}(1-\upsilon+\gamma)^2}.
\end{aligned}
\tag{7.87}
$$

Equations (7.85) and (7.86) can be rewritten as follows:

$$
\Delta(\gamma, \zeta) = \Psi(\gamma, \upsilon), \tag{7.88}
$$

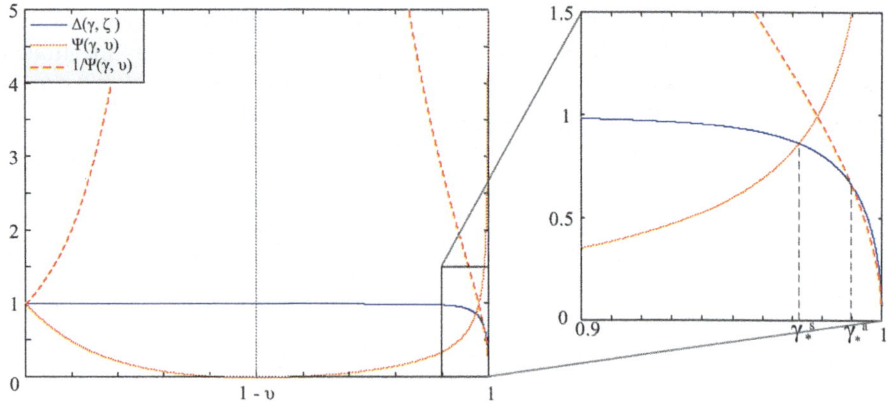

Fig. 7.3 Plots of $\Delta(\gamma, \zeta)$ and $\Psi(\gamma, \upsilon)$ with $\zeta = 5$ and $\upsilon = 0.5$

$$\Delta(\gamma, \zeta) = \frac{1}{\Psi(\gamma, \upsilon)}. \tag{7.89}$$

As previously mentioned, the condition for the existence of a nontrivial solution is that (7.88) or (7.89) is satisfied. Notably, $\Delta(\gamma, \zeta)$ depends on the slenderness ratio of the plate ζ, whereas $\Psi(\gamma, \upsilon)$ is a function relating to Poisson's number υ. The solutions γ_*^{s} and γ_*^{a} of (7.88) and (7.89), respectively, can be determined via the plots of $\Delta(\gamma, \zeta)$, $\Psi(\gamma, \upsilon)$, and $1/\Psi(\gamma, \upsilon)$ shown in Fig. 7.3, which describe the behaviors of these functions as functions of γ. In the range $\gamma \in [0, 1]$, (7.88) has one root $\gamma_*^{\mathrm{s}} = 0.9533$, whereas (7.89) has two roots: $\gamma_{*1}^{\mathrm{a}} = 0.9927$ and $\gamma_{*2}^{\mathrm{a}} = 1$. We observe that $\gamma_*^{\mathrm{s}} < \gamma_{*1}^{\mathrm{a}} < \gamma_{*2}^{\mathrm{a}}$. Due to the properties of functions $\Delta(\gamma, \zeta)$, $\Psi(\gamma, \upsilon)$, and $1/\Psi(\gamma, \upsilon)$, the root of (7.88) is always smaller than the roots of (7.89). Therefore, the critical speed υ_{cr}, which is the lowest speed at which the nontrivial equilibrium configuration exists, can be determined based on γ_*^{s}, that is,

$$\upsilon_{\mathrm{cr}} = \sqrt{\frac{D\pi^2}{\rho h l^2}\gamma_*^{\mathrm{s}2} + \frac{T_0}{\rho h}}. \tag{7.90}$$

Now, we investigate the influence of the plate parameters on γ_*^{s}. Figure 7.4 shows that the slenderness ratio $\zeta = b/l$ affects the trajectory of the function $\Delta(\gamma, \zeta)$, whereas the curvature of $\Psi(\gamma, \upsilon)$ is subjected to Poisson's number υ. The reduction of the slenderness ratio causes a decrease in the critical speed (i.e., Fig. 7.4a). In contrast, Fig. 7.4b reveals that the critical speed decreases as Poisson's number increases. The influence of ζ and υ on the value of γ_*^{s} is clearly shown in Fig. 7.5. Furthermore, the plate buckling mode is also significantly affected by the plate geometric and material properties. Figure 7.6 demonstrates the plate's buckling mode with $\zeta = 5$ and $\upsilon = 0.5$, wherein the red line represents the shape function $g(z)$, which is the slice of the buckling mode at $x = l/2$. If the slenderness ratio and Poisson's number are

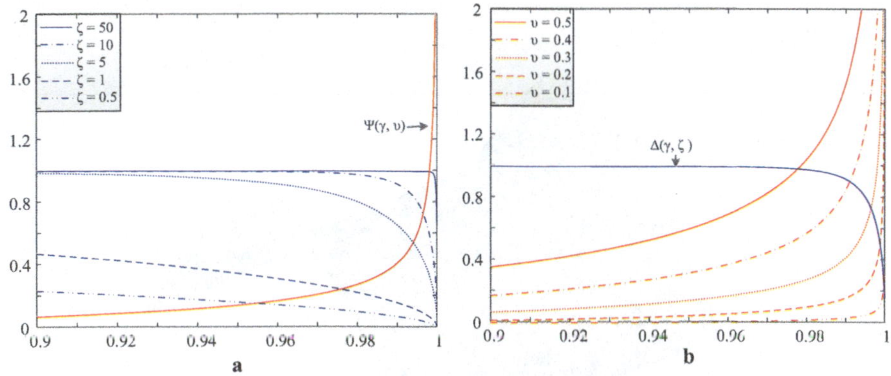

Fig. 7.4 Plot of $\Delta(\gamma, \zeta)$ and $\Psi(\gamma, \upsilon)$ with various values of the slenderness ratio and Poisson's number: **a** $\Delta(\gamma, \zeta)$ and **b** $\Psi(\gamma, \upsilon)$

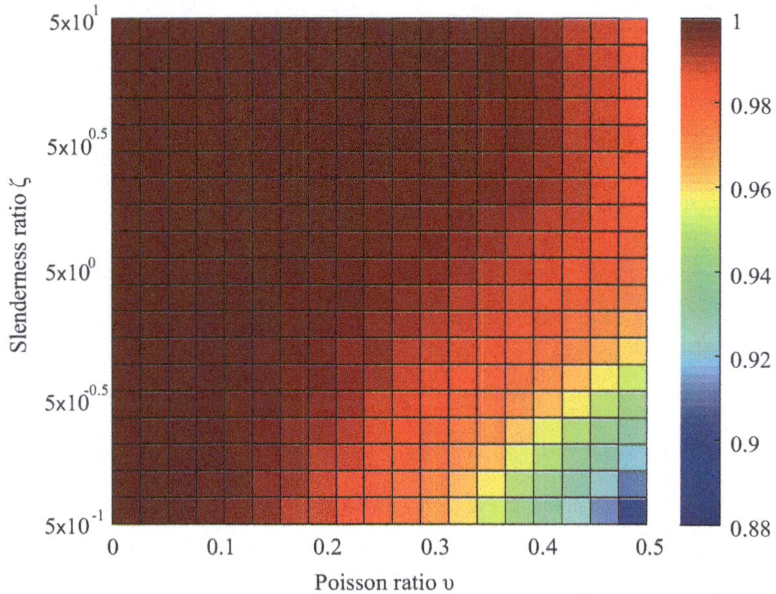

Fig. 7.5 Value of γ_*^s with variation in the slenderness ratio and Poisson's number (i.e., vertical axis is the logarithmic axis)

sufficiently small, the red line is almost straight. The curve of the red line increases as the values of the slenderness ratio and Poisson's number increase (i.e., Fig. 7.7).

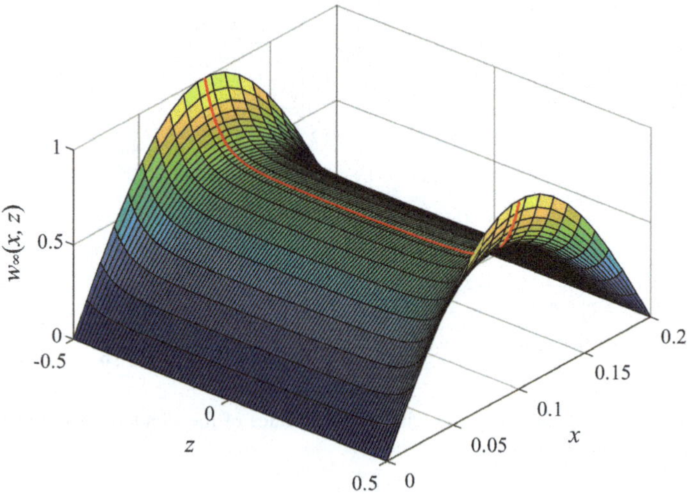

Fig. 7.6 Buckling mode of the axially moving isotropic plate with $\zeta = 5$ and $\upsilon = 0.5$

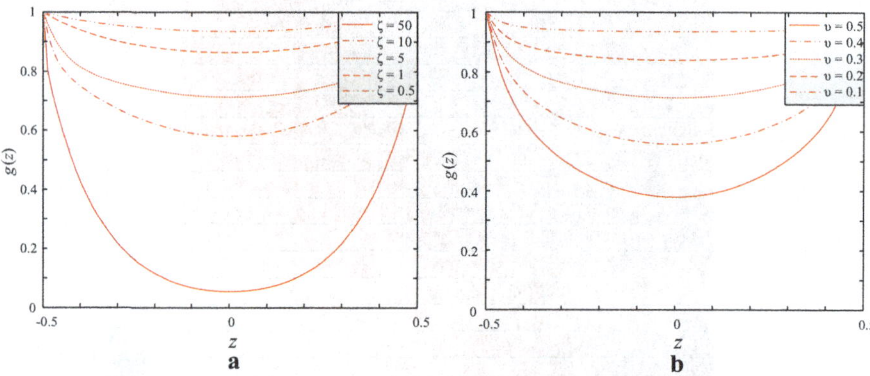

Fig. 7.7 Plot of $g(z)$ (i.e., red line in Fig. 7.6) with variation in **a** slenderness ratio and **b** Poisson's number

7.4.2 Equilibrium Solutions and Stability of an Orthotropic Plate

Let us consider the case of orthotropic plates. The governing equation of the lateral vibration is shown as follows:

$$\rho h\left(w_{tt} + 2v_0 w_{xt} + v_0^2 w_{xx}\right) \\ + D\left[w_{xxxx} + \left(2\chi\upsilon + \frac{Gh^3}{3D}\right)w_{xxzz} + \chi w_{zzzz}\right] - T_0 w_{xx} = 0, \tag{7.91}$$

and the boundary conditions are

$$w|_{x=0,l} = 0,$$
$$(w_{zz} + \upsilon w_{xx})|_{x=0,l} = 0, \tag{7.92}$$

$$(w_{zz} + \upsilon w_{xx})|_{z=\pm b} = 0,$$
$$\left[w_{zzz} + \left(\upsilon + \frac{4G}{\chi E}(1 - \chi \upsilon^2) \right) w_{xxz} \right] \bigg|_{z=\pm b} = 0. \tag{7.93}$$

The buckling problem of the orthotropic plate is formulated as an eigenvalue problem of the following PDE:

$$\left(\frac{\partial^4 w_\infty}{\partial x^4} + \vartheta \frac{\partial^4 w_\infty}{\partial x^2 \partial z^2} + \chi \frac{\partial^4 w_\infty}{\partial z^4} \right) + \frac{(\rho h v_0^2 - T_0)}{D} \frac{\partial^2 w_\infty}{\partial x^2} = 0 \tag{7.94}$$

where $\vartheta = 2\chi\upsilon + 4(1 - \chi\upsilon^2)G/E$. According to the boundary condition (7.92), we use the following representation:

$$w_\infty(x, z) = g(\overline{z}) \sin \frac{\pi x}{l} \tag{7.95}$$

where $\overline{z} = \pi z/l$ and $g(\overline{z})$ is an unknown function. By using this representation, the partial differential equation (7.91) and boundary conditions (7.93) can be rewritten as follows:

$$\chi \frac{\mathrm{d}^4}{\mathrm{d}\overline{z}^4} g(\overline{z}) - \vartheta \frac{\mathrm{d}^2}{\mathrm{d}\overline{z}^2} g(\overline{z}) + [1 - \gamma^2] g(\overline{z}) = 0, \tag{7.96}$$

$$\frac{\mathrm{d}^2}{\mathrm{d}\overline{z}^2} g(\overline{z}) - \upsilon g(\overline{z}) = 0, \quad \text{for } \overline{z} = \pm \frac{\zeta\pi}{2}, \tag{7.97}$$

$$\frac{\mathrm{d}^3}{\mathrm{d}\overline{z}^3} g(\overline{z}) - \left(\upsilon + \frac{4G}{\chi E}(1 - \chi\upsilon^2) \right) \frac{\mathrm{d}}{\mathrm{d}\overline{z}} g(\overline{z}) = 0,$$
$$\text{for } \overline{z} = \pm \frac{\zeta\pi}{2} \tag{7.98}$$

where $\gamma = \sqrt{l^2(\rho h v^2 - T_0)/\pi^2 D}$. The characteristic polynomial of (7.96) becomes

$$\chi r^4 - \vartheta r^2 + (1 - \gamma^2) = 0. \tag{7.99}$$

The solutions are

$$r = \pm\sqrt{\frac{\vartheta \pm \sqrt{\vartheta^2 - 4(1-\gamma^2)\chi}}{2\chi}}. \tag{7.100}$$

Now, we consider the range of γ^2 such that (7.99) has real roots, that is,

$$\begin{cases} \vartheta^2 - 4(1-\gamma^2)\chi \geq 0, \\ \vartheta - \sqrt{\alpha^2 - 4(1-\gamma^2)\chi} \geq 0, \end{cases} \Leftrightarrow \begin{cases} \gamma^2 \geq 1 - \frac{\vartheta^2}{4\chi}, \\ \gamma^2 \leq 1. \end{cases} \tag{7.101}$$

According to (7.101), the search of the lowest eigenvalue can be performed in the following range:

$$\max\left\{1 - \frac{\vartheta^2}{4\chi}, 0\right\} \leq \gamma^2 \leq 1. \tag{7.102}$$

We now consider the case that $\max\{1 - \vartheta^2/4\chi, 0\} < \gamma^2 \leq 1$. In this range, the characteristic polynomial of (7.96) has four distinct real roots, and thus, the general form of the solution of (7.96) can be represented in the following form:

$$g(\overline{z}) = A_1 \cosh(\alpha_o\overline{z}) + A_2 \sinh(\alpha_o\overline{z}) + A_3 \cosh(\beta_o\overline{z}) + A_4 \sinh(\beta_o\overline{z}), \tag{7.103}$$

where $\alpha_o = \sqrt{\left[\vartheta + \sqrt{\vartheta^2 - 4(1-\gamma^2)\chi}\right]/2\chi}$ and $\beta_o = \sqrt{\left[\vartheta - \sqrt{\vartheta^2 - 4(1-\gamma^2)\chi}\right]/2\chi}$; A_1, A_2, A_3, and A_4 are constants. By using a similar method applied in the previous section, we derive the zero determinant condition, wherein (7.96) has the nontrivial solution as follows:

$$\Delta(\gamma, \zeta) = \Psi(\gamma, \upsilon), \tag{7.104}$$

$$\text{or } \Delta(\gamma, \zeta) = \frac{1}{\Psi(\gamma, \upsilon)} \tag{7.105}$$

where

$$\Delta(\gamma, \zeta, \upsilon, G, E, \chi) = \tanh\left(\frac{\beta_o\zeta\pi}{2}\right)\coth\left(\frac{\alpha_o\zeta\pi}{2}\right), \tag{7.106}$$

$$\Psi(\gamma, \upsilon, G, E, \chi) = \frac{\alpha_o\left[\beta_o^2 - \upsilon\right]\left[\alpha_o^2 - \upsilon - \left(\frac{4G}{\chi E}(1 - \chi\upsilon^2)\right)\right]}{\beta_o\left[\alpha_o^2 - \upsilon\right]\left[\beta_o^2 - \upsilon - \left(\frac{4G}{\chi E}(1 - \chi\upsilon^2)\right)\right]}. \tag{7.107}$$

For the orthotropic case, $\Delta(\gamma, \zeta, \upsilon, G, E, \chi)$ and $\Psi(\gamma, \upsilon, G, E, \chi)$ depend on Poisson's number υ, the shear modulus G, Young's modulus E, and the orthotropic factor χ. Furthermore, Δ consists of the slenderness factor ζ. The functions $\Delta(\gamma, \zeta, \upsilon, G, E, \chi)$ and $\Psi(\gamma, \upsilon, G, E, \chi)$ can be simplified by using the geometric average approximation of the shear modulus (i.e., known as the Huber approximation), namely

$$G \approx G_{\mathrm{a}} = \frac{\sqrt{E_{\mathrm{X}} E_{\mathrm{Z}}}}{2(1 + \sqrt{\upsilon_{\mathrm{XZ}} \upsilon_{\mathrm{ZX}}})} = \frac{E\sqrt{\chi}}{2(1 + \upsilon\sqrt{\chi})}. \tag{7.108}$$

Substituting (7.108) into (7.106) and (7.107) yields

$$\Delta(\gamma, \zeta, \chi) = \tanh\left(\frac{\beta_{\mathrm{o}}\zeta\pi}{2}\right)\coth\left(\frac{\alpha_{\mathrm{o}}\zeta\pi}{2}\right), \tag{7.109}$$

$$\Psi(\gamma, \upsilon, \chi) = \frac{\alpha_{\mathrm{o}}\left[\beta_{\mathrm{o}}^2 - \upsilon\right]\left[\alpha_{\mathrm{o}}^2 - \frac{2}{\sqrt{\chi}} + \upsilon\right]}{\beta_{\mathrm{o}}\left[\alpha_{\mathrm{o}}^2 - \upsilon\right]\left[\beta_{\mathrm{o}}^2 - \frac{2}{\sqrt{\chi}} + \upsilon\right]}, \tag{7.110}$$

where

$$\alpha_{\mathrm{o}} = \sqrt{\frac{1+\gamma}{\sqrt{\chi}}} \text{ and } \beta_{\mathrm{o}} = \sqrt{\frac{1-\gamma}{\sqrt{\chi}}}. \tag{7.111}$$

With the approximation given by (7.108), $\Delta(\gamma, \zeta, \chi)$ and $\Psi(\gamma, \upsilon, \chi)$ do not depend on the shear modulus G and Young's modulus E. $\Delta(\gamma, \zeta, \chi)$ does not consist of Poisson's number. As mentioned in the isotropic case, the solution of (7.104), γ_*^{s}, is always smaller than the solution of (7.105). Therefore, we only consider the solution of (7.104), which is shown in Fig. 7.8. Once γ_*^{s} is determined, the critical speed at which the nontrivial equilibrium configuration exists can be derived by (7.90).

We consider the case wherein the characteristic polynomial of (7.96) has two double roots, i.e., $\gamma^2 = 1 - \vartheta^2/4\chi$. It follows that the general solution of (7.96) is expressed as follows:

$$g(\bar{z}) = A_1 \mathrm{e}^{\bar{z}\sqrt{\vartheta/2\chi}} + A_2 \mathrm{e}^{-\bar{z}\sqrt{\vartheta/2\chi}} + A_3 \bar{z} \mathrm{e}^{\bar{z}\sqrt{\vartheta/2\chi}} + A_4 \bar{z} \mathrm{e}^{-\bar{z}\sqrt{\vartheta/2\chi}}. \tag{7.112}$$

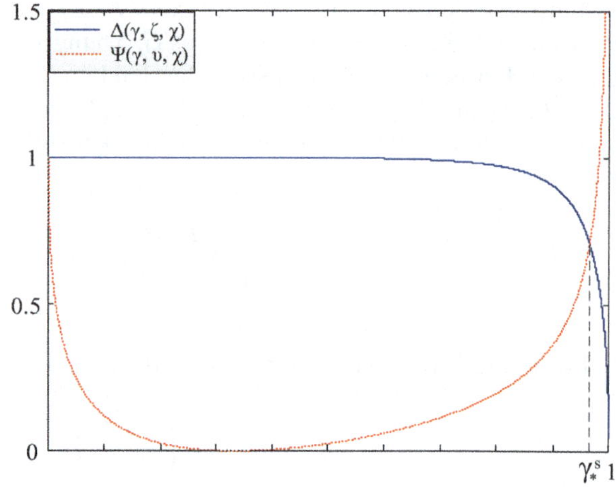

Fig. 7.8 Plots of $\Delta(\gamma, \zeta, \chi)$ and $\Psi(\gamma, \upsilon, \chi)$ with $\zeta = 10$, $\upsilon = 0.3$, and $\chi = 2$

Because the value of γ is fixed, the zero determinant condition does not hold for the general form above with the boundary conditions in (7.97) and (7.98). Consequently, we do not obtain a nontrivial solution for this case.

The remainder of this section investigates the influences of the orthotropic plate's geometric and material properties on the critical speed and the buckling mode. The impact of the slenderness ratio and Poisson's number on the trajectories of $\Delta(\gamma, \zeta, \chi)$ and $\Psi(\gamma, \upsilon, \chi)$, as shown in Fig. 7.9, is similar to the case of an axially moving isotropic plate. In addition, the increase or decrease of the orthotropic factor χ yields changes in both $\Delta(\gamma, \zeta, \chi)$ and $\Psi(\gamma, \upsilon, \chi)$ (i.e., Fig. 7.10). The values of γ_*^s with variation in the slenderness ratio, Poisson's number, and the orthotropic factor are revealed in Fig. 7.11. In Figs. 7.12 and 7.13, we observe that the buckling shape of the orthotropic plate is almost identical to that of the isotropic plate (i.e., Fig. 7.12) and significantly depends on the geometric and material properties of the plate (i.e., Fig. 7.13).

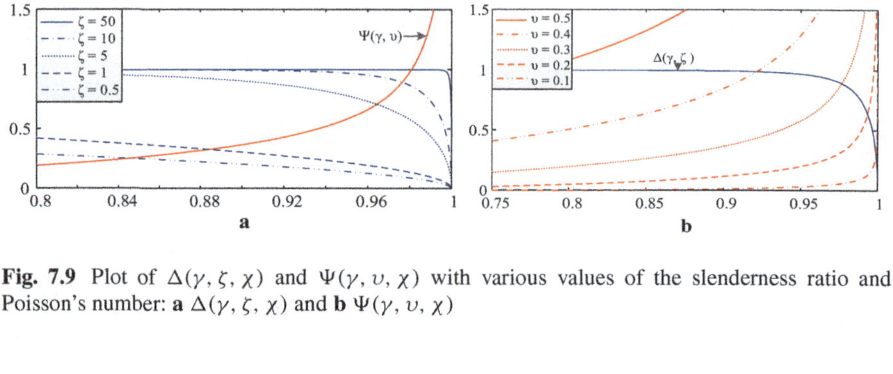

Fig. 7.9 Plot of $\Delta(\gamma, \zeta, \chi)$ and $\Psi(\gamma, \upsilon, \chi)$ with various values of the slenderness ratio and Poisson's number: **a** $\Delta(\gamma, \zeta, \chi)$ and **b** $\Psi(\gamma, \upsilon, \chi)$

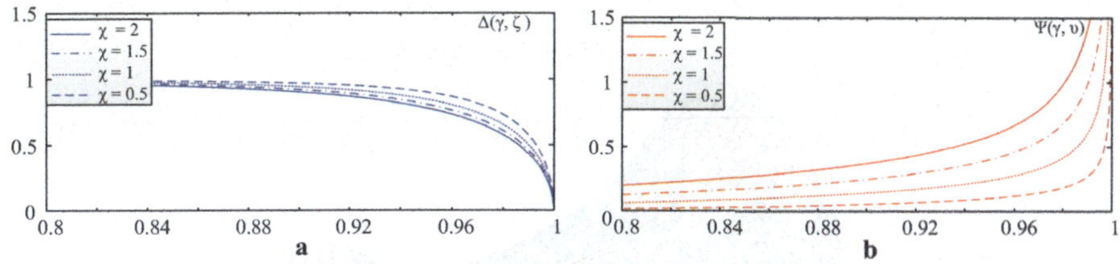

Fig. 7.10 Plot of $\Delta(\gamma, \zeta, \chi)$ and $\Psi(\gamma, \upsilon, \chi)$ with $\zeta = 5$, $\upsilon = 0.5$, and different orthotropic factors: **a** $\Delta(\gamma, \zeta, \chi)$ and **b** $\Psi(\gamma, \upsilon, \chi)$

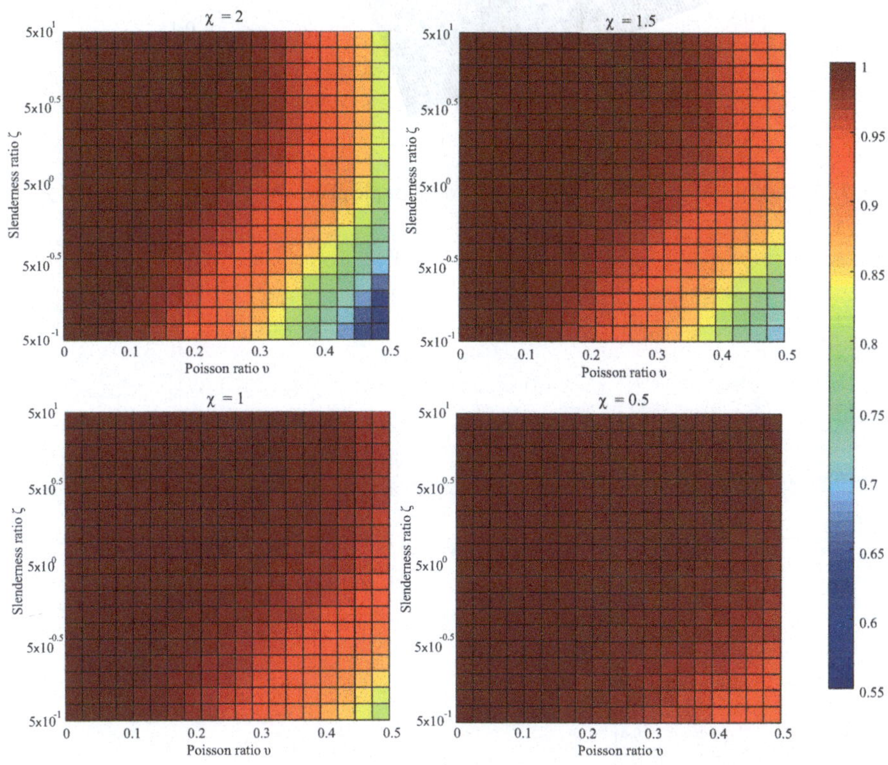

Fig. 7.11 Value of γ_*^s with variation in the slenderness ratio, Poisson's number, and the orthotropic factor (i.e., vertical axis is the logarithmic axis)

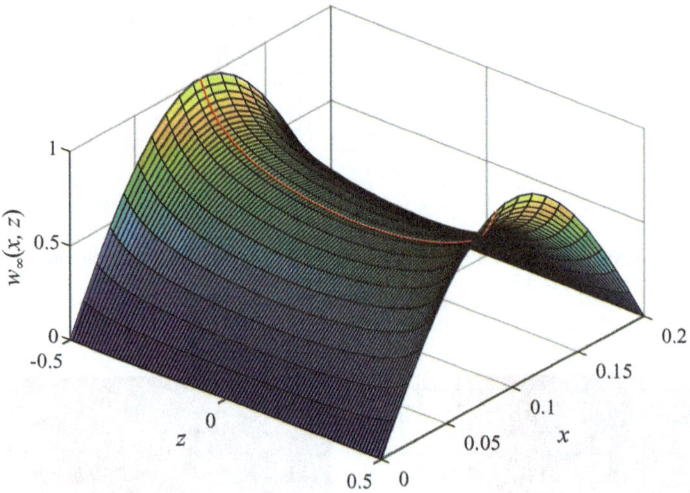

Fig. 7.12 Buckling mode of the axially moving orthotropic plate with $\zeta = 10$, $\upsilon = 0.3$, and $\chi = 2$

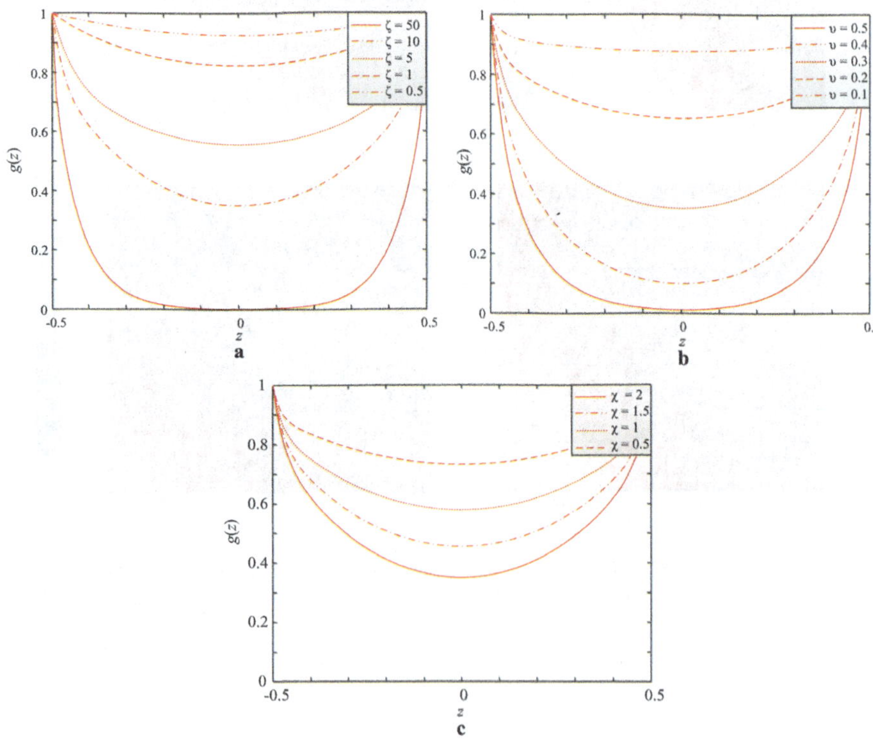

Fig. 7.13 Plot of $g(z)$ (i.e., red line in Fig. 7.12) with variation in **a** the slenderness ratio, **b** Poisson's ratio, and **c** the orthotropic factor

References

Abedi M, Asnafi A, Karami K (2014) To obtain approximate probability density functions for a class of axially moving viscoelastic plates under external and parametric white noise excitation. Nonlinear Dyn 78(3):1717–1727

Banichuk N, Barsuk A, Jeronen A, Tuovinen T, Neittaanmäki P (2020) Stability of axially moving materials. Springer, Berlin

Duan JB, Zhang DP, Wang WJ (2019) Flutter and divergence instability of axially-moving nanoplates resting on a viscoelastic foundation. Appl Sci-Basel 9(6):1097

Esmaeilzadeh M, Kadkhodayan M (2019) Numerical investigation into dynamic behaviors of axially moving functionally graded porous sandwich nanoplates reinforced with graphene platelets. Mater Res Express 6(10):1050b7

Hatami S, Azhari M, Saadatpour MM (2007) Free vibration of moving laminated composite plates. Compos Struct 80(4):609–620

Li C, Liu JJ, Cheng M, Fan XL (2017) Nonlocal vibrations and stabilities in parametric resonance of axially moving viscoelastic piezoelectric nanoplate subjected to thermo-electro-mechanical forces. Compos Pt B-Eng 116:153–169

Li HY, Li J, Lang TY, Zhu X (2018) Dynamics of an axially moving unidirectional plate partially immersed in fluid under two frequency parametric excitation. J Non-Linear Mech 99:31–39

Li HY, Li J, Liu YJ (2015) Internal resonance of an axially moving unidirectional plate partially immersed in fluid under foundation displacement excitation. J Sound Vib 358:124–141

Lin CC (1997) Stability and vibration characteristics of axially moving plates. Int J Solids Struct 34(24):3179–3190

Lin CC, Mote CD (1995) Equilibrium displacement and stress distribution in a two-dimensional, axially moving web under transverse loading. J Appl Mech-T Asme 62(3):772–779

Liu JJ, Li C, Fan XL, Tong LH (2017) Transverse free vibration and stability of axially moving nanoplates based on nonlocal elasticity theory. Appl Math Model 45:65–84

Marynowski K (2004) Nonlinear vibrations of an axially moving viscoelastic web with time-dependent tension. Chaos Soliton Fract 21(2):481–490

Marynowski K (2006) Two-dimensional rheological element in modelling of axially moving viscoelastic web. Eur J Mech a-Solid 25(5):729–744

Marynowski K (2012) Dynamic of axially moving orthotropic web. Springer, Berlin

Marynowski K, Kołakowski Z (1999) Dynamic behaviour of an axially moving thin orthotropic plate. J Appl Mech 37(1):109–128

Shin CH, Chung JT, Yoo HH (2006) Dynamic responses of the in-plane and out-of-plane vibrations for an axially moving membrane. J Sound Vib 297(3–5):794–809

Timoshenko SP, Gere JM (2009) Theory of elastic stability. Courier Corporation, New York

Ulsoy AG, Mote CD (1982) Vibration of wide band-saw blades. J Eng Ind-T Asme 104(1):71–78

Wang YQ, Guo XH, Sun Z, Li J (2014) Stability and dynamics of axially moving unidirectional plates partially immersed in a liquid. Int J Struct Stab Dyn 14(4):1450010

Wang YQ, Zu JW (2017) Instability of viscoelastic plates with longitudinally variable speed and immersed in ideal liquid. Int J Appl Mech 9(1):1750005

Yang XD, Chen LQ, Zu JW (2011) Vibrations and stability of an axially moving rectangular composite plate. J Appl Mech 78(1):011018

Yang XD, Zhang W, Chen LQ, Yao MH (2012) Dynamical analysis of axially moving plate by finite difference method. Nonlinear Dyn 67(2):997–1006

Zhang W, Lu SF, Yang XD (2014) Analysis on nonlinear dynamics of a deploying composite laminated cantilever plate. Nonlinear Dyn 76(1):69–93

Zhou YF, Wang ZM (2007) Transverse vibration characteristics of axially moving viscoelastic plate. Appl Math Mech-Engl 28(2):209–218

Zhou YF, Wang ZM (2019) Dynamic instability of axially moving viscoelastic plate. Eur. J Mech A-Solids 73:1–10

Chapter 8
Control of Axially Moving Plates and Membranes

Abstract This chapter presents the vibration suppression of axially moving plates/membranes by regulating the axial moving velocity. Numerical simulations are provided to illustrate the control performance.

Keywords Plate · Membrane · Axial velocity regulation · Iterative numerical techniques

Axially moving plates/membranes are two-dimensional models that involve two independent spatial variables, x and z. Their mathematical models are significantly complicated. Therefore, the design of a control law for axially moving plates/membranes based on the techniques presented in Chap. 4 can encounter mathematical problems. The studies on the control of axially moving plates/membranes are still very limited. One rare study on this topic was published by Nguyen and Hong (2012). In this work, the authors extended the control technique based on regulating the axial velocity to a two-dimensional model—an axially moving membrane. This chapter discusses vibration suppression of axially moving plates/membranes by determining the optimal velocity profile.

8.1 Control Based on Regulating Axial Velocity

The control of axially moving membranes based on regulating axial velocity is similar to the beam case. First, the ODE model of the membrane is derived. Based on this model, the minimizing problem of the upper bound of the vibration energy, wherein the axial velocity is used as an input, is stated. The iterative numerical technique is employed to solve this problem. Now, we consider the axially moving membrane described by (7.68) and (7.69):

© The Author(s), under exclusive license to Springer Nature Singapore Pte Ltd. 2022
K.-S. Hong et al., *Control of Axially Moving Systems*,
https://doi.org/10.1007/978-981-16-2915-0_8

$$\mathbf{M}_{\mathrm{w}}\ddot{\mathbf{q}}_{\mathrm{w}} + \mathbf{C}_{\mathrm{w}}\dot{\mathbf{q}}_{\mathrm{w}} + \mathbf{K}_{\mathrm{w}}(\mathbf{q}_{\mathrm{u}\eta})\mathbf{q}_{\mathrm{w}} = 0, \tag{8.1}$$

$$\mathbf{M}_{\mathrm{u}\eta}\ddot{\mathbf{q}}_{\mathrm{u}\eta} + \mathbf{C}_{\mathrm{u}\eta}\dot{\mathbf{q}}_{\mathrm{u}\eta} + \mathbf{K}_{\mathrm{u}\eta}\mathbf{q}_{\mathrm{u}\eta} = \mathbf{G}_{\mathrm{u}\eta}. \tag{8.2}$$

The ODE model of the lateral vibration is converted to the following state-space form with the state-space vector $\mathbf{x}_{\mathrm{w}} = \left[\, \dot{\mathbf{q}}_{\mathrm{w}}^{\mathrm{T}}\ \mathbf{q}_{\mathrm{w}}^{\mathrm{T}} \,\right]^{\mathrm{T}}$:

$$\dot{\mathbf{x}}_{\mathrm{w}} = \mathbf{f}_{\mathrm{w}}(\mathbf{x}_{\mathrm{w}}, v, \dot{v}) = \mathbf{A}_{\mathrm{w}}(t)\mathbf{x}_{\mathrm{w}}, \tag{8.3}$$

where

$$\mathbf{A}_{\mathrm{w}}(t) = \begin{bmatrix} -\mathbf{M}_{\mathrm{w}}^{-1}\mathbf{C}_{\mathrm{w}} & -\mathbf{M}_{\mathrm{w}}^{-1}\mathbf{K}_{\mathrm{w}} \\ \mathbf{0}_{N^2\times N^2} & \mathbf{I}_{N^2\times N^2} \end{bmatrix}. \tag{8.4}$$

The state-space model given by (8.3) can be rewritten as follows:

$$\begin{aligned} \dot{\mathbf{x}}_{\mathrm{w}} &= \mathbf{f}_{\mathrm{w}}(\mathbf{x}_{\mathrm{w}}, v, \dot{v}) = \mathbf{A}_{\mathrm{wc}}\mathbf{x}_{\mathrm{w}} + v\mathbf{B}_{\mathrm{w1}}\mathbf{x}_{\mathrm{w}} + v^2\mathbf{B}_{\mathrm{w2}}\mathbf{x}_{\mathrm{w}} + c_{\mathrm{e}}v\mathbf{B}_{\mathrm{w3}}\mathbf{x}_{\mathrm{w}} \\ &\quad + N_{\mathrm{X}}\mathbf{B}_{\mathrm{w4}}\mathbf{x}_{\mathrm{w}} + N_{\mathrm{Z}}\mathbf{B}_{\mathrm{w5}}\mathbf{x}_{\mathrm{w}} + N_{\mathrm{XZ}}\mathbf{B}_{\mathrm{w6}}\mathbf{x}_{\mathrm{w}}, \end{aligned} \tag{8.5}$$

where $\mathbf{A}_{\mathrm{wc}}$, $\mathbf{B}_{\mathrm{w1}}$, $\mathbf{B}_{\mathrm{w2}}$, $\mathbf{B}_{\mathrm{w3}}$, $\mathbf{B}_{\mathrm{w4}}$, $\mathbf{B}_{\mathrm{w5}}$, and $\mathbf{B}_{\mathrm{w6}}$ are defined as follows:

$$\mathbf{A}_{\mathrm{wc}} = \begin{bmatrix} -\mathbf{M}_{\mathrm{w}}^{-1}\mathbf{C}^{\mathrm{s}} & -\mathbf{M}_{\mathrm{w}}^{-1}(\mathbf{K}^{\mathrm{s}} + \mathbf{B}^{\mathrm{s}}) \\ \mathbf{0}_{N^2\times N^2} & \mathbf{I}_{N^2\times N^2} \end{bmatrix}, \tag{8.6}$$

$$\mathbf{B}_{\mathrm{w1}} = \begin{bmatrix} -\mathbf{M}_{\mathrm{w}}^{-1}\mathbf{B}^{\mathrm{s1}} & \mathbf{0}_{N^2\times N^2} \\ \mathbf{0}_{N^2\times N^2} & \mathbf{0}_{N^2\times N^2} \end{bmatrix}, \tag{8.7}$$

$$\mathbf{B}_{\mathrm{w2}} = \begin{bmatrix} \mathbf{0}_{N^2\times N^2} & \mathbf{M}_{\mathrm{w}}^{-1}\mathbf{B}^{\mathrm{s2}} \\ \mathbf{0}_{N^2\times N^2} & \mathbf{0}_{N^2\times N^2} \end{bmatrix}, \tag{8.8}$$

$$\mathbf{B}_{\mathrm{w3}} = \begin{bmatrix} \mathbf{0}_{N^2\times N^2} & \mathbf{M}_{\mathrm{w}}^{-1}\mathbf{B}^{\mathrm{s3}} \\ \mathbf{0}_{N^2\times N^2} & \mathbf{0}_{N^2\times N^2} \end{bmatrix}, \tag{8.9}$$

where $\mathbf{B}^{\mathrm{s}}$, $\mathbf{C}^{\mathrm{s}}$, $\mathbf{K}^{\mathrm{s}}$, $\mathbf{B}^{\mathrm{s1}}$, $\mathbf{B}^{\mathrm{s2}}$, and $\mathbf{B}^{\mathrm{s3}}$ are $N^2 \times N^2$ matrices whose structures are similar to matrix $\mathbf{M}_{\mathrm{w}}$. Their elements are defined as follows.

$$B_{mnij}^{\mathrm{s}} = \int_0^l (\varphi_m)_x(\varphi_i)_x \mathrm{d}x \int_0^b Z_n Z_j \, \mathrm{d}z, \tag{8.10}$$

$$C_{mnij}^{\mathrm{s}} = c_{\mathrm{e}} \int_0^l \varphi_m\varphi_i \, \mathrm{d}x \int_0^b Z_n Z_j \, \mathrm{d}z, \tag{8.11}$$

$$K^{s}_{mnij} = \rho h \dot{v} \int_0^l \varphi_m (\varphi_i)_x \, dx \int_0^b Z_n Z_j \, dz, \tag{8.12}$$

$$B^{s1}_{mnij} = 2\rho h \int_0^l \varphi_m (\varphi_i)_x \, dx \int_0^b Z_n Z_j \, dz, \tag{8.13}$$

$$B^{s2}_{mnij} = -\rho h \int_0^l \varphi_m (\varphi_i)_{xx} \, dx \int_0^b Z_n Z_j \, dz, \tag{8.14}$$

$$B^{s3}_{mnij} = \int_0^l \varphi_m (\varphi_i)_x \, dx \int_0^b Z_n Z_j \, dz. \tag{8.15}$$

The vibration energy of the membrane is derived as follows.

$$\begin{aligned} E(t) &= \frac{1}{2}\int_S \left\{\rho h (w_t + v w_x)^2\right\} dS + \frac{1}{2}\int_{SS} \left\{N_X w_x^2 + N_Z w_z^2 + 2N_{XZ} w_x w_z\right\} dS \\ &= \frac{1}{2}\rho h \zeta_1(t) + \frac{1}{2}\frac{Eh}{(1-\upsilon^2)}(\zeta_2(t) + \zeta_3(t)) + \frac{1}{2}\frac{Eh}{(1-\upsilon^2)}\zeta_4(t), \end{aligned} \tag{8.16}$$

where

$$\zeta_1(t) = \int_S \left\{\sum_{i=0}^{N}\sum_{j=0}^{N}\left[\dot{q}^w_{ij}\varphi_i Z_j + v q^w_{ij}(\varphi_i)_x Z_j\right]\right\}^2 dS, \tag{8.17}$$

$$\zeta_2(t) = \int_S \left\{\sum_{i=0}^{N}\sum_{j=0}^{N}\left[q^u_{ij}(X_i)_x Z_j + \upsilon q^\eta_{ij}\varphi_i (Z_j)_z\right]\left(\sum_{i=0}^{N}\sum_{j=0}^{N}\left[q^w_{ij}(\varphi_i)_x Z_j\right]\right)^2\right\} dS, \tag{8.18}$$

$$\zeta_3(t) = \int_A \left\{\sum_{i=0}^{N}\sum_{j=0}^{N}\left[q^\eta_{ij}\varphi_i (Z_j)_z + \upsilon q^u_{ij}(X_i)_x Z_j\right]\left(\sum_{i=0}^{N}\sum_{j=0}^{N}\left[q^w_{ij}\varphi_i (Z_j)_z\right]\right)^2\right\} dA, \tag{8.19}$$

$$\zeta_4(t) = \int_A \left\{\sum_{i=0}^{N}\sum_{j=0}^{N}\left[q^u_{ij} X_i (Z_j)_z + \upsilon q^\eta_{ij}(\varphi_i)_x Z_j\right]\left(\sum_{i=0}^{N}\sum_{j=0}^{N}\left[q^w_{ij}(\varphi_i)_x Z_j(z)\right]\right)\right.$$

$$\times\left(\sum_{i=0}^{N}\sum_{j=0}^{N}\left[q_{ij}^{w}\varphi_i(Z_j)_z\right]\right)\Bigg\}dA. \tag{8.20}$$

The Legendre polynomials and their spatial derivatives are bounded as follows:

$$|X_i(x)| < \sum_{k=0}^{P_i}\frac{(2i-2k)!}{2^i k!(i-k)!(i-2k)!}, \tag{8.21}$$

$$\left|Z_j(z)\right| < \sum_{k=0}^{P_j}\frac{(2j-2k)!}{2^j k!(j-k)!(j-2k)!}, \tag{8.22}$$

$$|(X_i(x))_x| < \sum_{k=0}^{P_i}\frac{(2i-2k)!}{2^i k!(i-k)!(i-2k)!}\frac{2(i-2k)}{l}, \tag{8.23}$$

$$\left|(Z_j(z))_z\right| < \sum_{k=0}^{P_j}\frac{(2j-2k)!}{2^j k!(j-k)!(j-2k)!}\frac{2(j-2k)}{b}. \tag{8.24}$$

Using inequalities (8.21)–(8.24), the vibration energy of the membrane is evaluated as

$$\begin{aligned} E(t) \le \frac{\rho hbl}{2}&\left\{\sum_{i=0}^{2N^2}\xi_1\left(1+\psi+(i)\frac{(i-N^2+1)\pi v}{l}\right)x_i\right\}^2 \\ &+\xi_3 bl\left\{\sum_{i=0}^{2N^2}\frac{\xi_1\psi+(i)(i-N^2+1)}{l}x_i\right\}^2 \\ &+\xi_4 bl\left\{\sum_{i=0}^{2N^2}\frac{\xi_2\psi+(i)(i-N^2+1)\pi}{l}x_i\right\}^2 \\ &+\xi_5 bl\left\{\sum_{i=0}^{2N^2}\xi_2\psi+(i)x_i\right\}^2, \end{aligned} \tag{8.25}$$

where positive constants ξ_1, ξ_2, ξ_3, ξ_4, and ξ_5 satisfy the following equations and inequalities.

$$\xi_1 = \sum_{k=0}^{P_j}\frac{(2j-2k)!}{2^j k!(j-k)!(j-2k)!} > \left|Z_j\right|, \tag{8.26}$$

$$\xi_2 = \sum_{k=0}^{P_j} \frac{(2j-2k)!}{2^j k!(j-k)!(j-2k)!} \frac{2(j-2k)}{b} > \left|(Z_j)_z\right|, \tag{8.27}$$

$$\xi_3 \geq \left| \frac{Eh}{2(1-\upsilon^2)} \sum_{i=0}^{N} \sum_{j=0}^{N} \left[q_{ij}^u (X_i)_x Z_j + \upsilon q_{ij}^\eta \varphi_i (Z_j)_z \right] \right|, \tag{8.28}$$

$$\xi_4 \geq \left| \frac{Eh}{2(1+\upsilon)} \sum_{i=0}^{N} \sum_{j=0}^{N} \left[q_{ij}^u X_i (Z_j)_z + \upsilon q_{ij}^\eta (\varphi_i)_x Z_j \right] \right|, \tag{8.29}$$

$$\xi_5 \geq \left| \frac{Eh}{2(1-\upsilon^2)} \sum_{i=0}^{N} \sum_{j=0}^{M} \left[q_{ij}^\eta \varphi_i (Z_j)_z + \upsilon q_{ij}^u (X_i)_x Z_j \right] \right|. \tag{8.30}$$

$\psi_{\mathrm{H}}(i)$ is the Heaviside step function given by

$$\psi_{\mathrm{H}}(i) = \frac{1}{2} \left\{ \operatorname{sgn}(i - N^2) + \left| \operatorname{sgn}(i - N^2) \right| \right\}. \tag{8.31}$$

According to (8.25), the upper bound of the vibration energy is derived, that is

$$\overline{E}_{\mathrm{w}}(t) = \frac{1}{2} \mathbf{x}_{\mathrm{w}}^{\mathrm{T}} \mathbf{Q}_{\mathrm{w}}(t) \mathbf{x}_{\mathrm{w}}. \tag{8.32}$$

The $2N^2 \times 2N^2$ matrix $\mathbf{Q}_{\mathrm{w}}(t)$ is positive definite and symmetric, whose elements are defined as follows:

$$\begin{aligned} q_{ij} &= \frac{\rho hbl}{4} \xi_1^2 \left(1 + \psi_{\mathrm{H}}(i) \frac{(i - N^2 + 1)}{l} \pi v \right) \left(1 + \psi_{\mathrm{H}}(j) \frac{(j - N^2 + 1)}{l} \pi v \right) \\ &+ \frac{b\pi^2}{2l} \left(\xi_3 \xi_1^2 + \xi_4 \xi_2^2 \right) \psi_{\mathrm{H}}(i) \psi_{\mathrm{H}}(j) \left(i - N^2 + 1 \right) \left(j - N^2 + 1 \right) \\ &+ \frac{bl}{2} \xi_5 \xi_2^2 \psi_{\mathrm{H}}(i) \psi_{\mathrm{H}}(j). \end{aligned} \tag{8.33}$$

From the bound of the vibration energy given by (8.32) and state Eq. (8.3), we establish an optimal problem that is similar to the beam case. Consider the problem of minimizing the functional

$$\overline{E}_{\mathrm{w}}(t_{\mathrm{f}}) = \frac{1}{2} \mathbf{x}_{\mathrm{w}}^{\mathrm{T}}(t_{\mathrm{f}}) \mathbf{Q}_{\mathrm{w}}(t_{\mathrm{f}}) \mathbf{x}_{\mathrm{w}}(t_{\mathrm{f}}) \tag{8.34}$$

subjected to the differential constraint given by (8.3) and the initial and final conditions

$$\mathbf{x}_{\mathrm{w}}(0) = \mathbf{x}_{\mathrm{w}0}, \tag{8.35}$$

$$\mathbf{x}_{\mathrm{w}}(t_{\mathrm{f}}) = 0, \tag{8.36}$$

where $v(t)$ is the control input subjected to constraints $v(t_{\mathrm{f}}) = 0$ and $v(t) > 0$. The combination of the final condition and (8.34) yields the following cost function:

$$J = \frac{1}{2}\mathbf{x}_{\mathrm{w}}^{\mathrm{T}}(t_{\mathrm{f}})[\mathbf{Q}_{\mathrm{w}}(t_{\mathrm{f}}) + \mathbf{W}]\mathbf{x}_{\mathrm{w}}(t_{\mathrm{f}}), \tag{8.37}$$

where $\mathbf{W}$ is a $2N^2 \times 2N^2$ weight coefficient matrix. To apply the conjugate gradient algorithm, we define a Hamiltonian function: $H{=}\boldsymbol{\lambda}^{\mathrm{T}}\mathbf{f}_{\mathrm{w}}$, where $\boldsymbol{\lambda}$ is the vector of adjoint variables described by the following differential equation:

$$\dot{\boldsymbol{\lambda}} = -\frac{\partial H}{\partial \mathbf{x}_{\mathrm{w}}} = -\boldsymbol{\lambda}^{\mathrm{T}}\mathbf{A}_{\mathrm{w}}, \tag{8.38}$$

$$\boldsymbol{\lambda}(t_{\mathrm{f}}) = [\mathbf{Q}_{\mathrm{w}}(t_{\mathrm{f}}) + \mathbf{W}]\mathbf{x}_{\mathrm{w}}(t_{\mathrm{f}}). \tag{8.39}$$

According to (8.5), the gradient is derived as follows:

$$g(t) = \frac{\partial H}{\partial v} = \boldsymbol{\lambda}^{\mathrm{T}}(\mathbf{B}_{\mathrm{w}1} + 2v\mathbf{B}_{\mathrm{w}2} + c\mathbf{B}_{\mathrm{w}3})\mathbf{x}_{\mathrm{w}} \tag{8.40}$$

Let $v_i(t)$ be the i-th approximation to the optimal velocity profile. The corresponding gradient $g_i(t)$ is obtained after the following steps: (i) Solve the ODE of the longitudinal and transverse vibrations (8.1) forwards with $v(t) = v_i(t)$ to determine $\mathbf{q}_{\mathrm{u}\eta}$, and then solve the state-space Eq. (8.1) forwards with $v_i(t)$ and $\mathbf{q}_{\mathrm{u}\eta}$; (ii) solve the adjoint Eq. (8.38) backward by using the boundary condition (8.39); and (iii) computing the gradient $g_i(t)$ by using (8.40). The corresponding gradient direction $p_i(t)$ and $(i+1)$-th approximation of the velocity profile are updated by

$$p_i(t) = -g_i(t) + \beta_{i-1}p_{i-1}(t), \tag{8.41}$$

$$v_{i+1}(t) = v_i(t) + \alpha_i p_i(t), \tag{8.42}$$

where α_i is the step size of the i-th iteration and

$$\beta_{i-1} = \frac{\int_0^{t_{\mathrm{f}}} g_i(t)g_i(t)\mathrm{d}t}{\int_0^{t_{\mathrm{f}}} g_{i-1}(t)g_{i-1}(t)\mathrm{d}t}. \tag{8.43}$$

The procedure for determining the optimal velocity profile by the conjugate gradient algorithm is similar to the beam case (i.e., the entire procedure summarized in Sect. 4.1.3). For example, a numerical simulation is performed to verify the control strategies.

Example 8.1 An axially moving membrane is described by (8.1)-(8.2), wherein $E = 0.1 \times 10^9$ Pa, $l = 4$ m, $b = 0.5$ m, $h = 1.8 \times 10^{-3}$ m, $T_0 = 10^4$ N, $\rho = 1{,}522$ kg/m^2, and $v_0 = 5$ m/s. The initial condition is $w(x, 0) = 0.04 \sin(\pi x/l)Z_0(z)$. The axial velocity is 5 m/s at the initial time $t = 0$; then, the axial velocity reaches the zero value at time $t_f = 5$ s. The conventional velocity profile is constructed with a slope, wherein the deceleration is 1, whereas the optimal velocity is generated by using the conjugate gradient algorithm. The simulation procedure can be performed similarly to Example 4.2. We can observe the significant effect of the proposed control strategies via the vibration response and the vibration energy shown in Fig. 8.1 and Fig. 8.2, respectively.

This control technique can be further applied to the plate case. Fundamentally, the structure of the ODE models of an axially moving plate and an axially moving membrane are similar. Therefore, we can easily extend this method to an axially moving plate. Control by regulating the axial velocity is the only published method that handles the vibrations of axially moving membranes/plates. This method is an ODE-based control method; therefore, it has a severe limitation: spillovers due to the residual modes. Additionally, this method is an open-loop control method. It required the exact values of system parameters; thus, the method is unable to deal with the systems with uncertainties. Due to these disadvantages, the implementation of this algorithm in practice is still challenging.

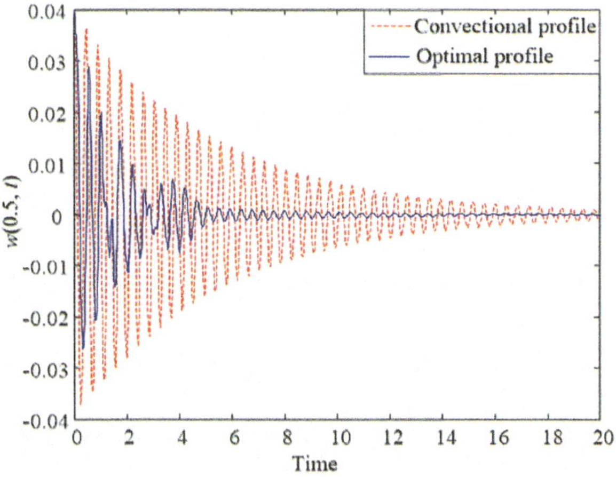

Fig. 8.1 Lateral vibration of the membrane in $x = 0.5\,l$ and $z = 0.5\,b$

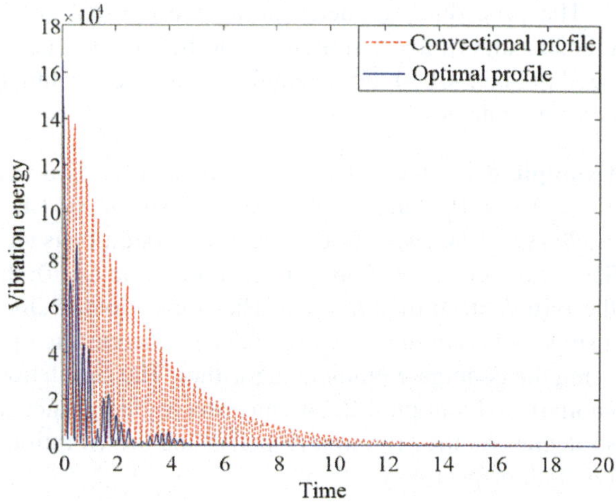

Fig. 8.2 Vibration energy of the membrane during the deceleration period

Reference

Nguyen QC, Hong K-S (2012) Transverse vibration control of axially moving membranes by regulation of axial velocity. IEEE Trans Control Syst Technol 20(4):1124–1131

Chapter 9
Conclusions and Future Directions

Abstract Concluding remarks for individual chapters are made. Also, future research directions regarding the dynamics and control of axially moving systems are discussed.

Keywords Concluding remarks · Future research direction

This book comprehensively discusses the dynamics analysis and control system designs of axially moving systems. According to the discussions in the previous chapters, several concluding remarks are drawn. Additionally, the outline of future research directions for the dynamics and vibration control of axially moving systems is presented.

9.1 Concluding Remarks

Although investigations on axially moving systems (AMSs) have been performed for more than six decades, this subject is still an exciting research field. Fundamentally, the axially moving system is characterized by a continuous material under tension. Depending on the system's flexibility and geometric parameters, we categorized AMSs into four models: string, beam, belt, and plate models. This book developed the dynamic models of many cases of AMSs based on the extended Hamilton principle and then proposed various control methods.

Chapter 2 begins with the linear dynamic model of an axially moving string [i.e., refer to (2.14)]. This model is the most fundamental model for describing an axially moving elastic material, wherein only an initial tension is considered. Later, dynamic models of many cases of axially moving strings were also established. Remarkably, the kinetic energy of axially moving strings, which contains the motions of material elements i.e., (2.3), is similar in most cases. In contrast, the potential energy that depends on the axial tension and material's stress–strain relation is different for each case. For instance, the potential energy of the strings with varying length is subjected to a spatiotemporal axial force (2.23). For the case of strings with large amplitude vibration, the potential energy, i.e., (2.20), consists of the influence of axial

© The Author(s), under exclusive license to Springer Nature Singapore Pte Ltd. 2022
K.-S. Hong et al., *Control of Axially Moving Systems*,
https://doi.org/10.1007/978-981-16-2915-0_9

deformation. Also, the equations that govern the moving viscoelastic string were also derived. In these cases, the stress–strain relation is described by viscoelastic models, such as the Kelvin–Voigt model, standard linear solid model, and Burgers model. The PDEs that describe system vibrations of the string can be transformed into a low-dimensional system of ODEs by using the Galerkin method [i.e., refer to (2.62)]. These ODEs are convenient for dynamics analysis and control design. Furthermore, we discussed stability regions, bifurcation, and chaotic dynamics of axially moving strings in Sect. 2.4.

The dynamics of the axially moving beam model comprise the main topic discussed in Chap. 3. Unlike the string model, the potential energy of the axially moving beam accounts for the energy due to the bending stiffness [i.e., Euler–Bernoulli beam, (3.7)], further addresses the rotary inertia effect [i.e., Rayleigh beam, (3.38)], and also considers shear deformation [i.e., Timoshenko beam, (3.40)–(3.41)]. The dynamics of the Bernoulli beam and Rayleigh beam are described by one fourth-order PDE. In contrast, the Timoshenko beam's mathematical model includes two second-order PDEs consisting of two state variables: lateral vibration and angle of rotation of the cross-section. In Chap. 3, fluid-conveying pipes, a special type of axially moving systems, were also discussed. Next, the dynamic models of axially moving nanoscale beams (3.109) and axially moving composite beams (3.118) were derived. We characterized the stress–strain relation of the nanoscale materials by considering the size-dependent effect (3.101), whereas one of the laminated composite materials is attributed to staking many different materials (3.114). According to the Galerkin truncation method, the ODE model of an Euler–Bernoulli beam was derived in Sect. 3.3 [i.e., refer (3.128)]. Additionally, we performed stability analysis of an axially moving beam by using the multiple scales method. In Sect. 3.4, bifurcation and chaotic dynamics of an Euler–Bernoulli beam and a Timoshenko beam were investigated.

Another type of AMS, axially moving belts, is discussed in Chap. 5. The belt model considers the longitudinal inertia force; therefore, its dynamic model describes both lateral vibration and longitudinal vibration. The early mathematical model of the translating belt, (5.15)–(5.16), was developed based on the closed-form strain (5.5). Simple models (5.21), (5.22), and (5.27) were subsequently established based on the different approximate strains (5.19). These models have an essential meaning in dynamics analysis and control design. Additionally, the equation of motion of axially moving viscoelastic belts was given in (5.28)–(5.29), whereas the belt model based on the Timoshenko beam theory was shown in (5.35)–(5.37). Furthermore, the governing equations that describe the lateral, longitudinal, and transverse vibrations of axially moving systems, (5.41)–(5.43), reveal that the dynamics of the lateral and transverse vibration of a one-dimensional model are almost equivalent. The remainder of this chapter discussed the ODE model and dynamic analysis of axially moving belts. The approximate model based on the Galerkin method of an axially moving belt was shown in (5.73) and (5.74). We discussed the dynamical issues of axially moving belts, including stability, bifurcation, and chaos, in Sect. 5.4.

The last type of AMS, axially moving plates, differs in the number of independent spatial variables compared with the three foregoing types. The plate is a two-dimensional model whose vibrations depend on both x and z. Chapter 7 discusses the dynamics of axially moving plates in detail. First, the equations of motion of an axially moving orthotropic plate, (7.14)–(7.16), were derived by using the extended Hamilton principle. Based on these equations, the mathematical model of the moving membrane, wherein the bending moment is disregarded, was presented in (7.23)–(7.25). Dynamic models of viscoelastic plates using the two-dimensional standard linear solid model and laminated composite plates were also established (i.e., refer to (7.33) and (7.40)–(7.42), respectively). The ODE models of axially moving plates/membranes were subsequently developed based on the extended Galerkin method. This method only requires the basis functions to satisfy the geometric boundary conditions. Accordingly, two ODE models for moving plates/membranes with different geometric boundary conditions were developed [i.e., refer to (7.54) and (7.68)–(7.69)]. Section 7.4 focuses on analyzing the equilibrium solutions and the stability of the moving isotropic and orthotropic plates. The simulation results further reveal that the critical speed and buckling mode substantially depend on the slenderness ratio, Poisson's number, and the orthotropic factor.

We discussed various control methods of axially moving systems in Chap. 4 (string, beam), Chap. 6 (belt), and Chap. 8 (plate). Fundamentally, the control methods for the string and beam (i.e., Chap. 4) and the belt (i.e., Chap. 6) are similar. These control methods include control based on the ODE model and control based on the PDE model. Concerning the ODE-based control techniques, the control law was designed based on the approximate model. This design is convenient for implementing the conventional control methods developed for ODE systems, such as model-based feedback control, sliding mode control, and control by regulating the axial velocity. Unfortunately, ODE-based control techniques can cause spillover problems. Therefore, control techniques using the PDE model, such as the transfer function-based method, wave cancelation method, and boundary control based on the Lyapunov method, have been developed to overcome this drawback. The most prominent technique is boundary control using Lyapunov stability analysis. The main trait of this control technique is to analyze an energy-like Lyapunov function. Hence, this technique allows nonlinear axially moving systems and hybrid systems to be addressed. Additionally, the boundary control approach does not interfere with the operation of the system. For these reasons, boundary control based on the Lyapunov method is regarded as a practical control solution for the vibration control of AMSs. Contrary to the diversity of control techniques for one-dimensional models (i.e., string, beam, and belt), the reported works lack studies on vibration control of two-dimensional models (i.e., plate and membrane). A rarely reported control algorithm is the control technique by regulating axial velocity (i.e., refer to Chap. 8). This algorithm is open-loop control, designed based on the ODE model of the membrane/plate. Even if the simulation results showed that this algorithm is effective in vibration suppression, implementing this algorithm for actual systems is still challenging due to the disadvantages of open-loop and ODE-based control.

9.2 Future Directions

Based on the discussions on axially moving systems in this book, the following specific aspects outline future research concerning the dynamics and vibration control of axially moving systems.

1. *Dynamic analysis of axially moving composite materials and axially moving nanoscale materials*: The dynamics of an axially moving system substantially depend on the materials' stress–strain relations. Using different elasticity/viscoelasticity theories to consider an axially moving system generates different dynamic models. Several types of materials with complex constitutive relations have been used in axially moving systems, such as laminated composite materials and nanoscale materials. Thus far, studies on axially moving composite materials and nanoscale materials are still limited despite research endeavors in recent years. Due to their broad applications, the dynamic behaviors of axially moving composite materials and nanoscale materials must be investigated in future studies.
2. *Nonideal boundary conditions*: Most of the published studies on axially moving systems with nonideal boundaries only concentrate on investigating linear vibrations. There is a lack of works concerning the effects of nonideal boundary conditions on complex dynamics, such as bifurcation and chaotic motions. In the future, this aspect must be further explored.
3. *Vibration control of axially moving Timoshenko beams*: Most control strategies for axially moving beams have been developed based on mathematical models using the Euler–Bernoulli beam theory. Vibration control of moving beams, wherein the Timoshenko beam theory is used to establish a dynamic model, has not been investigated. Since the Timoshenko beam theory accounts for the influences of both shear deformation and rotational inertia, it is superior for describing the dynamics of thick and short beams and the high modes of vibration. The control scheme for axially moving Timoshenko beams should be investigated in the future to undertake these cases.
4. *Vibration control of axially moving plates and membranes*: Another uncharted field concerning the control of axially moving systems is vibration suppression of axially moving plates and membranes. The current control strategy for the moving membrane presented in Chap. 8 is an open-loop and ODE-based control method. The disadvantage of this method is that it required the exact values of system parameters; as a result, the method is unable to handle systems in the presence of uncertainties. In the future, we need to pursue novel closed-loop control approaches for axially moving plates and membranes.
5. *Control of hybrid axially moving systems*: Most studies on the control of AMSs usually consider a single AMS. In the industry, an axially moving system rarely operates in isolation. For example, the motion of an axially moving cable in a container crane depends on a trolley's translational motion, and the dynamics of roll-to-roll systems consist of the dynamics of many axially moving systems

and rollers. In the future, the vibration control problem of these hybrid systems needs to receive more attention.

6. *Influence of harsh environments*: Many axially moving systems operate in harsh environments; for instance, fluid-conveying pipes in the ocean are subjected to sea currents. Environmental factors such as the sea current or high temperatures can cause changes in the dynamic behaviors of AMSs. Therefore, existing control schemes need to extend to AMSs, wherein environmental factors are considered further.
7. *Experimental studies*: Measuring the vibration of flexible systems is always problematic from both the technical point of view and economic point of view. These problems are significant barriers to experimental studies on axially moving systems. Thus far, almost all studies on the dynamic behaviors and control of axially moving systems have been verified via numerical simulation. To demonstrate the applicability of these studies, further advanced experimental research is needed.

Appendix A

Derivation of the Closed-loop Transfer Function of Eq. (4.71)

The closed-loop transfer function of axially moving string in (4.71) is expressed as follows:

$$
\begin{aligned}
G_c(x,\xi,s) &= \frac{G_0(x,x_a,s)G(s)G_0(x_s,\xi,s)}{1-G_0(x_s,x_a,s)G(s)} + G_0(x,\xi,s) \\
&= \frac{G_0(x,x_a,s)G(s)G_0(x_s,\xi,s) + G_0(x,\xi,s) - G_0(x,\xi,s)G_0(x_s,x_a,s)G(s)}{1-G_0(x_s,x_a,s)G(s)} \\
&= \frac{G_0(x,\xi,s) + G(s)[G_0(x,x_a,s)G_0(x_s,\xi,s) - G_0(x,\xi,s)G_0(x_s,x_a,s)]}{1-G_0(x_s,x_a,s)G(s)} \\
&= \frac{\frac{N_0(x,\xi,s)}{D_0(s)} + \alpha\frac{N(s)}{D(s)}e^{-t_d s}\left[\frac{N_0(x,x_a,s)}{D_0(s)}\frac{N_0(x_s,\xi,s)}{D_0(s)} - \frac{N_0(x,\xi,s)}{D_0(s)}\frac{N_0(x_s,x_a,s)}{D_0(s)}\right]}{1-\frac{N_0(x_s,x_a,s)}{D_0(s)}\alpha\frac{N(s)}{D(s)}e^{-t_d s}} \\
&= \frac{N_0(x,\xi,s)D(s) + \alpha\frac{N(s)}{D_0(s)}e^{-t_d s}[N_0(x,x_a,s)N_0(x_s,\xi,s) - N_0(x,\xi,s)N_0(x_s,x_a,s)]}{D_0(s)D(s) - \alpha N_0(x_s,x_a,s)N(s)e^{-t_d s}} \\
&= \frac{N_c(x,\xi,s)}{D_c(s)}.
\end{aligned}
\tag{A.1}
$$

Note that $N_c(x,\xi,s)$ is analytic in s for all x and ξ. It can be verified by substituting (4.64) and (4.65) into $N_c(x,\xi,s)$. Therefore, the eigenvalues of the closed-loop system can be obtained by solving the equation $D_c(s) = 0$.

© The Editor(s) (if applicable) and The Author(s), under exclusive license to Springer Nature Singapore Pte Ltd. 2022
K.-S. Hong et al., *Control of Axially Moving Systems*,
https://doi.org/10.1007/978-981-16-2915-0

Appendix B

Criterion 4.1 (Stability for a Linear String)

The axially moving linear string with the closed-loop transfer function in (4.71) is asymptotically stable if all roots of $D_c(s) = 0$ lie in C_-, where C_- is the left-hand s-plane $\{s \in C_- : \mathrm{Re}(s) < 0\}$.

Let us consider the roots of $D_c(s) = 0$ when $\alpha = 0$. In this case, we observe that the roots of $D_c(s) = 0$ include p_k^0 and the roots of $D(s) = 0$. p_k^0 lie on the imaginary axis, whereas the roots of $D(s) = 0$ lie in C_- due to the stable controller. We denote the root locus of the kth vibration mode that begins from p_k^0 by $p_k^c(\alpha)$. To assure the system's asymptotical stability, we must establish the condition of the controller's transfer function such that all root loci initially move toward C_- as α is increased. Furthermore, the critical value α_{cr} of the control gain, wherein the roots remain in C_- as $\alpha \in (0, \alpha_{cr})$, is also determined.

The direction of $p_k^c(\alpha)$ at $\alpha = 0^+$ is derived based on the sign of the real part of $\left.\frac{\mathrm{d}p_k^c(\alpha)}{\mathrm{d}\alpha}\right|_{\alpha=0^+}$. According to (4.72), we obtain

$$\left.\frac{\mathrm{d}p_k^c(\alpha)}{\mathrm{d}\alpha}\right|_{\alpha=0^+} = \frac{N_0(x_s, x_a, p_k^0)N(p_k^0)\mathrm{e}^{-t_d p_k^0}}{\left.\frac{\mathrm{d}D_0(p_k^0)}{\mathrm{d}s}\right|_{\alpha=0^+} D(p_k^0)}$$
$$= -\frac{\mathrm{i}}{k\pi}\sin(k\pi x_s)\sin(k\pi x_a)H(p_k^0)\mathrm{e}^{-\mathrm{i}k\pi\theta}. \qquad \text{(B.1)}$$

All root loci initially move toward C_- as α begins to increase from zero if $\mathrm{Re}\left(\left.\frac{\mathrm{d}p_k^c(\alpha)}{\mathrm{d}\alpha}\right|_{\alpha=0^+}\right) < 0$. If α continuously increases, the root loci can cross the imaginary before moving into the right-hand s-plane. We must determine the critical value α_{cr} such that the first root locus intersects the imaginary axis as α approaches to α_{cr}. Suppose that the root loci cross the imaginary axis at $\pm\mathrm{i}\beta$, $\beta > 0$. Let

© The Editor(s) (if applicable) and The Author(s), under exclusive license
to Springer Nature Singapore Pte Ltd. 2022
K.-S. Hong et al., *Control of Axially Moving Systems*,
https://doi.org/10.1007/978-981-16-2915-0

$$\zeta = v_0(x_a - x_s) + t_d(1 - v_0^2),$$
$$\xi_{min} = \min(x_a, x_s),\ \xi_{max} = \max(x_a, x_s),$$
$$H(i\beta) = \frac{N(i\beta)}{D(i\beta)} = A(\beta)e^{i\Theta(\beta)},\ A(\beta) = \sqrt{\mathrm{Re}(H(i\beta))^2 + \mathrm{Im}(H(i\beta))^2},$$
$$\Theta(\beta) = \tan^{-1}\left(\frac{\mathrm{Im}(H(i\beta))}{\mathrm{Re}(H(i\beta))}\right). \tag{B.2}$$

The control gain α can be determined by substituting (B.2), (4.64), and (4.65) into (4.72), namely

$$\begin{aligned}
\alpha &= \frac{D_0(i\beta)D(i\beta)}{N_0(x_s, x_a, i\beta)N(i\beta)}e^{t_d i\beta} \\
&= \frac{D_0(i\beta)}{N_0(x_s, x_a, i\beta)A(\beta)e^{i\Theta(\beta)}}e^{i\beta\left(\frac{\zeta}{(1-v_0^2)} - v_0\frac{(x_a - x_s)}{(1-v_0^2)}\right)} \\
&= \frac{\sinh\left(i\beta(1-v_0^2)^{-1}\right)e^{i\left(\frac{\zeta\beta}{(1-v_0^2)} - \Theta(\beta)\right)}e^{-i\beta v_0\frac{(x_a - x_s)}{(1-v_0^2)}}}{\frac{1}{i\omega}\sinh\left(i\beta\xi_{min}(1-v_0^2)^{-1}\right)\sinh\left(i\beta(1-\xi_{max})(1-v_0^2)^{-1}\right)e^{-i\beta v_0\frac{(x_a - x_s)}{(1-v_0^2)}}A(\beta)} \\
&= \frac{\beta\sin\left(\beta(1-v_0^2)^{-1}\right)}{\sin\left(\xi_{min}\beta(1-v_0^2)^{-1}\right)\sin\left((1-\xi_{max})\beta(1-v_0^2)^{-1}\right)A(\beta)}e^{i\left(\frac{\zeta\beta}{(1-v_0^2)} - \Phi(\beta)\right)}.
\end{aligned} \tag{B.3}$$

Because α is a positive real number, (B.3) leads to the following condition

$$\begin{aligned}
&\sin[\zeta(1-v_0^2)^{-1}\beta - \Theta(\omega)] = 0 \\
&\Rightarrow \tan\left(\zeta(1-v_0^2)^{-1}\beta\right) = \tan(\Theta(\beta)) = \frac{\mathrm{Im}(H(i\beta))}{\mathrm{Re}(H(i\beta))}.
\end{aligned} \tag{B.4}$$

Let the solutions of equation (B.4) be β_n, which satisfy $0 < \beta_1 < \beta_2 < \ldots < \beta_n < \ldots$. The values of the control gain α, wherein the root loci cross the imaginary axis, are calculated as follows.

$$\alpha_n = \frac{\beta_n\sin\left(\beta_n(1-v_0^2)^{-1}\right)}{\sin\left(\xi_{min}\beta_n(1-v_0^2)^{-1}\right)\sin\left((1-\xi_{max})\beta_n(1-v_0^2)^{-1}\right)A(\beta_n)}e^{i\left(\frac{\zeta\beta_n}{(1-v_0^2)} - \Theta(\beta_n)\right)}. \tag{B.5}$$

The critical value of the control gain is $\alpha_{cr} = \inf_n\{\alpha_n\}$. It should be noted that α_{cr} is either a finite value or infinite value. If (B.4) has no positive real roots, $\alpha_{cr} \to \infty$. According to the above results, Criterion 4.1 for the stability of a linear string is made.

Appendix C

Proof of Lemmas 4.6 and 4.7

Proof of Lemma 4.6
Using Lemma 4.2, we obtain the following inequality

$$
\begin{aligned}
V_3 &\leq \zeta_2 \rho A l \int_0^l \left[w_x^2 + (w_t + v_0 w_x)^2\right] \mathrm{d}x \\
&\leq \frac{\zeta_2 \rho A l}{\min\{\rho A, T_0\}} V_1.
\end{aligned}
\tag{C.1}
$$

It follows that

$$
\left(\zeta_1 - \frac{\zeta_2 \rho A l}{\min\{\rho A, T_0\}}\right) V_1 \leq V_1 + V_3 \leq \left(\zeta_1 + \frac{\zeta_2 \rho A l}{\min\{\rho A, T_0\}}\right) V_1. \tag{C.2}
$$

Furthermore, we obtain

$$
\lambda_1 (V_1 + V_2 + V_4 + V_5) \leq V \leq \lambda_2 (V_1 + V_2 + V_4 + V_5), \tag{C.3}
$$

where

$$
\begin{aligned}
\lambda_1 &= \min\{1, \zeta_1 - \zeta_2 \rho A l / \min\{\rho A, T_0\}\}, \\
\lambda_2 &= \max\{1, \zeta_1 + \zeta_2 \rho A l / \min\{\rho A, T_0\}\}.
\end{aligned}
\tag{C.4}
$$

Note that ζ_2 is chosen to satisfy $0 < \zeta_2 < \min\{\rho A, T_0\}/\rho A l$.

© The Editor(s) (if applicable) and The Author(s), under exclusive license
to Springer Nature Singapore Pte Ltd. 2022
K.-S. Hong et al., *Control of Axially Moving Systems*,
https://doi.org/10.1007/978-981-16-2915-0

Proof of Lemma 4.7

The time rates of change of $V_1(t)$, $V_2(t)$, $V_3(t)$, $V_4(t)$, and $V_5(t)$ are expressed as follows:

$$\dot{V}_1 = \zeta_1 T_0 \left[w_x w_t + v_0 w_x^2 \right] \Big|_{x=0}^{x=l},$$
$$\dot{V}_2 = m_a \frac{Dw(l,t)}{Dt} \frac{\partial}{\partial t} \frac{Dw(l,t)}{Dt},$$
$$\dot{V}_3 = \frac{1}{2}\zeta_2 \rho A l \frac{Dw(l,t)}{Dt}^2 + \frac{1}{2}\zeta_2 T_0 l w_x(l,t)^2 - \frac{\zeta_2}{2}\int_0^l \left[\rho A (w_t + v_0 w_x)^2 + T_0 w_x^2\right] dx,$$
$$\dot{V}_4 = \frac{1}{2}\left[\dot{\hat{\boldsymbol{\theta}}}^{\mathrm{T}} \mathbf{B} \tilde{\boldsymbol{\theta}} + \tilde{\boldsymbol{\theta}}^{\mathrm{T}} \mathbf{B} \dot{\hat{\boldsymbol{\theta}}} \right],$$
$$\dot{V}_5 = \beta_4 \dot{\hat{d}}_b \tilde{d}_b. \tag{C.5}$$

From (C.5) and the control law given by (4.150), the time rate of change of $V(t)$ is derived as

$$\begin{aligned}
\dot{V} &= \zeta_1 T_0 \left[w_x w_t + v_0 w_x^2 \right] \Big|_{x=0}^{x=l} + \frac{Dw(l,t)}{Dt}\left[-d(t) - \mathrm{sgn}\left(\frac{Dw(l,t)}{Dt}\right)\hat{d}_b - \mathbf{w}^{\mathrm{T}}\tilde{\boldsymbol{\theta}} + \alpha_2 w_t(l,t) \right. \\
&\quad \left. - \alpha_3 w_x(l,t) - \alpha_1 \frac{Dw(l,t)}{Dt} \right] + \frac{1}{2}\zeta_2 \rho A l \left[\frac{Dw(l,t)}{Dt}\right]^2 + \frac{1}{2}\zeta_2 T_0 l w_x(l,t)^2 \\
&\quad - \frac{\zeta_2}{2}\int_0^l \left[\rho A (w_t + v_0 w_x)^2 + T_0 w_x^2\right] dx + \frac{1}{2}\left[\dot{\hat{\boldsymbol{\theta}}}^{\mathrm{T}} \mathbf{B} \tilde{\boldsymbol{\theta}} + \tilde{\boldsymbol{\theta}}^{\mathrm{T}} B \dot{\hat{\boldsymbol{\theta}}} \right] + \beta_4 \dot{\hat{d}}_b \tilde{d}_b \\
&= \zeta_1 T_0 w_x(l,t) w_t(l,t) + \zeta_1 T_0 v_0 w_x(l,t)^2 - \zeta_1 T_0 v_0 w_x(0,t)^2 + \frac{Dw(l,t)}{Dt}[-d(t) \\
&\quad - \mathrm{sgn}\left(\frac{Dw(l,t)}{Dt}\right)\hat{d}_b] - \frac{Dw(l,t)}{Dt}\mathbf{w}^{\mathrm{T}}\tilde{\boldsymbol{\theta}} + \alpha_2 \frac{Dw(l,t)}{Dt} w_t(l,t) - \alpha_3 \frac{Dw(l,t)}{Dt} w_x(l,t) \\
&\quad - \alpha_1 \left[\frac{Dw(l,t)}{Dt}\right]^2 + \frac{1}{2}\zeta_2 \rho A l \left[\frac{Dw(l,t)}{Dt}\right]^2 + \frac{1}{2}\zeta_2 T_0 l w_x(l,t)^2 - \frac{\zeta_2}{\zeta_1} V_1 + \dot{\hat{\boldsymbol{\theta}}}^{\mathrm{T}} \mathbf{B} \tilde{\boldsymbol{\theta}} + \beta_4 \dot{\hat{d}}_b \tilde{d}_b \\
&= 2\zeta_1 T_0 w_x(l,t) w_t(l,t) + \zeta_1 T_0 v_0 w_x(l,t)^2 - \zeta_1 T_0 v_0 w_x(0,t)^2 + \alpha_2 w_t(l,t)^2 - \alpha_3 v_0 w_x(l,t)^2 \\
&\quad + \frac{1}{2}\zeta_2 T_0 l w_x(l,t)^2 - \zeta_1 T_0 w_x(l,t) w_t(l,t) + \alpha_2 v_0 w_x(l,t) w_t(l,t) - \alpha_3 w_x(l,t) w_t(l,t) \\
&\quad - \alpha_1 \left[\frac{Dw(l,t)}{Dt}\right]^2 + \frac{1}{2}\zeta_2 \rho A l \left[\frac{Dw(l,t)}{Dt}\right]^2 - \frac{\zeta_2}{\zeta_1} V_1 + \dot{\hat{\boldsymbol{\theta}}}^{\mathrm{T}} \mathbf{B} \tilde{\boldsymbol{\theta}} - \frac{Dw(l,t)}{Dt}\mathbf{w}^{\mathrm{T}}\tilde{\boldsymbol{\theta}} \\
&\quad + \beta_4 \dot{\hat{d}}_b \tilde{d}_b + \frac{Dw(l,t)}{Dt}\left[-d(t) - \mathrm{sgn}\left(\frac{Dw(l,t)}{Dt}\right)\hat{d}_b \right] \\
&= -\zeta_1 \frac{T_0}{v_0} w_t(l,t)^2 + \zeta_1 \frac{T_0}{v_0} l \left[\frac{Dw(l,t)}{Dt}\right]^2 - \zeta_1 T_0 v_0 w_x(0,t)^2 + \frac{1}{2}\alpha_2 w_t(l,t)^2 \\
&\quad - \alpha_3 v_0 w_x(l,t)^2 + \frac{1}{2}\zeta_2 T_0 l w_x(l,t)^2 - \frac{1}{2}\alpha_2 v_0^2 w_x(l,t)^2 - (\zeta_1 T_0 + \alpha_3) w_x(l,t) w_t(l,t) \\
&\quad + \frac{1}{2}\alpha_2 \left[\frac{Dw(l,t)}{Dt}\right]^2 - \alpha_1 \left[\frac{Dw(l,t)}{Dt}\right]^2 + \frac{1}{2}\zeta_2 \rho A l \left[\frac{Dw(l,t)}{Dt}\right]^2 - \frac{\zeta_2}{\zeta_1} V_1
\end{aligned}$$

$$+\left[\dot{\hat{\boldsymbol{\theta}}}^{\mathrm{T}}-\frac{\mathrm{D}w(l,t)}{\mathrm{D}t}\mathbf{w}^{\mathrm{T}}\mathbf{B}\right]\mathbf{B}^{-1}\tilde{\boldsymbol{\theta}}+\beta_4\dot{\hat{d}}_{\mathrm{b}}\tilde{d}_{\mathrm{b}}+\frac{\mathrm{D}w(l,t)}{\mathrm{D}t}\left[-d(t)-\mathrm{sgn}\left(\frac{\mathrm{D}w(l,t)}{\mathrm{D}t}\right)\hat{d}_{\mathrm{b}}\right]. \quad (C.6)$$

Using the inequality $-w_x(l,t)w_t(l,t)\le \zeta_3 w_x(l,t)^2/2+w_x(l,t)^2/2\zeta_3$, wherein ζ_3 is a positive constant, the following inequality is obtained.

$$\begin{aligned}\dot{V}=&-\frac{\zeta_2}{\zeta_1}V_1-\zeta_1T_0v_0w_x(0,t)^2-\frac{1}{\zeta_3}\left(2\zeta_3\zeta_1\frac{T_0}{v_0}-\zeta_1T_0-\alpha_3-\zeta_3\alpha_2\right)w_t(l,t)^2\\&-\frac{1}{2}(2\alpha_3v_0-\zeta_2T_0l+\alpha_2v_0^2-\zeta_1\zeta_3T_0-\zeta_3\alpha_3)w_x(l,t)^2\\&-\left(\alpha_1-\frac{1}{2}\alpha_2-\frac{1}{2}\zeta_2\rho Al-\zeta_1\frac{T_0}{v_0}l\right)\left[\frac{\mathrm{D}w(l,t)}{\mathrm{D}t}\right]^2\\&+\left[\dot{\hat{\boldsymbol{\theta}}}^{\mathrm{T}}-\frac{\mathrm{D}w(l,t)}{\mathrm{D}t}\mathbf{w}^{\mathrm{T}}\mathbf{B}\right]\mathbf{B}^{-1}\tilde{\boldsymbol{\theta}}+\beta_4\dot{\hat{d}}_{\mathrm{b}}\tilde{d}_{\mathrm{b}}\\&+\frac{\mathrm{D}w(l,t)}{\mathrm{D}t}\left[-d(t)-\mathrm{sgn}\left(\frac{\mathrm{D}w(l,t)}{\mathrm{D}t}\right)\hat{d}_{\mathrm{b}}\right].\end{aligned}\quad (C.7)$$

Substituting the adaptation law given by (4.151) into (C.7) yields

$$\begin{aligned}\dot{V}=&-\frac{\zeta_2}{\zeta_1}V_1-\zeta_1T_0v_0w_x(0,t)^2-\frac{1}{\zeta_3}\left(2\zeta_3\zeta_1\frac{T_0}{v_0}-\zeta_1T_0-\alpha_3-\zeta_3\alpha_2\right)w_t(l,t)^2\\&-\frac{1}{2}(2\alpha_3v_0-\zeta_2T_0l+\alpha_2v_0^2-\zeta_1\zeta_3T_0-\zeta_3\alpha_3)w_x(l,t)^2\\&-\left(\alpha_1-\frac{1}{2}\alpha_2-\frac{1}{2}\zeta_2\rho Al-\zeta_1\frac{T_0}{v_0}l\right)\left[\frac{\mathrm{D}w(l,t)}{\mathrm{D}t}\right]^2\\&-\left[d_{\mathrm{b}}+d(t)\mathrm{sgn}\left(\frac{\mathrm{D}w(l,t)}{\mathrm{D}t}\right)\right]\left|\frac{\mathrm{D}w(l,t)}{\mathrm{D}t}\right|-\frac{\beta_4}{2}\xi_2\hat{d}_{\mathrm{b}}^2-\frac{\xi_1}{2}\hat{\boldsymbol{\theta}}^{\mathrm{T}}\mathbf{B}^{-1}\hat{\boldsymbol{\theta}}-\frac{\beta_4}{2}\xi_2\tilde{d}_{\mathrm{b}}^2\\&-\frac{\xi_1}{2}\tilde{\boldsymbol{\theta}}^{\mathrm{T}}\mathbf{B}^{-1}\tilde{\boldsymbol{\theta}}+\frac{1}{2}\left[\beta_4\xi_2d_{\mathrm{b}}^2+\xi_1\boldsymbol{\theta}^{\mathrm{T}}\mathbf{B}^{-1}\boldsymbol{\theta}\right].\end{aligned}\quad (C.8)$$

If α_1, α_2, α_3, ζ_1, ζ_2, and ζ_3 satisfy

$$\begin{aligned}&2\zeta_3\zeta_1\frac{T_0}{v_0}-\zeta_1T_0-\alpha_3-\zeta_3\alpha_2\ge 0,\\&2\alpha_3v_0-\zeta_2T_0l+\alpha_2v_0^2-\zeta_1\zeta_3T_0-\zeta_3\alpha_3\ge 0,\\&\alpha_1-\frac{1}{2}\alpha_2-\frac{1}{2}\zeta_2\rho Al-\zeta_1\frac{T_0}{v_0}l\ge 0,\end{aligned}\quad (C.9)$$

then

$$\dot{V}\le -\lambda_3(V_1+V_2+V_4+V_5)+\delta, \quad (C.10)$$

where

$$\lambda_3 = \min\left\{\frac{\zeta_2}{2\zeta_1}, \frac{1}{2}\left(\alpha_1 - \frac{1}{2}\alpha_2 - \frac{1}{2}\zeta_2\rho Al - \zeta_1\frac{T_0}{v_0}l\right), \xi_1, \xi_2\right\},$$

$$\varepsilon = \frac{1}{2}\left[\beta_4\xi_2 d_b^2 + \xi_1\boldsymbol{\theta}^{\mathrm{T}}\mathbf{B}^{-1}\boldsymbol{\theta}\right]. \tag{C.11}$$

Appendix D

Expansion of (7.13)

$$\frac{\partial L_{\mathrm{p1}}}{\partial w} = \frac{\partial L_{\mathrm{p1}}}{\partial u} = \frac{\partial L_{\mathrm{p1}}}{\partial \eta} = 0, \tag{D.1}$$

$$\frac{\partial}{\partial x}\frac{\partial L_{\mathrm{p1}}}{\partial w_x} = \rho h(v_0 w_{xt} + v_0^2 w_{xx}) - (N_{\mathrm{X}} w_x)_x - (N_{\mathrm{XZ}} w_z)_x, \tag{D.2}$$

$$\frac{\partial}{\partial z}\frac{\partial L_{\mathrm{p1}}}{\partial w_z} = -(N_{\mathrm{Z}} w_z)_z - (N_{\mathrm{XZ}} w_x)_z, \tag{D.3}$$

$$\frac{\partial}{\partial t}\frac{\partial L_{\mathrm{p1}}}{\partial w_t} = \rho h(w_{tt} + v_0 w_{xt}), \tag{D.4}$$

$$\frac{\partial^2}{\partial x^2}\frac{\partial L_{\mathrm{p2}}}{\partial w_{xx}} = M_{\mathrm{X}xx}, \tag{D.5}$$

$$\frac{\partial^2}{\partial z^2}\frac{\partial L_{\mathrm{p2}}}{\partial w_{zz}} = M_{\mathrm{Z}zz}, \tag{D.6}$$

$$\frac{\partial^2}{\partial x \partial z}\frac{\partial L_{\mathrm{p2}}}{\partial w_{xz}} = 2M_{\mathrm{XZ}xz}, \tag{D.7}$$

$$\frac{\partial}{\partial x}\frac{\partial L_{\mathrm{p1}}}{\partial u_x} = \rho h(v_0 u_{xt} + v_0^2 u_{xx}) - N_{\mathrm{X}x}, \tag{D.8}$$

$$\frac{\partial}{\partial z}\frac{\partial L_{\mathrm{p1}}}{\partial u_z} = -N_{\mathrm{XZ}z}, \tag{D.9}$$

$$\frac{\partial}{\partial t}\frac{\partial L_{\mathrm{p1}}}{\partial u_t} = \rho h(u_{tt} + v_0 u_{xt}), \tag{D.10}$$

© The Editor(s) (if applicable) and The Author(s), under exclusive license
to Springer Nature Singapore Pte Ltd. 2022
K.-S. Hong et al., *Control of Axially Moving Systems*,
https://doi.org/10.1007/978-981-16-2915-0

$$\frac{\partial}{\partial x}\frac{\partial L_{\mathrm{p1}}}{\partial \eta_x} = \rho h(v_0\eta_{xt} + v_0^2\eta_{xx}) - N_{\mathrm{XZ}x}, \tag{D.11}$$

$$\frac{\partial}{\partial z}\frac{\partial L_{\mathrm{p1}}}{\partial \eta_z} = -N_{\mathrm{Z}z}, \tag{D.12}$$

$$\frac{\partial}{\partial t}\frac{\partial L_{\mathrm{p1}}}{\partial \eta_t} = \rho h(\eta_{tt} + v_0\eta_{xt}). \tag{D.13}$$

Substituting the above terms into (7.13) yields

$$\begin{aligned}
0 = &\int_{t_1}^{t_2}\int_S \{-\rho h(w_{tt} + 2v_0 w_{xt} + v_0^2 w_{xx}) + (N_{\mathrm{X}} w_x)_x + (N_{\mathrm{XZ}} w_z)_x + (N_{\mathrm{Z}} w_z)_z \\
&+(N_{\mathrm{XZ}} w_x)_z + M_{\mathrm{X}xx} + M_{\mathrm{Z}zz} + 2M_{\mathrm{XZ}xz}\}\delta w \mathrm{d}S\mathrm{d}t \\
&+\int_{t_1}^{t_2}\int_S (-\rho h(u_{tt} + 2v_0 u_{xt} + v_0^2 u_{xx}) + N_{\mathrm{X}x} + N_{\mathrm{XZ}z})\delta u \mathrm{d}S\mathrm{d}t \\
&+\int_{t_1}^{t_2}\int_S (-\rho h(\eta_{tt} + 2v_0 \eta_{xt} + v_0^2 \eta_{xx}) + N_{\mathrm{XZ}x} + N_{\mathrm{Z}z})\delta \eta \mathrm{d}S\mathrm{d}t \\
&+\int_{t_1}^{t_2}\int_0^b (\rho h v_0(w_t + v_0 w_x) - N_{\mathrm{X}} w_x - N_{\mathrm{XZ}} w_z - M_{\mathrm{X}x} - 2M_{\mathrm{XZ}z})\delta w|_{x=0}^{x=l}\mathrm{d}z\mathrm{d}t \\
&+\int_{t_1}^{t_2}\int_0^b (\rho h v_0(u_t + v_0 u_x) - N_{\mathrm{X}})\delta u|_{x=0}^{x=l}\mathrm{d}z\mathrm{d}t \\
&+\int_{t_1}^{t_2}\int_0^b (\rho h v_0(\eta_t + v_0 \eta_x) - N_{\mathrm{XZ}})\delta \eta|_{x=0}^{x=l}\mathrm{d}z\mathrm{d}t \\
&+\int_{t_1}^{t_2}\int_0^l (-N_{\mathrm{Z}} w_z - N_{\mathrm{XZ}} w_z - M_{\mathrm{Z}z} - 2M_{\mathrm{XZ}x})\delta w|_{z=0}^{z=b}\mathrm{d}x\mathrm{d}t \\
&+\int_{t_1}^{t_2}\int_0^l N_{\mathrm{XZ}}\delta u|_{z=0}^{z=b}\mathrm{d}x\mathrm{d}t + \int_{t_1}^{t_2}\int_0^l N_{\mathrm{Z}}\delta \eta|_{z=0}^{z=b}\mathrm{d}x\mathrm{d}t \\
&+\int_S \rho h(w_t + v_0 w_x)\delta w|_{t=t_1}^{t=t_2}\mathrm{d}S + \int_S \rho h(u_t + v_0 u_x)\delta u|_{t=t_1}^{t=t_2}\mathrm{d}S \\
&+\int_S \rho h(\eta_t + v_0 \eta_x)\delta \eta|_{t=t_1}^{t=t_2}\mathrm{d}S
\end{aligned}$$

$$
+\int_{t_1}^{t_2}\int_{0}^{b} M_{\mathrm{X}}\delta w_x|_{x=0}^{x=l}\mathrm{d}z\mathrm{d}t+\int_{t_1}^{t_2}\int_{0}^{l} M_{\mathrm{Z}}\delta w_z|_{z=0}^{z=l}\mathrm{d}x\mathrm{d}t
$$

$$
+\int_{t_1}^{t_2}\left(2M_{\mathrm{XZ}}\delta w|_{x=0}^{x=l}\right)\Big|_{z=0}^{z=b}\mathrm{d}t. \tag{D.14}
$$

Then, the equations of motion of an axially moving orthotropic plate (7.14)–(7.16) can be derived based on (D.14).

Appendix E

Elements of Mass Matrices, Damping Matrices, and Stiffness Matrices in (7.68) and (7.69)

The state vectors of (7.68) and (7.76) are given by

$$\mathbf{q}_{\mathrm{w}} = \left[q_{00}^{\mathrm{w}}, q_{10}^{\mathrm{w}}, \ldots, q_{i0}^{\mathrm{w}}, \ldots, q_{N0}^{\mathrm{w}}, q_{01}^{\mathrm{w}}, q_{11}^{\mathrm{w}}, \ldots, q_{i1}^{\mathrm{w}}, \ldots, \right.$$
$$\left. q_{N1}^{\mathrm{w}}, \ldots, q_{0j}^{\mathrm{w}}, q_{1j}^{\mathrm{w}}, \ldots, q_{ij}^{\mathrm{w}}, \ldots, q_{Nj}^{\mathrm{w}}, \ldots, q_{NN}^{\mathrm{w}}\right]^{\mathrm{T}}, \tag{E.1}$$

$$\mathbf{q}_{\mathrm{u}\eta} = \begin{bmatrix} \mathbf{q}_{\mathrm{u}} \\ \mathbf{q}_{\eta} \end{bmatrix}, \tag{E.2}$$

where

$$\mathbf{q}_{\mathrm{u}} = \left[q_{00}^{\mathrm{u}}, q_{10}^{\mathrm{u}}, \ldots, q_{i0}^{\mathrm{u}}, \ldots, q_{N0}^{\mathrm{u}}, q_{01}^{\mathrm{u}}, q_{11}^{\mathrm{u}}, \ldots, q_{i1}^{\mathrm{u}}, \ldots, \right.$$
$$\left. q_{N1}^{\mathrm{u}}, \ldots, q_{0j}^{\mathrm{u}}, q_{1j}^{\mathrm{u}}, \ldots, q_{ij}^{\mathrm{u}}, \ldots, q_{Nj}^{\mathrm{u}}, \ldots, q_{NN}^{\mathrm{u}}\right]^{\mathrm{T}}, \tag{E.3}$$

$$\mathbf{q}_{\eta} = \left[q_{00}^{\eta}, q_{10}^{\eta}, \ldots q_{i0}^{\eta}, \ldots, q_{N0}^{\eta}, q_{01}^{\eta}, q_{11}^{\eta}, \ldots, q_{i1}^{\eta}, \ldots, \right.$$
$$\left. q_{N1}^{\eta}, \ldots, q_{0j}^{\eta}, q_{1j}^{\eta}, \ldots, q_{ij}^{\eta}, \ldots, q_{Nj}^{\eta}, \ldots, q_{NN}^{\eta}\right]^{\mathrm{T}}. \tag{E.4}$$

The mass, damping, and stiffness matrices in (7.68) are defined as follows:

© The Editor(s) (if applicable) and The Author(s), under exclusive license to Springer Nature Singapore Pte Ltd. 2022
K.-S. Hong et al., *Control of Axially Moving Systems*,
https://doi.org/10.1007/978-981-16-2915-0

$$\mathbf{M}_{\mathrm{w}} = \begin{bmatrix} M^{\mathrm{w}}_{0000} & M^{\mathrm{w}}_{0010} & \cdots & M^{\mathrm{w}}_{00ij} & \cdots & M^{\mathrm{w}}_{00NN} \\ M^{\mathrm{w}}_{1000} & M^{\mathrm{w}}_{1010} & \cdots & M^{\mathrm{w}}_{10ij} & \cdots & M^{\mathrm{w}}_{10NN} \\ \vdots & \vdots & \ddots & & & \vdots \\ M^{\mathrm{w}}_{mn00} & M^{\mathrm{w}}_{mn10} & & \ddots & & M^{\mathrm{w}}_{mnij} \\ \vdots & \vdots & & & \ddots & \vdots \\ M^{\mathrm{w}}_{NN00} & M^{\mathrm{w}}_{NN10} & \cdots & M^{\mathrm{w}}_{mnij} & \cdots & M^{\mathrm{w}}_{NNNN} \end{bmatrix}, \tag{E.5}$$

$$M^{\mathrm{w}}_{mnij} = \rho h \int_0^l \varphi_m \varphi_i \mathrm{d}x \int_0^l Z_n Z_j \mathrm{d}z, \tag{E.6}$$

$$\mathbf{C}_{\mathrm{w}} = \begin{bmatrix} C^{\mathrm{w}}_{0000} & C^{\mathrm{w}}_{0010} & \cdots & C^{\mathrm{w}}_{00ij} & \cdots & C^{\mathrm{w}}_{00NN} \\ C^{\mathrm{w}}_{1000} & C^{\mathrm{w}}_{1010} & \cdots & C^{\mathrm{w}}_{10ij} & \cdots & C^{\mathrm{w}}_{10NN} \\ \vdots & \vdots & \ddots & & & \vdots \\ C^{\mathrm{w}}_{mn00} & C^{\mathrm{w}}_{mn10} & & \ddots & & C^{\mathrm{w}}_{mnij} \\ \vdots & \vdots & & & \ddots & \vdots \\ C^{\mathrm{w}}_{NN00} & C^{\mathrm{w}}_{NN10} & \cdots & C^{\mathrm{w}}_{mnij} & \cdots & C^{\mathrm{w}}_{NNNN} \end{bmatrix}, \tag{E.7}$$

$$C^{\mathrm{w}}_{mnij} = 2\upsilon\rho h \int_0^l \varphi_m (\varphi_i)_x \mathrm{d}x \int_0^l Z_n Z_j \mathrm{d}z + c \int_0^l \varphi_m \varphi_i \mathrm{d}x \int_0^l Z_n Z_j \mathrm{d}z, \tag{E.8}$$

$$\mathbf{K}_{\mathrm{w}} = \begin{bmatrix} K^{\mathrm{w}}_{0000} & K^{\mathrm{w}}_{0010} & \cdots & K^{\mathrm{w}}_{00ij} & \cdots & K^{\mathrm{w}}_{00NN} \\ K^{\mathrm{w}}_{1000} & K^{\mathrm{w}}_{1010} & \cdots & K^{\mathrm{w}}_{10ij} & \cdots & K^{\mathrm{w}}_{10NN} \\ \vdots & \vdots & \ddots & & & \vdots \\ K^{\mathrm{w}}_{mn00} & K^{\mathrm{w}}_{mn10} & & \ddots & & K^{\mathrm{w}}_{mnij} \\ \vdots & \vdots & & & \ddots & \vdots \\ K^{\mathrm{w}}_{NN00} & K^{\mathrm{w}}_{NN10} & \cdots & K^{\mathrm{w}}_{mnij} & \cdots & K^{\mathrm{w}}_{NNNN} \end{bmatrix}, \tag{E.9}$$

$$\begin{aligned} K^{\mathrm{w}}_{mnij} &= (c\upsilon + \rho h \dot{\upsilon}) \int_0^l \varphi_m (\varphi_i)_x \mathrm{d}x \int_0^l Z_n Z_j \mathrm{d}z + \rho h \upsilon^2 \int_0^l \varphi_m (\varphi_i)_{xx} \mathrm{d}x \int_0^l Z_n Z_j \mathrm{d}z \\ &+ \frac{Eh}{1-\upsilon^2} \sum_{k=0}^{N} \sum_{r=0}^{N} q^{u}_{kr} \xi^{\mathrm{u}}_{mnijkr} + \frac{Eh}{1-\upsilon^2} \sum_{k=0}^{N} \sum_{r=0}^{N} q^{\eta}_{kr} \xi^{\eta}_{mnijkr}, \end{aligned} \tag{E.10}$$

$$\xi^{\mathrm{u}}_{mnijkr} = \int_0^l (\varphi_m)_x (\varphi_i)_x (X_k)_x \mathrm{d}x \int_0^l Z_n Z_j Z_r \mathrm{d}z$$

$$+\frac{1+\upsilon}{2}\int_{0}^{l}\varphi_m(\varphi_i)_x X_k \mathrm{d}x\int_{0}^{l}(Z_n)_z Z_j (Z_r)_z \mathrm{d}z$$

$$+\frac{1+\upsilon}{2}\int_{0}^{l}(\varphi_m)_x\varphi_i X_k \mathrm{d}x\int_{0}^{l}Z_n (Z_j)_z (Z_r)_z \mathrm{d}z$$

$$+\upsilon\int_{0}^{l}\varphi_m\varphi_i (X_k)_z \mathrm{d}x\int_{0}^{l}(Z_n)_z (Z_j)_z Z_r \mathrm{d}z, \tag{E.11}$$

$$\xi^{\eta}_{mnijkr}=\upsilon\int_{0}^{l}(\varphi_m)_x(\varphi_i)_x\varphi_k \mathrm{d}x\int_{0}^{l}Z_n Z_j (Z_r)_z \mathrm{d}z$$

$$+\frac{1+\upsilon}{2}\int_{0}^{l}\varphi_m(\varphi_i)_x(\varphi_k)_x \mathrm{d}x\int_{0}^{l}(Z_n)_z Z_j Z_r \mathrm{d}z$$

$$+\frac{1+\upsilon}{2}\int_{0}^{l}(\varphi_m)_x\varphi_i(\varphi_k)_x \mathrm{d}x\int_{0}^{l}Z_n (Z_j)_z Z_r \mathrm{d}z$$

$$+\int_{0}^{l}\varphi_m\varphi_i\varphi_k \mathrm{d}x\int_{0}^{l}(Z_n)_z (Z_j)_z (Z_r)_z \mathrm{d}z, \tag{E.12}$$

where c is the damping coefficient. The mass matrix of (7.69) is given as follows:

$$\mathbf{M}_{\mathrm{u}\eta}=\begin{bmatrix}\mathbf{M}_{\mathrm{u}} & \mathbf{0}_{N^2\times N^2}\\ \mathbf{0}_{N^2\times N^2} & \mathbf{M}_{\eta}\end{bmatrix} \tag{E.13}$$

where

$$\mathbf{M}_{\mathrm{u}}=\begin{bmatrix} M^{\mathrm{u}}_{0000} & M^{\mathrm{u}}_{0010} & \cdots & M^{\mathrm{u}}_{00ij} & \cdots & M^{\mathrm{u}}_{00NN}\\ M^{\mathrm{u}}_{1000} & M^{\mathrm{u}}_{1010} & \cdots & M^{\mathrm{u}}_{10ij} & \cdots & M^{\mathrm{u}}_{10NN}\\ \vdots & \vdots & \ddots & & & \vdots\\ M^{\mathrm{u}}_{mn00} & M^{\mathrm{u}}_{mn10} & & \ddots & & M^{\mathrm{u}}_{mnij}\\ \vdots & \vdots & & & \ddots & \vdots\\ M^{\mathrm{u}}_{NN00} & M^{\mathrm{u}}_{NN10} & \cdots & M^{\mathrm{u}}_{mnij} & \cdots & M^{\mathrm{u}}_{NNNN}\end{bmatrix}, \tag{E.14}$$

$$M^{\mathrm{u}}_{mnij}=\rho h\int_{0}^{l}X_m X_i \mathrm{d}x\int_{0}^{b}Z_n Z_j \mathrm{d}z, \tag{E.15}$$

$$
\mathbf{M}_{\eta} = \begin{bmatrix} M^{\eta}_{0000} & M^{\eta}_{0010} & \cdots & M^{\eta}_{00ij} & \cdots & M^{\eta}_{00NN} \\ M^{\eta}_{1000} & M^{\eta}_{1010} & \cdots & M^{\eta}_{10ij} & \cdots & M^{\eta}_{10NN} \\ \vdots & \vdots & \ddots & & & \vdots \\ M^{\eta}_{mn00} & M^{\eta}_{mn10} & & \ddots & & M^{\eta}_{mnij} \\ \vdots & \vdots & & & \ddots & \vdots \\ M^{\eta}_{NN00} & M^{\eta}_{NN10} & \cdots & M^{\eta}_{mnij} & \cdots & M^{\eta}_{NNNN} \end{bmatrix}, \tag{E.16}
$$

$$
M^{\eta}_{mnij} = \rho h \int_0^l \varphi_m \varphi_i \mathrm{d}x \int_0^b Z_n Z_j \mathrm{d}z. \tag{E.17}
$$

The damping matrix of (7.69) is given as follows:

$$
\mathbf{C}_{\mathrm{u}\eta} = \begin{bmatrix} \mathbf{C}_{\mathrm{u}} & \mathbf{0}_{N^2 \times N^2} \\ \mathbf{0}_{N^2 \times N^2} & \mathbf{C}_{\eta} \end{bmatrix} \tag{E.18}
$$

where

$$
\mathbf{C}_{\mathrm{u}} = \begin{bmatrix} C^{\mathrm{u}}_{0000} & C^{\mathrm{u}}_{0010} & \cdots & C^{\mathrm{u}}_{00ij} & \cdots & C^{\mathrm{u}}_{00NN} \\ C^{\mathrm{u}}_{1000} & C^{\mathrm{u}}_{1010} & \cdots & C^{\mathrm{u}}_{10ij} & \cdots & C^{\mathrm{u}}_{10NN} \\ \vdots & \vdots & \ddots & & & \vdots \\ C^{\mathrm{u}}_{mn00} & C^{\mathrm{u}}_{mn10} & & \ddots & & C^{\mathrm{u}}_{mnij} \\ \vdots & \vdots & & & \ddots & \vdots \\ C^{\mathrm{u}}_{NN00} & C^{\mathrm{u}}_{NN10} & \cdots & C^{\mathrm{u}}_{mnij} & \cdots & C^{\mathrm{u}}_{NNNN} \end{bmatrix}, \tag{E.19}
$$

$$
C^{\mathrm{u}}_{mnij} = 2 v \rho h \int_0^l X_m (X_i)_x \mathrm{d}x \int_0^b Z_n Z_j \mathrm{d}z, \tag{E.20}
$$

$$
\mathbf{C}_{\eta} = \begin{bmatrix} C^{\eta}_{0000} & C^{\eta}_{0010} & \cdots & C^{\eta}_{00ij} & \cdots & C^{\eta}_{00NN} \\ C^{\eta}_{1000} & C^{\eta}_{1010} & \cdots & C^{\eta}_{10ij} & \cdots & C^{\eta}_{10NN} \\ \vdots & \vdots & \ddots & & & \vdots \\ C^{\eta}_{mn00} & C^{\eta}_{mn10} & & \ddots & & C^{\eta}_{mnij} \\ \vdots & \vdots & & & \ddots & \vdots \\ C^{\eta}_{NN00} & C^{\eta}_{NN10} & \cdots & C^{\eta}_{mnij} & \cdots & C^{\eta}_{NNNN} \end{bmatrix}, \tag{E.21}
$$

$$
C^{\eta}_{mnij} = 2 v \rho h \int_0^l \varphi_m (\varphi_i)_x \mathrm{d}x \int_0^b Z_n Z_j \mathrm{d}z. \tag{E.22}
$$

The stiffness matrix of (7.69) is given as follows:

$$\mathbf{K}_{\mathrm{u}\eta}=\begin{bmatrix}\mathbf{K}_{\mathrm{u}} & \overline{\mathbf{K}}_{\mathrm{u}\eta}\\ \overline{\mathbf{K}}_{\eta\mathrm{u}} & \mathbf{K}_{\eta}\end{bmatrix} \tag{E.23}$$

where

$$\mathbf{K}_{\mathrm{u}}=\begin{bmatrix} K^{\mathrm{u}}_{0000} & K^{\mathrm{u}}_{0010} & \cdots & K^{\mathrm{u}}_{00ij} & \cdots & K^{\mathrm{u}}_{00NN}\\ K^{\mathrm{u}}_{1000} & K^{\mathrm{u}}_{1010} & \cdots & K^{\mathrm{u}}_{10ij} & \cdots & K^{\mathrm{u}}_{10NN}\\ \vdots & \vdots & \ddots & & & \vdots\\ K^{\mathrm{u}}_{mn00} & K^{\mathrm{u}}_{mn10} & & \ddots & & K^{\mathrm{u}}_{mnij}\\ \vdots & \vdots & & & \ddots & \vdots\\ K^{\mathrm{u}}_{NN00} & K^{\mathrm{u}}_{NN10} & \cdots & K^{\mathrm{u}}_{mnij} & \cdots & K^{\mathrm{u}}_{NNNN}\end{bmatrix}, \tag{E.24}$$

$$\begin{aligned}K^{\mathrm{u}}_{mnij}&=\upsilon^2\rho h\int_0^l X_m(X_i)_{xx}\mathrm{d}x\int_0^b Z_nZ_j\mathrm{d}z+\dot{\upsilon}\rho h\int_0^l X_m(X_i)_x\mathrm{d}x\int_0^b Z_nZ_j\mathrm{d}z\\ &+\frac{Eh}{1-\upsilon^2}\int_0^l (X_m)_x(X_i)_x\mathrm{d}x\int_0^b Z_nZ_j\mathrm{d}z\\ &+\frac{Eh}{2(1+\upsilon)}\int_0^l X_mX_i\mathrm{d}x\int_0^b (Z_n)_z(Z_j)_z\mathrm{d}z,\end{aligned} \tag{E.25}$$

$$\overline{\mathbf{K}}_{\mathrm{u}\eta}=\begin{bmatrix} \overline{K}^{\mathrm{u}\eta}_{0000} & \overline{K}^{\mathrm{u}\eta}_{0010} & \cdots & \overline{K}^{\mathrm{u}\eta}_{00ij} & \cdots & \overline{K}^{\mathrm{u}\eta}_{00NN}\\ \overline{K}^{\mathrm{u}\eta}_{1000} & \overline{K}^{\mathrm{u}\eta}_{1010} & \cdots & \overline{K}^{\mathrm{u}\eta}_{10ij} & \cdots & \overline{K}^{\mathrm{u}\eta}_{10NN}\\ \vdots & \vdots & \ddots & & & \vdots\\ \overline{K}^{\mathrm{u}\eta}_{mn00} & \overline{K}^{\mathrm{u}\eta}_{mn10} & & \ddots & & \overline{K}^{\mathrm{u}\eta}_{mnij}\\ \vdots & \vdots & & & \ddots & \vdots\\ \overline{K}^{\mathrm{u}\eta}_{NN00} & \overline{K}^{\mathrm{u}\eta}_{NN10} & \cdots & \overline{K}^{\mathrm{u}\eta}_{mnij} & \cdots & \overline{K}^{\mathrm{u}\eta}_{NNNN}\end{bmatrix}, \tag{E.26}$$

$$\begin{aligned}K^{\mathrm{u}\eta}_{mnij}&=\frac{Eh\upsilon}{1-\upsilon^2}\int_0^l (X_m)_x\varphi_i\mathrm{d}x\int_0^b Z_n(Z_j)_z\mathrm{d}z\\ &+\frac{Eh}{2(1+\upsilon)}\int_0^l X_m(\varphi_i)_x\mathrm{d}x\int_0^b (Z_n)_zZ_j\mathrm{d}z,\end{aligned} \tag{E.27}$$

$$\overline{\mathbf{K}}_{\eta u} = \begin{bmatrix} \overline{K}^{\eta u}_{0000} & \overline{K}^{\eta u}_{0010} & \cdots & \overline{K}^{\eta u}_{00ij} & \cdots & \overline{K}^{\eta u}_{00NN} \\ \overline{K}^{\eta u}_{1000} & \overline{K}^{\eta u}_{1010} & \cdots & \overline{K}^{\eta u}_{10ij} & \cdots & \overline{K}^{\eta u}_{10NN} \\ \vdots & \vdots & \ddots & & & \vdots \\ \overline{K}^{\eta u}_{mn00} & \overline{K}^{\eta u}_{mn10} & & \ddots & & \overline{K}^{\eta u}_{mnij} \\ \vdots & \vdots & & & \ddots & \vdots \\ \overline{K}^{\eta u}_{NN00} & \overline{K}^{\eta u}_{NN10} & \cdots & \overline{K}^{\eta u}_{mnij} & \cdots & \overline{K}^{\eta u}_{NNNN} \end{bmatrix}, \tag{E.28}$$

$$K^{\eta u}_{mnij} = \frac{Eh\upsilon}{1-\upsilon^2}\int_0^l \varphi_m (X_i)_x \mathrm{d}x \int_0^b (Z_n)_z Z_j \mathrm{d}z + \frac{Eh}{2(1+\upsilon)}\int_0^l (\varphi_m)_x X_i \mathrm{d}x \int_0^b Z_n (Z_j)_z \mathrm{d}z, \tag{E.29}$$

$$\mathbf{K}_{\eta} = \begin{bmatrix} K^{\eta}_{0000} & K^{\eta}_{0010} & \cdots & K^{\eta}_{00ij} & \cdots & K^{\eta}_{00NN} \\ K^{\eta}_{1000} & K^{\eta}_{1010} & \cdots & K^{\eta}_{10ij} & \cdots & K^{\eta}_{10NN} \\ \vdots & \vdots & \ddots & & & \vdots \\ K^{\eta}_{mn00} & K^{\eta}_{mn10} & & \ddots & & K^{\eta}_{mnij} \\ \vdots & \vdots & & & \ddots & \vdots \\ K^{\eta}_{NN00} & K^{\eta}_{NN10} & \cdots & K^{\eta}_{mnij} & \cdots & K^{\eta}_{NNNN} \end{bmatrix}, \tag{E.30}$$

$$K^{\eta}_{mnij} = \dot{\upsilon}\rho h \int_0^l \varphi_m (\varphi_i)_x \mathrm{d}x \int_0^b Z_n Z_j \mathrm{d}z + \upsilon^2 \rho h \int_0^l \varphi_m (\varphi_i)_{xx} \mathrm{d}x \int_0^b Z_n Z_j \mathrm{d}z + \frac{Eh}{1-\upsilon^2}\int_0^l \varphi_m \varphi_i \mathrm{d}x \int_0^b (Z_n)_z (Z_j)_z \mathrm{d}z + \frac{Eh}{2(1+\upsilon)}\int_0^l (\varphi_m)_x (\varphi_i)_{xi} \mathrm{d}x \int_0^b Z_n Z_j \mathrm{d}z. \tag{E.31}$$

The force vector in (7.69) is given as follows.

$$\mathbf{G}_{u\eta} = \left[\, \mathbf{G}\ \mathbf{0}_{1\times N^2} \,\right]^{\mathrm{T}} \tag{E.32}$$

where

$$\mathbf{G} = \left[\, G_{00}\ G_{10}\ \ldots\ G_{mn}\ \ldots\ G_{NN} \,\right], \tag{E.33}$$

$$G_{mn} = \left((T_0 + \rho h l \dot{\upsilon}) X_m(l) - T_0 X_m(0) - \rho h \dot{\upsilon} \int_0^l X_m \mathrm{d}x \right) \int_0^l Z_n \mathrm{d}x. \tag{E.34}$$

Appendix F

MATLAB Codes

Example 4.1

"*control_string*" function code

```
function                [ddq1,ddq2,ddq3,ddq4]              =
controlled_string(q1,q2,q3,q4,dq1,dq2,dq3,dq4)
%--MODEL BASED FEEDBACK CONTROL--&&
```

© The Editor(s) (if applicable) and The Author(s), under exclusive license to Springer Nature Singapore Pte Ltd. 2022
K.-S. Hong et al., *Control of Axially Moving Systems*,
https://doi.org/10.1007/978-981-16-2915-0

```
%%---System Parameters----&&
L=1;         % string's length
v=9;         % axial velocity
rho = 10;    % mass density
A = 0.01;    % cross-section area
T0 = 10;     % tension
n=4;         % number of simulated modes
%%
%---Coefficient Matrix---%
M=zeros(n,n);
C=zeros(n,n);
K=zeros(n,n);
for i=1:n
    for j=1:n
      %---Mass matrix----
        if i==j
            M(j,i)=L/2;
        else
            M(j,i)=0;
        end
      %---Damping matrix---
        if i==j
            C(j,i)=0;
        else
            if mod(i+j,2)==0
                C(j,i)=0;
            end
            if mod(i+j,2)==1
                C(j,i)=2*i*j/(j^2-i^2);
            end
        end
      %---Stiffness matrix---
        if i==j
            K(j,i)=(-pi^2*i^2/(2*L))*(rho*A*v^2-T0);
        else
```

```
                K(j,i)=0;
            end
        end
    end
    %%
    %---Control law----
    % Positions of actuator: a1=L/4; a2=3*L/8;
    % Positions of sensor: s1=5L/8; s2=3L/4;
    a=sin(pi/4);
    b=sin(3*pi/8);
    c=sin(9*pi/8);
    case_flag=2; %Setting spillover cases: 1/No spillover,
    2/Only control spillover, 3/Only observation spillover, 4/
    Both two spillover
    if case_flag==1 %---No Spillover---
        G1=[1 0;0 1;0 0;0 0];
        G2=[G1' zeros(2,4);zeros(2,4) G1'];
        end
    if case_flag==2  %---Control spillover--
        G1=[1 0;0 1;(a^2-c)/(a^2-b) (a*c-a*b)/(a^2-b);1/(a^2-b)
    -a/(a^2-b)];
        G1_0=[1 0;0 1;0 0;0 0];
        G2=[G1_0' zeros(2,4);zeros(2,4) G1_0'];
        end
    if case_flag==3 %---Observation spillover--
        G1=[1 0;0 1;0 0;0 0];
        G1_0=[1 0;0 1;(a^2-c)/(a^2-b) (a*c-a*b)/(a^2-b);1/(a^2-
    b) -a/(a^2-b)];
        G2=[G1_0' zeros(2,4);zeros(2,4) G1_0'];
       end
    if case_flag==4 %---Control and Observation spillover---
        G1=[1 0;0 1;(a^2-c)/(a^2-b) (a*c-a*b)/(a^2-b);1/(a^2-b)
    -a/(a^2-b)];
        G2=[G1' zeros(2,4);zeros(2,4) G1'];
        end
    %--Control gain matrix
    G=[pi^2*v^2/L^2 0 (2*pi/L)*sqrt(T0/(rho*A)) -4*v/(3*L); 0
    4*pi^2*v^2/L^2 4*v/(3*L) (4*pi/L)*sqrt(T0/(rho*A))];
    %%
    Y=[q1;q2;q3;q4;dq1;dq2;dq3;dq4];
    Mat_A=[zeros(n,n) eye(n);-inv(M)*K -inv(M)*C];
    Mat_B=[zeros(n,n);inv(M)];
    dY=(Mat_A-Mat_B*G1*G*G2)*Y;
    ddq1=dY(5);
    ddq2=dY(6);
    ddq3=dY(7);
    ddq4=dY(8);
```

Example 4.2

Matlab code

```
clear
clc

CG_flag=0;
t0 = 0; tf = 5;
step_adj =10;
max_iteration = 1000;                    % Maximum number of
iterations
t_segment = 201;
Tv = linspace(t0, tf, t_segment);   % Discretize time

%Initial velocity profile
v_initial=4;
v_slope=v_initial/tf;
v=v_initial-v_slope*Tv;
dv=-v_slope+0.*Tv;
g=0;

%Beam parameter
[EI,rho,A,L,d,T,a,b,c,n]=systemparameter();

options = odeset('RelTol', 1e-4, 'AbsTol',1e-4*ones(1,2*n));

%% PARAMETRIC MATRICES
for i=1:n
  for j=1:n
    if i==j
        M(i,j)=L/2;
        Cs(i,j)=b*L/2;
        Ks(i,j)=a^2*(i*pi/L)^4+(i*pi/L)^2*c^2;
        Bs1(i,j)=0;
        Bs2(i,j)=-(i*pi/L)^2;
        Bs3(i,j)=0;
    else
        M(i,j)=0;
        Cs(i,j)=0;
        Ks(i,j)=0;
        Bs2(i,j)=0;
       if mod(i+j,2)==0
        Bs1(i,j)=0;
        Bs3(i,j)=0;
       else
        Bs1(i,j)=4*i*j/(j^2-i^2);
        Bs3(i,j)=2*i*j/(j^2-i^2);
       end
    end
  end
```

```
end
Ac=[-inv(M)*Cs  -inv(M)*Ks
     eye(n)      zeros(n)];
B1=[-inv(M)*Bs1    zeros(n)
     zeros(n)      zeros(n)];
B2=[zeros(n)      -inv(M)*Bs2
    zeros(n)       zeros(n)];
B3=[zeros(n)      -inv(M)*Bs3
    zeros(n)       zeros(n)];
for i=1:2*n
    psy(i)=0.2*(sign(i-n)+abs(sign(i-n)));
end
W=diag([10^5,10^5,10^5,10^5,10^5,2*10^5,4*10^5,10^5]);
% Weighting coefficient matrix

%% INTERATIVE NUMERICAL PROCEDUCE
initx=zeros(1,2*n);
initx(n+1)=0.04;
TspanX=linspace(t0, tf, t_segment);
TspanLamda=linspace(tf, t0, t_segment);
for i = 1:max_iteration
   [Tx,X]  =  ode45(@(t,x)  stateEq(t,x,v,dv,Tv),  TspanX,
initx, options);
   x1 = X(:,1); x2 = X(:,2); x3 = X(:,3);x4 = X(:,4);x5 =
X(:,5);x6 = X(:,6);x7 = X(:,7);x8 = X(:,8);
   Q = MatrixQ(n,rho*A,L,EI,T,psy,v,t_segment);
   initLamda=[Q+W]*X(t_segment,:)';
   [Tlamda,Lamda]             =             ode45(@(t,lamda)
costateEq(t,lamda,v,dv,Tv),      TspanLamda,      initLamda,
options);
   for k=1:2*n
    Lamda(:,k) = interp1(Tlamda,Lamda(:,k),Tx);
   end
   g_old=g;
   g = deltaH(Lamda,X,v,B1,B2,B3,t_segment,b);
   g_Norm = g*g';
   if i==1
    p=-g;
   else
    p_old=g;
    beta=(g*g')/(g_old*g_old');
    p=-g+CG_flag*beta*p_old;
   end
   J(i) = 0.5*X(t_segment,:)*(Q+W)*X(t_segment,:)';
   step=5*J(i)/g_Norm;
   v_old = v;

   % Find step size alpha to minimize the cost function J
   if J(i)<10^3
```

```
        for j=1:15
          stepScan(j)=(2^(j-1))*10^-6;
          stepScan(1)=0;
          v                                                      =
AdjControl(p,Tx,v_old,Tv,stepScan(j),t_segment,v_initial);
          dv=AdjAcceleration(v,t_segment);
          [Tx,X] = ode45(@(t,x) stateEq(t,x,v,dv,Tv), TspanX,
initx, options);
          Q = MatrixQ(n,rho*A,L,EI,T,psy,v,t_segment);
          Jscan(j)                                               =
0.5*X(t_segment,:)*(Q+W)*X(t_segment,:)';
              if Jscan(j)>3*J(i)
                  break
              end
        end
        [J(i) index]=min(Jscan);
        step= stepScan(index);
        if step==0
          stepScan=linspace(0,512*10^-6,100);
          for j=1:length(stepScan)
          stepScan(1)=0;
          v                                                      =
AdjControl(p,Tx,v_old,Tv,stepScan(j),t_segment,v_initial);
          dv=AdjAcceleration(v,t_segment);
          [Tx,X] = ode45(@(t,x) stateEq(t,x,v,dv,Tv), TspanX,
initx, options);
          Q = MatrixQ(n,rho*A,L,EI,T,psy,v,t_segment);
          Jscan(j)                                               =
0.5*X(t_segment,:)*(Q+W)*X(t_segment,:)';
         end
          [J(i) index]=min(Jscan);
          step= stepScan(index);
          if step==0
              step=0.001;
          end
       end
     end
     v = AdjControl(p,Tx,v_old,Tv,step,t_segment,v_initial);
     dv=AdjAcceleration(v,t_segment);

     [i J(i) 1000000*step]
end
w45=x5*sin(pi*0.45)+x6*sin(2*pi*0.45)+x7*sin(3*pi*0.45)+x8*
4*sin(2*pi*0.45);
% plot the state variables & cost for each iteration
figure(1);
plot(Tv,v);
hold on
figure(2);
```

```
plot(J,'x-');
hold on
xlabel('Iteration number');
ylabel('J');

%%
figure(3);
subplot(4,2,1)
x = linspace(0,10);
plot(Tv,x5)
title('Subplot 1: disp model 1')

subplot(4,2,2)
plot(Tv,x1)
title('Subplot 2: velo 1')

subplot(4,2,3)
plot(Tv,x6)
title('Subplot 3: dis_mode 1')

subplot(4,2,4)
plot(Tv,x2)
title('Subplot 4: vel_mode 3)')
subplot(4,2,5)
plot(Tv,x7)
title('Subplot 3: dis_mode 1')

subplot(4,2,6)
plot(Tv,x3)
title('Subplot 4: vel_mode 3)')
subplot(4,2,7)
plot(Tv,x8)
title('Subplot 3: dis_mode 1')

subplot(4,2,8)
plot(Tv,x4)
title('Subplot 4: vel_mode 3)')

%% SUB_FUNCTIONs
% State equations
function dx = stateEq(t,x,v,dv,Tv)
v = interp1(Tv,v,t);
dv = interp1(Tv,dv,t);
A = MatrixA(v,dv);
dx=A*x;
end
% Matrix A
function A = MatrixA(v,dv)
[EI,rho,A,L,d,T,a,b,c,n]=systemparameter();
```

```
for i=1:n
  for j=1:n
    if i==j
        M(i,j)=L/2;
        C(i,j)=b*L/2;
        K(i,j)=a^2*(i*pi/L)^4+(i*pi/L)^2*(c^2-v^2);
    else
       M(i,j)=0;
       if mod(i+j,2)==0
        C(i,j)=0;
        K(i,j)=0;
       else
        C(i,j)=4*i*j*v/(j^2-i^2);
        K(i,j)=2*i*j*(b*v+dv)/(j^2-i^2);
       end
    end
  end
end
A=[-inv(M)*C   -inv(M)*K
    eye(n)      zeros(n)];
end
% Matrix Q
function Q = MatrixQ(n,rhoA,L,EI,T,psy,v,t_segment)
for i=1:2*n
  for j=1:2*n

Q1(i,j)=rhoA*(1+i*pi*v(t_segment)*psy(i)/L)*(1+j*pi*v(t_seg
ment)*psy(j)/L);
    Q2(i,j)=T*(psy(i)+i*pi/L)*(psy(j)+j*pi/L);
    Q3(i,j)=EI*(psy(i)+(i*pi/L)^2)*(psy(j)+(j*pi/L)^2);
  end
end
Q=L*(Q1+Q2+Q3);
end

% Costate equations
function dp = costateEq(t,p,v,dv,Tv)
v = interp1(Tv,v,t);
dv = interp1(Tv,dv,t);
A = MatrixA(v,dv);
dp=-A'*p;
end

% Computinmg the gradient
function g = deltaH(P,X,v,B1,B2,B3,t_segment,b)
for i=1:t_segment
    g(i)=P(i,:)*(B1+2*v(i)*B2+b*B3)*X(i,:)';
end
end
```

```
% Adjusting the velocity profile
function                       v_new                          =
AdjControl(p,tx,v,tv,step,t_segment,v_initial)

p = interp1(tx,p,tv);
v_new = v + step*p;
v_new(1)=v_initial;
v_new(t_segment)=0;
for i=1:t_segment
    if v_new(i)<0
       v_new(i)=0;
    end
end
end

% Computing the accelerate profile
function dv_new = AdjAcceleration(v_new,t_segment)
dv_new(1)=0;
for i=2:t_segment
    dv_new(i)=v_new(i)-v_new(i-1);
end
end

% Setting the beam parameteters
function [EI,rho,A,L,d,T,a,b,c,n]=systemparameter()
E=69*10^7;
I=5.2e-10;
EI=E*I;
rho=2700;
A=2.5e-4;
L=4;
d=0.3;
T=300;
a=sqrt(EI/(rho*A));
b=d/(rho*A);
c=sqrt(T/(rho*A));
n=4;
end
```

Example 4.3

```
clc
clear
%% SYSTEM PARAMETERS (Rubber string)
v=0.5;                  %Axial velocity (m/s)
L=1;                    %String length (m)
T=1.53;                 %Initial tension (N)
rho=1530;               %Mass density (kg/m3)
A=0.001;                %Cross-section area (m2)
Time=15;                %Simulation time (sec)
%% SIMULATION CASES
case_flag=2;
%Setting cases: 1/Colocation case, 2/Non-colocation case,
3/Non-colocation and boundary control case
if case_flag==1
    xa=14; xs=xa; % Location of actuator and sensor
end
if case_flag==2
    xa=12; xs=20-xa;
end
if case_flag==3
    xs=19;
end
%% FINITE DIFFERENTIAL METHOD
dx=0.05;                   %Spatial step
dt=0.00005;                %Temporal step
nx=fix(L/dx);              %Number of spatial point
nt=fix(Time/dt);    %Number of temporal point
x=linspace(0,L,nx);
% Field variable
yn=zeros(nx,1);     % yn: displacement vector at the current
time  (t=i)
ym=yn;             % ym: displacement vector at the previous
time (t=i-1)
yp=yn;               % yp: displacement vector at the later
time (t=i+1)
w=zeros(nx,nt);     % Lateral vibration
% Initial conditions
for i=1:nx
yn(i)=0.05*sin((i-1)*pi/(nx-1));
end
yp(:)=yn(:);
% Boundary conditions
yn(1)=0;
yn(nx)=0;
% Supplementary parameters and variables
k1=v*sqrt(rho*A/T);
k2=(1-v^2*rho*A/T);
ys=zeros(1,nt);
% Solving equation using a FDM-based loop
control_flag=1;
if control_flag==1
for j=2:nt-1
    w(:,j)=yn;
    ys(j)=yn(xs);
    ym=yn; yn=yp;
```

```
 if case_flag==3 % Boundary control based on wave cancelation
method
     for i=2:nx-1
        yxx=(yn(i+1)-2*yn(i)+yn(i-1))/dx^2;
        yxt=(yn(i+1)-yn(i-1)-ym(i+1)+ym(i-1))/(2*dx*dt);
        yp(i)=2*yn(i)-ym(i)+dt^2*(-k1*yxt+k2*yxx);
    end
    Time_delay=fix(((1-xs/nx)/(1+k1))/dt);
    if (j-2)<Time_delay
      yst=0;
    else
      yst=(ys(j-Time_delay)-ys(j-Time_delay-1))/dt;
    end
    yp(nx)=yp(nx-1)-1*(dx*yst);
 else % Pointwise control based on transfer function method
    for i=2:nx-1
        yxx=(yn(i+1)-2*yn(i)+yn(i-1))/dx^2;
        yxt=(yn(i+1)-yn(i-1)-ym(i+1)+ym(i-1))/(2*dx*dt);
        if i==xa
            if xs==xa
                yst=(yn(xs)-ym(xs))/dt;
                yp(i)=2*yn(i)-ym(i)+dt^2*(-k1*yxt+k2*yxx-
40*yst);
            else
                Time_delay=fix((1-k1*(xa-xs)/nx)/(1-
k1^2)/dt);
                if (j-2)<Time_delay
                   yst=0;
                else
                   yst=(ys(j-Time_delay)-ys(j-Time_delay-
1))/dt;
                end
                yp(i)=2*yn(i)-ym(i)+dt^2*(-
k1*yxt+k2*yxx+1*yst);
            end
        else
            yp(i)=2*yn(i)-ym(i)+dt^2*(-k1*yxt+k2*yxx);
        end
    end
 end
end
else
for j=2:nt-1
    w(:,j)=yn;
    ys(j)=yn(xs);
    ym=yn; yn=yp;
     for i=2:nx-1
        yxx=(yn(i+1)-2*yn(i)+yn(i-1))/dx^2;
        yxt=(yn(i+1)-yn(i-1)-ym(i+1)+ym(i-1))/(2*dx*dt);
        yp(i)=2*yn(i)-ym(i)+dt^2*(-k1*yxt+k2*yxx);
     end
end
end
%% TO SHOW THE VIBRATION AT x=0.5 m
wm=w(fix(nx/2),:);
tm=linspace(0,Time,nt);
plot(tm,wm)
xlabel('Time (sec)')
ylabel('w(0.5,t)')
axis([0 15 -0.1 0.1]);
```

Example 4.4

```
clc
clear
%% SYSTEM PARAMETERS
v=1;                %Axial velocity (m/s)
L=1;                %String length (m)
T=30;               %Initial tension (N)
rho=1530;             %Mass density (kg/m3)
A=0.001;             %Cross-section area (m2)
Time=10;            %Simulation time (sec)
%% FINITE DIFFERENTIAL METHOD
dx=0.05;               %Spatial step
dt=0.00005;            %Temporal step
nx=fix(L/dx);          %Number of spatial point
nt=fix(Time/dt);       %Number of temporal point
x=linspace(0,L,nx);

% Field variable
yn=zeros(nx,1);      % yn: spatial vector at the current time
(t=i)
ym=yn;                  % ym: spatial vector at the previous
time (t=i-1)
yp=yn;                 % yp: spatial vector at the later time
(t=i+1)
w=zeros(nx,nt);       % lateral vibration

%Initial conditions
for i=1:nx
yn(i)=0.05*sin((i-1)*pi/(nx-1));
end
yp(:)=yn(:);

% Boundary conditions
yn(1)=0;
yn(nx)=0;
```

```
% Supplementary parameters
k1=v;
k2=(T/(rho*A)-v^2);

for j=2:nt-1
    w(:,j)=yn;
    ym=yn; yn=yp; %Save current and previous arrays
    for i=2:nx-1
    yxx=(yn(i+1)-2*yn(i)+yn(i-1))/dx^2;
    yxt=(yn(i+1)-yn(i-1)-ym(i+1)+ym(i-1))/(dx*dt);
    yp(i)=2*yn(i)-ym(i)+dt^2*(-k1*yxt+k2*yxx);
    end
%       yp(nx)=yn(nx)-0.9*(dt/dx)*(yn(nx)-yn(nx-1));
end

%%FIGURE%%
wm=w(fix(nx/2),:);
sizet=size(wm);
tm=linspace(0,Time,sizet(2));
plot(tm,wm)
axis([0 10 -0.06 0.06]);
```

Example 4.5

```
"Dynamic model" function code

function [ddy, ddq1, ddq2] = Dynamic_model(u,  L, dL, ddL,
y, q1, q2, dy, dq1, dq2)
%% SYSTEM PARAMETERS
rho=90;       % Density mass (kg/m3)
me=1000;      % Payload mass (kg)
Mass=10^4;    % Trolley mass (kg)
g=9.8;        % Gravitional acceleration (m/s2)

%% GALERKIN DECOMPOSITION
phi=[sqrt(2);-sqrt(2)];  %Mode shape at x=L(t)
%-------Integral coefficients of mode shape-------
A=[1/2  -3/4
   3/4   1/2];
B=[pi^2/12+1/2    15/8
    15/8          3*pi^2/4+1/2];
C=[pi^2/8+1/2     3/2
    3/2           9*pi^2/8+1/2];
D=[pi^2/4     0
    0       9*pi^2/4];
Y=[2*sqrt(2)/pi; 2*sqrt(2)/(3*pi)];
J=(sqrt(2)*pi/(2*L^(3/2)))*(rho*dL^2+(me*g+rho*L)*(ddL-g));
```

```
%---------------Supplementary matrices----------------
Ms=[rho+2*me/L     -2*me/L
    -2*me/L      rho+2*me/L];
Gs=[-2*rho*(dL/L)*(0.5-A(1,1))       2*rho*(dL/L)*A(1,2)
    2*rho*(dL/L)*A(2,1)                        -2*rho*(dL/L)*(0.5-
A(2,2))];
Cs=[-2*me*dL/L^2        2*me*dL/L^2
     2*me*dL/L^2      -2*me*dL/L^2];
Ks=[(1/4)*rho*(dL/L)^2-rho*(dL/L)^2*B(1,1)+rho*((g-
ddL)/L)*C(1,1)+me*((g-
ddL)/L^2)*D(1,1)+2*me*((3/4)*dL^2/L^3-0.5*ddL/L^2)          -
rho*(dL/L)^2*B(1,2)+rho*((g-ddL)/L)*C(1,2)+me*((g-
ddL)/L^2)*D(1,2)-2*me*((3/4)*dL^2/L^3-0.5*ddL/L^2)
     -rho*(dL/L)^2*B(2,1)+rho*((g-ddL)/L)*C(2,1)+me*((g-
ddL)/L^2)*D(2,1)-2*me*((3/4)*dL^2/L^3-0.5*ddL/L^2)
(1/4)*rho*(dL/L)^2-rho*(dL/L)^2*B(2,2)+rho*((g-
ddL)/L)*C(2,2)+me*((g-
ddL)/L^2)*D(2,2)+2*me*((3/4)*dL^2/L^3-0.5*ddL/L^2)];
Hs=[(rho*(dL/L)^2-ddL/L)*(0.5-A(1,1))        -(rho*(dL/L)^2-
ddL/L)*A(1,2)
    -(rho*(dL/L)^2-ddL/L)*A(2,1)                (rho*(dL/L)^2-
ddL/L)*(0.5-A(2,2))];
% ---------------Coefficient matrices --------------
Mo=[Mass                                        0         0
    rho*Y(1)*sqrt(L)+me*phi(1)/sqrt(L)     Ms(1,1)  Ms(1,2)
    rho*Y(2)*sqrt(L)+me*phi(2)/sqrt(L)    Ms(2,1)  Ms(2,2)];
Co=[rho*dL         0                      0
    0       Cs(1,1)+Gs(1,1)         Cs(1,2)+Gs(1,2)
    0       Cs(2,1)+Gs(2,1)         Cs(2,2)+Gs(2,2)];
Ko=[0             J              3*J
    0       Ks(1,1)+Hs(1,1)    Ks(1,2)+Hs(1,2)
    0       Ks(2,1)+Hs(2,1)    Ks(2,2)+Hs(2,2)];
Fo=[u;0;0];
X=[y;q1;q2];
dX=[dy;dq1;dq2];
ddX=-inv(Mo)*(Co*dX+Ko*X-Fo);
ddy=ddX(1);
ddq1=ddX(2);
ddq2=ddX(3);
```

Example 4.6

```
clc
clear
%% SYSTEM PARAMETERS (Rubber string)
v=0.7;                  %Axial velocity (m/s)
L=1;                    %String length (m)
T=30;                   %Initial tension (N)
rho=1530;               %Mass density (kg/m3)
A=0.001;                %Cross-section area (m2)
pA=rho*A;
ca=1.5;
ma=2;
Time=15;                %Simulation time (sec)
%% FINITE DIFFERENTIAL METHOD
dx=0.05;                %Spatial step
dt=0.00005;             %Temporal step
nx=fix(L/dx);           %Number of spatial point
nt=fix(Time/dt);        %Number of temporal point
x=linspace(0,L,nx);

% Field variable
yn=zeros(nx,1);        % yn: displacement vector at the current
time  (t=i)
ym=yn;                 % ym: displacement vector at the previous
time (t=i-1)
yp=yn;                  % yp: displacement vector at the later
time (t=i+1)
w=zeros(nx,nt);         % lateral vibration
d=zeros(1,nt);
d_bound=50;
for j=1:nt
    d(j)=d_bound*sin(45*pi*j*dt);
end
% Initial conditions
for i=1:nx
yn(i)=0.05*sin((i-1)*pi/(nx-1));
end
yp(:)=yn(:);

% Boundary conditions
yn(1)=0;
yn(nx)=0;
theta_est=zeros(3,nt);
pA_hat=1;
ma_hat=2;
ca_hat=0.01;
theta_est(:,1)=[pA_hat ma_hat ca_hat]';
d_bound_est=zeros(1,nt);

% Supplementary parameters
k1=v;
k2=(T/pA-v^2);

kb1=(ca-pA*v);
kb2=(T-v^2*pA);
```

```
% Control gain
a1=225;a2=25;a3=15;
g1=178;g2=155;
b4=35;
test=zeros(1,nt);
for j=2:nt
    w(:,j)=yn;
    ym=yn; yn=yp; %Save current and previous arrays
    for i=2:nx-1
        yxx=(yn(i+1)-2*yn(i)+yn(i-1))/dx^2;
        yxt=(yn(i+1)-yn(i-1)-ym(i+1)+ym(i-1))/(2*dx*dt);
        yp(i)=2*yn(i)-ym(i)+dt^2*(-k1*yxt+k2*yxx);
    end
%  Boundary control law
   ytn=(yn(nx)-ym(nx))/dt;
   yxn=(yp(nx)-yp(nx-1))/dx;
   yxtn=(yn(nx)-yn(nx-1)-ym(nx)+ym(nx-1))/(dx*dt);
%  yp(nx)=exp(-alpha1*j*dt);
   vector_w=[v*(ytn+v*yxn) v*yxtn -ytn];
   theta=[pA ma ca]';
   B=[10 0 0;0 10 0;0 0 10];
   theta_est(:,j)=theta_est(:,j-1)+dt*(-g1*theta_est(:,j-
1)+(ytn+v*yxn)*B*vector_w');
   d_bound_est(j)=d_bound_est(j-1)+dt*(-g2*d_bound_est(:,j-
1)+abs(ytn+v*yxn)/b4);
   f=-vector_w*theta_est(:,j)+T*yxn-a1*(ytn+v*yxn)+a2*ytn-
a3*yxn-sign(ytn+v*yxn)*d_bound_est(j);
%  f=0;
   yp(nx)=2*yn(nx)-ym(nx)+dt^2*(-kb1*ytn-kb2*yxn+f-
d(j))/ma;
end

%%FIGURE%%
wm=w(fix(nx/2),:);
ss=size(wm);
tm=linspace(0,Time,ss(2));
plot(tm,wm)
xlabel('Time (sec)')
ylabel('w(0.5,t)')
axis([0 15 -0.06 0.06]);
```

Example 6.1

```
clc
clear

%% SYSTEM PARAMETERS (Aluminium string)
v=2;                  %Axial velocity (m/s)
L=20;                 %String length (m)
T0=9800*10^3;         %Initial tension (N)
rho=7850;             %Mass density (kg/m3)
E=200*10^9;           %Young modulus (N/m2)
A=1.5*0.005;          %Cross-section area (m2)
EA=E*A;
pA=rho*A;
Time=5;     %Simulation time (sec)
%% FINITE DIFFERENTIAL METHOD
dx=0.2;                     %Spatial step
dt=0.00001;                 %Temporal step
nx=fix(L/dx);               %Number of spatial point
nt=fix(Time/dt);            %Number of temporal point
x=linspace(0,L,nx);

% Field variable
% Lateral vibrations
yn=zeros(nx,1);      % yn: spatial vector at the current time
(t=i)
ym=yn;                 % ym: spatial vector at the previous
time (t=i-1)
yp=yn;                % yp: spatial vector at the later time
(t=i+1)
w=zeros(nx,nt);      % lateral vibration
% Longitudinal vibration
zn=zeros(nx,1);       % zn: spatial vector  at the current
time  (t=i)
zm=zn;                 % zm: spatial vector at the previous
time (t=i-1)
zp=zn;                % zp: spatial vector at the later time
(t=i+1)
u=zeros(nx,nt);      % longitudinal vibration

%Initial conditions
for i=1:nx
yn(i)=0.02*sin((i-1)*pi/(nx-1));
zn(i)=0.01*sin((i-1)*pi/(nx-1));
end
yp(:)=yn(:);
yp(:)=yn(:);
zp(:)=zn(:);
zp(:)=zn(:);
% Boundary conditions
yn(1)=0;
yn(nx)=0;
zn(1)=0;
zn(nx)=0;
% Control gain
K1=0.1;
```

```
K2=0.01;
for j=3:nt-1
    w(:,j)=yn;
    u(:,j)=zn;
    ym=yn; yn=yp; zm=zn; zn=zp; %Save current and previous
arrays
    %Belt span
    for i=2:nx-1
    yx=(yn(i+1)-yn(i-1))/(2*dx);
    yxx=(yn(i+1)-2*yn(i)+yn(i-1))/dx^2;
    yxt=(yn(i+1)-yn(i-1)-ym(i+1)+ym(i-1))/(2*dx*dt);
    zx=(zn(i+1)-zn(i-1))/(2*dx);
    zxx=(zn(i+1)-2*zn(i)+zn(i-1))/dx^2;
    zxt=(zn(i+1)-zn(i-1)-zm(i+1)+zm(i-1))/(2*dx*dt);
    zp(i)=2*zn(i)-zm(i)+dt^2*(EA*(zxx+yxx*yx)/pA-2*v*zxt-
v^2*zxx);
    yp(i)=2*yn(i)-
ym(i)+dt^2*(T0*yxx/pA+EA*(yxx*zx+yx*zxx+(3/2)*yx^2*yxx)/pA-
(2*v*yxt+v^2*yxx));
    end
    %Right boundary
    ynx=(yp(nx)-yp(nx-1))/dx;
    znx=(zp(nx)-zp(nx-1))/dx;
    fc=(T0+EA*(znx+ynx^2/2))*ynx;
    yp(nx)=0;
    yp(nx)=yn(nx)-K1*(fc+K2*10^-18 *fc^3)*10^-9;
end

%%FIGURE%%
wm=w(fix(nx/2),:);
um=u(fix(nx/2),:);
sizet=size(wm);
tm=linspace(0,Time,sizet(2));
plot(tm,wm)
```

Index

© The Editor(s) (if applicable) and The Author(s), under exclusive license to Springer Nature Singapore Pte Ltd. 2022
K.-S. Hong et al., *Control of Axially Moving Systems*,
https://doi.org/10.1007/978-981-16-2915-0

The manufacturer's authorised representative in the EU is Springer Nature Customer Service Centre GmbH, Europaplatz 3, 69115 Heidelberg, Germany. If you have any concerns regarding our products, please contact ProductSafety@springernature.com

Printed and bound by CPI Group (UK) Ltd, Croydon, CR0 4YY
30/09/2026
02225329-0012